Bulk Crystal and Thin Film Formation

Bulk Crystal and
Thin Film Formation

Edited by **Sharon Levine**

NYRESEARCH
P R E S S

New York

Published by NY Research Press,
23 West, 55th Street, Suite 816,
New York, NY 10019, USA
www.nyresearchpress.com

Bulk Crystal and Thin Film Formation
Edited by Sharon Levine

International Standard Book Number: 978-1-63238-067-8 (Hardback)

Printed in the United States of America.

Contents

Preface

In present-day research and development, materials manufacturing crystal growth is referred to as a method to solve a broad spectrum of technological tasks in the fabrications of materials with stipulated properties. This all-inclusive book enables a reader to achieve insight into essential characteristics of the field, including formation of thin films, crystallization of proteins, rise of bulk organic crystals, low-dimensional structures, and other organic compounds.

The information contained in this book is the result of intensive hard work done by researchers in this field. All due efforts have been made to make this book serve as a complete guiding source for students and researchers. The topics in this book have been comprehensively explained to help readers understand the growing trends in the field.

I would like to thank the entire group of writers who made sincere efforts in this book and my family who supported me in my efforts of working on this book. I take this opportunity to thank all those who have been a guiding force throughout my life.

Editor

Part 1

Growth of Thin Films and Low-Dimensional Structures

Controlled Growth of C-Oriented AlN Thin Films: Experimental Deposition and Characterization

Manuel García-Méndez

Centro de Investigación en Ciencias Físico-Matemáticas,
FCFM de la UANL Manuel L. Barragán S/N, Cd. Universitaria,
México

1. Introduction

Nowadays, the science of thin films has experienced an important development and specialization. Basic research in this field involves a controlled film deposition followed by characterization at atomic level. Experimental and theoretical understanding of thin film processes have contributed to the development of relevant technological fields such as microelectronics, catalysis and corrosion.

The combination of materials properties has made it possible to process thin films for a variety of applications in the field of semiconductors. Inside that field, the nitrides *III-IV* semiconductor family has gained a great deal of interest because of their promising applications in several technology-related issues such as photonics, wear-resistant coatings, thin-film resistors and other functional applications (Moreira et al., 2011; Morkoç, 2008).

Aluminium nitride (*AlN*) is an *III-V* compound. Its more stable crystalline structure is the hexagonal würzite lattice (see figure 1). Hexagonal *AlN* has a high thermal conductivity (260 $Wm^{-1}K^{-1}$), a direct band gap (E_g=5.9-6.2 eV), high hardness (2 x 10^3 kgf mm^{-2}), high fusion temperature (2400°C) and a high acoustic velocity. *AlN* thin films can be used as gate dielectric for ultra large integrated devices (ULSI), or in GHz-band surface acoustic wave devices due to its strong piezoelectricity (Chaudhuri et al., 2007; Chiu et al., 2007; Jang et al., 2006; Kar et al., 2006; Olivares et al., 2007; Prinz et al., 2006). The performance of the *AlN* films as dielectric or acoustical/electronic material directly depends on their properties at microstructure (grain size, interface) and surface morphology (roughness). Thin films of *AlN* grown at a *c-axis* orientation (preferential growth perpendicular to the substrate) are the most interesting ones for applications, since they exhibit properties similar to monocrystalline *AlN*. A high degree of *c-axis* orientation together with surface smoothness are essential requirements for *AlN* films to be used for applications in surface acoustic wave devices (Jose et al., 2010; Moreira et al., 2011).

On the other hand, the oxynitrides MeN_xO_y (Me=metal) have become very important materials for several technological applications. Among them, aluminium oxynitrides may have promising applications in diferent technological fields. The addition of oxygen into a growing *AlN* thin film induces the production of ionic metal-oxygen bonds inside a matrix of covalent metal-nitrogen bond. Placing oxygen atoms inside the würzite structure of AlN can produce important modifications in their electrical and optical properties of the films, and thereby changes in their thermal conductivity and piezoelectricity features are produced too (Brien & Pigeat, 2008; Jang et al., 2008). Thus, the addition of oxygen would allow to tailor the properties of the AlN_xO_y films between those of pure aluminium oxide (Al_2O_3) and nitride (AlN), where the concentration of Al, N and O can be varied depending on the specific application being pursued (Borges et al., 2010; Brien & Pigeat, 2008; Ianno et al., 2002; Jang et al., 2008). Combining some of their advantages by varying the concentration of Al, N and O, aluminium oxynitride films ($AlNO$) can produce applications in corrosion protective coatings, optical coatings, microelectronics and other technological fields (Borges et al., 2010; Erlat et al., 2001; Xiao & Jiang, 2004). Thus, the study of deposition and growth of AlN films with the addition of oxygen is a relevant subject of scientific and technological current interest.

Thin films of AlN (pure and oxidized) can be prepared by several techniques: chemical vapor deposition (CVD) (Uchida et al., 2006; Sato el at., 2007; Takahashi et al., 2006), molecular beam epitaxy (MBE) (Brown et al., 2002; Iwata et al., 2007), ion beam assisted deposition (Lal et al., 2003; Matsumoto & Kiuchi, 2006) or direct current (DC) reactive magnetron sputtering.

Among them, reactive magnetron sputtering is a technique that enables the growth of c-$axis$ AlN films on large area substrates at a low temperature (as low as 200°C or even at room temperature). Deposition of AlN films at low temperature is a "must", since a high-substrate temperature during film growth is not compatible with the processing steps of device fabrication. Thus, reactive sputtering is an inexpensive technique with simple instrumentation that requires low processing temperature and allows fine tuning on film properties (Moreira et al., 2011).

In a reactive DC magnetron process, molecules of a reactive gas combine with the sputtered atoms from a metal target to form a compound thin film on a substrate. Reactive magnetron sputtering is an important method used to prepare ceramic semiconducting thin films. The final properties of the films depend on the deposition conditions (experimental parameters) such as substrate temperature, working pressure, flow rate of each reactive gas (Ar, O_2, N_2), power source delivery (voltage input), substrate-target distance and incidence angle of sputtered particles (Ohring, 2002). Reactive sputtering can successfully be employed to produce AlN thin films of good quality, but to achieve this goal requires controlling the experimental parameters while the deposition process takes place.

In this chapter, we present the procedure employed to grow AlN and $AlNO$ thin-films by DC reactive magnetron sputtering. Experimental conditions were controlled to get the growth of c-$axis$ oriented films.

The growth and characterization of the films was mainly explored by way of a series of examples collected from the author´s laboratory, together with a general reviewing of what already has been done. For a more detailed treatment of several aspects, references to highly-respected textbooks and subject-specific articles are included.

One of the most important properties of any given thin film system relies on its crystalline structure. The structural features of a film are used to explain the overall film properties, which ultimately leads to the development of a specific coating system with a set of required properties. Therefore, analysis of films will be concerned mainly with structural characterization.

Crystallographic orientation, lattice parameters, thickness and film quality were characterized through X-ray Diffraction (XRD) and UV-Visible spectroscopy (UV-Vis). Chemical indentification of phases and elemental concentration were characterized through X-ray photoelectron spectroscopy (*XPS*). From these results, an analysis of the interaction of oxygen into the *AlN* film is described. For a better understanding of this process, theoretical calculations of Density of States (DOS) are included too.

The aim of this chapter is to provide from our experience a step wise scientific/technical guide to the reader interested in delving into the fascinating subject of thin film processing.

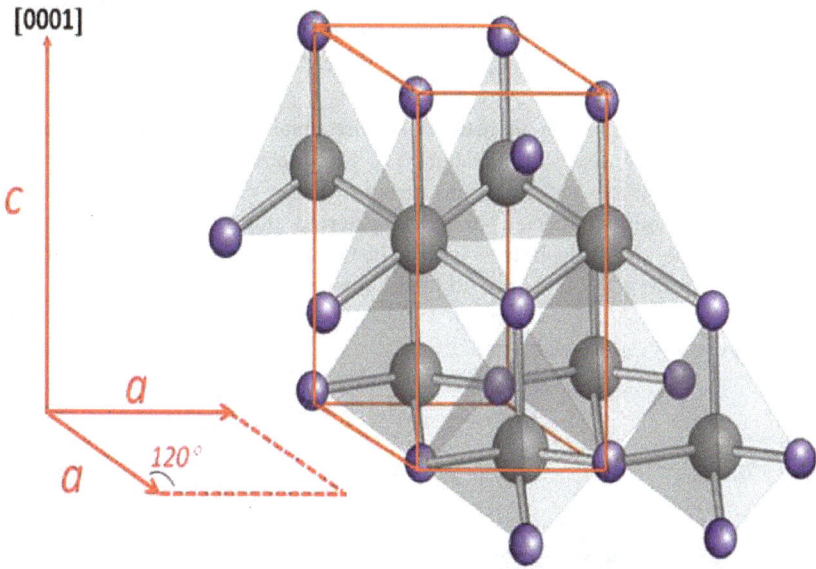

Fig. 1. Würzite structure of *AlN*. Hexagonal *AlN* belongs to the space group 6mm with lattice parameters c=4.97 Å and a=3.11 Å.

2. Deposition and growth of AlN films

The sputtering process consists in the production of ions within generated plasma, on which the ions are accelerated and directed to a target. Then, ions strike the target and material is ejected or sputtered to be deposited in the vicinity of a substrate. The plasma generation and sputtering process must be performed in a closed chamber environment, which must be maintained in vacuum. To generate the plasma gas particles (usually argon) are fed into the chamber. In *DC* sputtering, a negative potential U is applied to the target (cathode). At critical applied voltage, the initially insulating gas turns to electrical conducting medium. Then, the positively charged Ar^+ ions are accelerated toward the cathode. During ionization, the cascade reaction goes as follows:

$$e^- + Ar \rightarrow 2e^- + Ar^+$$

where the two additional (secondary) electrons strike two more neutral ions that cause the further gas ionization. The gas pressure "P" and the electrode distance "d" determine the breakdown voltage "V_B" to set the cascade reaction, which is expressed in terms of a product of pressure and inter electrode spacing:

$$V_B = \frac{APd}{\ln(Pd) + B} \tag{1}$$

where A and B are constants. This result is known as Paschen´s Law (Ohring, 2002).

In order to increase the ionization rate by emitted secondary electrons, a ring magnet (magnetron) below the target can be used. Hence, the electrons are trapped and circulate over the surface target, depicting a cycloid. Thus, the higher sputter yield takes place on the target area below this region. An erosion zone (trace) is "carved" on the target surface with the shape of the magnetic field.

Equipment description: Films under investigation were obtained by *DC* reactive magnetron sputtering in a laboratory deposition system. The high vacuum system is composed of a pirex chamber connected to a mechanic and turbomolecular pump. Inside the chamber the magnetron is placed and connected to a DC external power supply. In front of the magnetron stands the substrate holder with a heater and thermocouple integrated. The distance target-substrate is about 5 cm and target diameter 1". The power supply allows to control the voltage input (*Volts*) and an external panel display readings of current (*Amperes*) and sputtering power (*Watts*) (see Figure 2).

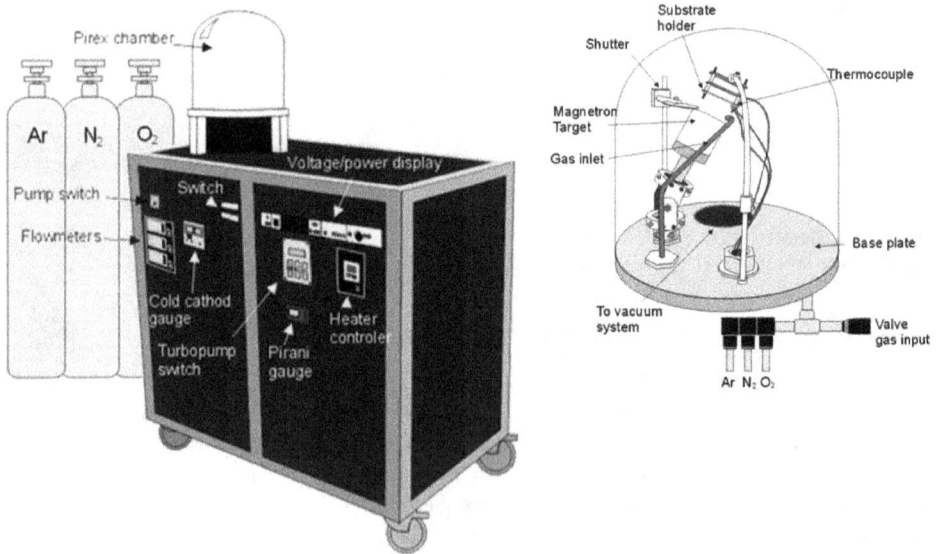

Fig. 2. Schematic diagram of the equipment utilized for film fabrication.

Deposition procedure: A disc of *Al* (2.54 cm diameter, 0.317 cm thick, 99.99% purity) was used as a target. Films were deposited on silica and glass substrates that were ultrasonically cleaned in an acetone bath. For deposition, the sputtering chamber was pumped down to a base pressure below 1×10^{-5} Torr. When the chamber reached the operative base pressure, the *Al* target was cleaned *in situ* with *Ar+* ion bombardment for 20 minutes at a working pressure of 10 mTorr (20 *sccm* gas flow). A shutter is placed between the target and the substrate throughout the cleaning process. The Target was systematically cleaned to remove any contamination before each deposition.

Sputtering discharge gases of *Ar*, N_2 and O_2 (99.99 % purity) were admitted separately and regulated by individual mass flow controllers. A constant gas mixture of *Ar* and N_2 was used in the sputtering discharge to grow *AlN* films; a gas mixture of *Ar*, N_2 and O_2 was used to grow *AlNO* films.

A set of eight films were prepared: four samples on glass substrates (*set 1*) and four samples on silica substrates (*set 2*). From *set 1*, two samples correspond to *AlN* (15 min of deposition time, labeled *S1* and *S2*) and two to *AlNO* (10 min of deposition time, labeled *S3* and *S4*). From *set 2*, three samples correspond to *AlN* (10 min of deposition time, labeled *S5*, *S6* and *S7*) and one to *AlNO* (10 min of deposition time, labeled *S8*). All samples were deposited using an *Ar* flow of 20 *sccm*, an N_2 flow of 1 *sccm* and an O_2 flow of 1 *sccm*. In all samples (excluding the ones grown at room temperature.), the temperature was supplied during film deposition.

Tables 1 (a) (*set 1*) and 1 (b) (set 2) summarize the experimental conditions of deposition. Calculated optical thickness by formula 4 is included in the far right column.

Sample	*(Film) °C - time*	*V (Volts)*	*P (Watts)*	*Thickness (nm)*
S1	*(AlN)* RT-15 min	360	120	980
S2	*(AlN)* 100°C-15 min	360	130	970
S3	*(AlNO)* RT-10 min	360	190	820
S4	*(AlNO)* 120°C-10 min	360	185	940

Table 1a. Deposition parameters for *DC* sputtered films grown on glass substrates (*set 1*)

Sample	*(Film) °C - time*	*V (Volts)*	*P (Watts)*	*Thickness (nm)*
S5	*(AlN)*RT-10 min	340	100	630
S6	*(AlN)*100°C-10 min	330	110	630
S7	*(AlN)*200°C-10 min	340	120	730
S8	*(AlNO)*RT-10 min	380	140	490

Table 1b. Deposition parameters for *DC* sputtered films grown on silica substrates (*set 2*).

3. Structural characterization

XRD measurements were obtained using a Philips X'Pert diffractometter equipped with a copper anode $K\alpha$ radiation, λ =1.54 Å. High resolution *theta/2Theta* scans (Bragg-Brentano geometry) were taken at a step size of 0.005°. Transmission spectra were obtained with a UV- Visible double beam Perkin Elmer 350 spectrophotometer.

Figure 3 (a) and (b) display the *XRD* patterns of the films deposited on glass (*set 1*) and silica (*set 2*) substrates, respectively.

The diffraction pattern of films displayed in figure 3 match with the standard *AlN* würzite spectrum (JCPDS card 00-025-1133, a=3.11 Å, c=4.97 Å) (Powder Diffraction file, 1998). The highest intensity of the *(002)* reflection at $2\theta \approx 35.9^0$ indicates an oriented growth along the *c-axis* perpendicular to substrate.

From *set 1*, it can be observed that the intensity of *(002)* diffraction peak is the highest in *S2*. In this case, the temperature of 100^0C increased the crystalline ordering of film. In *S3* and *S4* the intensity of *(002)* diffraction and grain size are very similar for both samples, which shows that applied temperature on *S4* had not effect in improving its crystal ordering.

From *set 2*, it can be observed that the intensity of *(002)* diffraction peak is the highest in *S5*. Generally, temperature gives atoms an extra mobility, allowing them to reach the lowest thermodynamically favored lattice positions hence, the crystal size becomes larger and the crystallinity of the film improves. However, the temperature applied to *S6* and *Ss* makes no effect to improve their crystallinity. In this case, a substrate temperature higher than 100°C can trigger a re-sputtering of the atoms that arrive at the substrate's surface level and crystallinity of films experiences a downturn.

From *set 1* and *set 2*, *S2* and *S5*, respectively, were the ones that presented the best crystalline properties. A temperature ranging from RT to 100°C turned out to be the critical experimental factor to get a highly oriented crystalline growth.

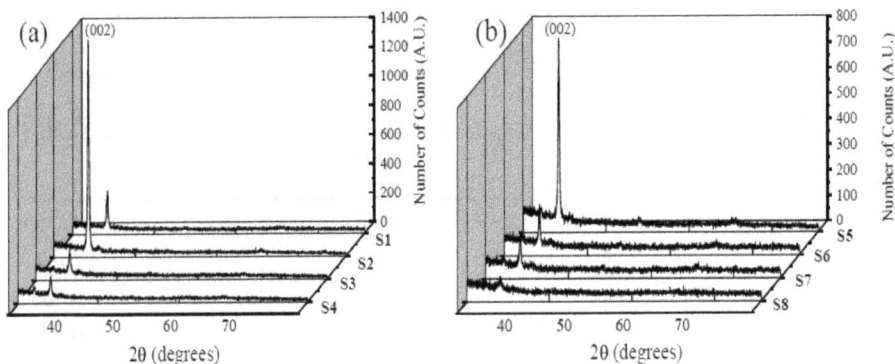

Fig. 3. *XRD* patterns of films deposited on (a) glass and (b) silica substrates.

In terms of the role of oxygen, for *S3*, *S4* and *S8*, the presence of alumina (γ-Al_2O_3: *JCPDS file 29-63*) or spinel (γ-*AlON*: *JCPDS files 10-425 and 18-52*) compounds in the diffraction patterns was not detected. However, it is known from thermodynamic that elemental aluminium reacts more favorably with oxygen than nitrogen: it is more possible to form Al_2O_3 by gaseous phase reaction of $Al+(3/2)O_2$ than AlN of $Al+(1/2)N$ since $\Delta G(Al_2O_3)=-1480\ KJ/mol$ and $\Delta G(AlN)=-253\ KJ/mol$ (Borges et al., 2010; Brien & Pigeat, 2007). Therefore, the existence of Al_2O_3 or even spinel $AlNO$ phases in samples cannot be discarded, but maybe in such a small proportions as to be detected by *XRD*.

S1, *S2* and *S5* show a higher crystalline quality than *S3*, *S4* and *S8*. For these last samples, the extra O_2 introduced to the chamber promotes the oxidation of the target-surface (target poisoning). In extreme cases when the target is heavily poisoned, oxidation can cause an arcing of the magnetron system. Formation of aluminium oxide on the target can act as an electrostatic shell, which in turn can affect the sputtering yield and the kinetic energy of species which impinge on substrate with a reduction of the sputtering rate: The lesser energy of species reacting on substrate, the lesser crystallinity of films.

Also, the oxygen can enter in to the *AlN* lattice through a mechanism involving a vacancy creation process by substituting a nitrogen atom in the weakest *Al-N bond* aligned parallel to [0001] direction. During the process, the mechanism of ingress of oxygen into the lattice is by diffussion (Brien & Pigeat, 2007; Brien & Pigeat, 2008; Jose et al., 2010). On the other hand, the ionic radius of oxygen (r_O=0.140 nm) is almost ten times higher than that of nitrogen (r_N=0.01-0.02 nm) (Callister, 2006). Thus, the oxygen causes an expansion of the crystal lattice through point defects. As the oxygen content increases, the density of point defects increases and the stacking of hexagonal *AlN* arrangement is disturbed . It has been reported that the *Al* and *O* atoms form octahedral atomic configurations that eventually become planar defects. These defects usually lie in the basal *{001}* planes (Brien & Pigeat, 2008; Jose et al., 2010).

As was mentioned, during the deposition of thin films, the oxygen competes with the nitrogen to form an oxidized *Al-compound*. The resulting films are then composed of separated phases of *AlN* and Al_xO_y domains. The presence of Al_xO_y domains provokes a disruption in the preferential growth of the film.

For example, in *S4*, the applied temperature of 120⁰C can promote an even more efficient diffusive ingress of oxygen into the *AlN* lattice and such temperature was not a factor contributing to improve crystallinity. In *S3* and *S8*, oxygen by itself was the factor that provoked a film´s low crystalline growth.

By using the Bragg angle (θ_b) as variable that satisfies the Bragg equation:

$$2d_{hkl}Sen\,\theta_b = n\lambda \qquad (2)$$

and the formula applied for hexagonal systems:

$$\frac{1}{d_{hkl}^2} = \frac{4}{3}\left(\frac{h^2 + hk + k^2}{a^2}\right) + \frac{l^2}{c^2} \qquad (3)$$

the length of the lattice parameters "a" and "c" can then be obtained from the experimental data.

As films crystallized in a hexagonal würzite structure, *XRD* patterns were processed with a software program in order to obtain the lattice parameters "*a*" and "*c*". The *AlN* würzite structure from the *JCPDS* database (*PDF file 00-025-1133, c= 4.97 Å, a=3.11 Å*) was taken as a reference (Powder Diffraction File, 1998). For the fitting, input parameters of *(h k l)* planes with their corresponding *theta-angle* are given. By using the Bragg formula and the equation of distance between planes (for a hexagonal lattice), the lattice parameters are then calculated by using a multiple correlation analysis with a least squares minimization. The 2θ angles were set fixed while lattice parameters were allowed to fit. Calculated lattice parameters "*a*" and "*c*" and grain size "*L*" by formula (4) are included in Table 2.

	a	*c*	*c/a*	*L*
S1	3.11	4.99	1.60	21
S2	3.11	4.98	1.60	23
S3	3.13	5.0	1.59	21
S4	3.14	5.0	1.59	20
S5	3.13	4.98	1.59	24
S6	3.13	4.99	1.59	20
S7	3.11	4.99	1.60	17
S8	3.11	5.0	1.60	17

Table 2. Lattice parameters "*a*" (nm) and "*c*" (nm) obtained from *XRD* measurements.

The average grain size "*L*" is obtained through the Debye-Scherrer formula (Patterson, 1939):

$$L = \frac{K\lambda}{B\cos\theta_b} \tag{4}$$

where K is a dimensionless constant that may range from 0.89 to 1.30 depending on the specific geometry of the scattering object.

For a perfect two dimenssional lattice, when every point on the lattice produces a spherical wave, the numerical calculations give a value of K=0.89. A cubic three dimensional crystal is best described by K=0.94 (Patterson, 1939).

The measure of the peak width, the full width at half maximum (FWHM) for a given θ_b is denoted by B (for a gaussian type curve).

From table 2, it can be observed that the calculated lattice parameters differ slightly from the ones reported from the *JCPDS* database, mainly the "*c*" value, particularly for *S3*, *S4* and *S8*. Introduction of oxygen into the *AlN* matrix along the {001} planes also modifies the lattice parameters. As expected, the "*c*" value is the most affected.

The quality of samples can also be evaluated from UV-Visible spectroscopy (Guo et al., 2006). By analysing the measured *T vs λ* spectra at normal incidence, the absorption coefficient (α) and the film thickness can be obtained.

If the thickness of the film is uniform, interference effects between substrate and film (because of multiple reflexions from the substrate/film interface) give rise to oscillations. The number of oscillations is related to the film thickness. The appearence of these oscillations on analized films indicates uniform thickness. If the thickness "*t*" were not uniform or slightly tappered, all interference effects would be destroyed and the *T vs λ* spectrum would look like a smooth curve (Swanepoel, 1983).

Oscillations are useful to calculate the thickness of films using the formula (Swanepoel, 1983; Zong et al., 2006):

$$t = \frac{1}{2n\left(\frac{1}{\lambda_2} - \frac{1}{\lambda_1}\right)} \quad (5)$$

Where t is the thickness of film, n the refractive index, λ_1 and λ_2 are the wavelength of two adjacent peaks. Calculated optical thickness of samples using the above mentioned formula, are included in Tables 1(a) and (b).

Regarding the absorbance (α), a T vs λ curve can be divided (grossly) into four regions. In the transparent region $\alpha=0$ and the transmitance is a function of n and t through multiple reflexions. In the region of weak absorption α is small and the transmission starts to reduce. In the region of medium absorption the transmission experiences the effect of absoption even more. In the region of strong absorption the transmission decreases abruptly. This last region is also named the absorption edge.

Near the absorption edge, the absorption coefficient can expressed as:

$$h\nu\alpha = \beta(h\nu\text{-}E_g)^{\gamma} \quad (6)$$

where $h\nu$ is the photon energy, E_g the optical band gap and γ is the parameter measuring the type of band gap (direct or indirect) (Guerra et al., 2011; Zong et al., 2006).

Thus, the optical band gap is determined by applying the Tauc model and the Davis and Mott model in the high absorbance region. For AlN films, the transmittance data provide the best linear curve in the band edge region, taking n=1/2, implying that the transition is direct in nature (for indirect transition n=2). Band gap is obtained by plotting $(\alpha h\nu)^2$ vs $h\nu$ by extrapolating the linear part of the absorption edge to find the intercept with the energy axis. By using UV-Vis measurements for $AlNO$ films on glass sustrates, authors of ref. (Jang et al., 2008) found band gap values between 6.63 to 6.95 eV, depending the Ar:O ratio.

From our measurements, figure 4 displays the optical spectra (T vs λ curve) graphs. The oscillations detected in the curves attest the high quality in homogeneity of deposited films. All the samples have oscillation regardless their degree of crystallinity. An important feature to note is that curves present differences in the "sharpness", at the onset of the strong absorption zone. These differences are attributed to deposition conditions, where final density of films, presence of deffects and thickness, modify the shape of the curve at the band edge.

A FESEM micrograph cross-section of S2 is displayed on figure 5. From figure, it is possible to identify a well defined substrate/film interface and a section of film with homogeneous thickness. Together with micrographs, in-situ EDAX analyses were conducted in two specific regions of the film. An elemental analysis by EDAX allows to distinguish the differences in elemental concentration depending on the analized zone. In the film zone , an elemental concentration of Al (54.7 %) and N (45.2 %) was detected, as expected for AlN film. Conversely, in the substrate zone, elemental concentration of Si and O with traces of Ca, Na, Mg was detected, as expected for glass.

At this stage, we can establish that during the sputtering process, the oxygen diffuses in to the growing AlN films. Then, the oxygen attaches to available Al, forming Al_xO_y phases. Dominions of these phases, contained in the whole film, can induce defects. These defects are piled up along the c-axis. From X-ray diffractograms, a low and narrow intensity at the (0002) reflection indicates low crystallographic ordering. By calculating lattice parameters "a" and "c" and evaluating how far their obtained values deviate from the JCPDF standard (mainly the "c" distance), also provides evidence about the degree of crystalline disorder. In films, a low crystallographic ordering does not imply a disruption in the homogeneity, as was already detected by UV-Visible measurements. A more detailed analysis concerning the identification and nature of the phases contained in films were performed with a spectroscopic technique.

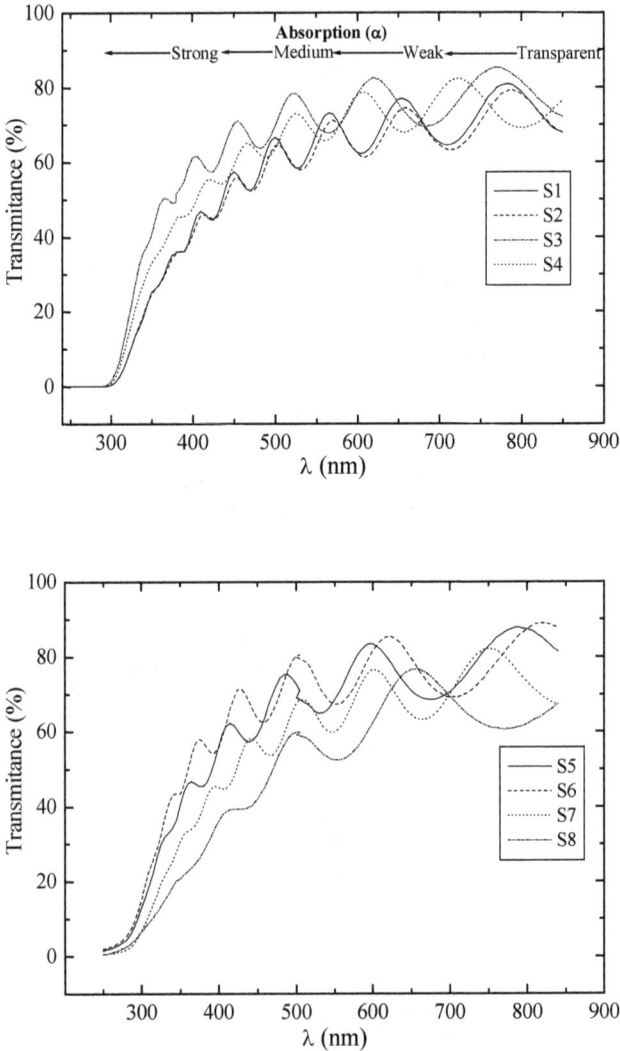

Fig. 4. Optical transmission spectra of deposited films.

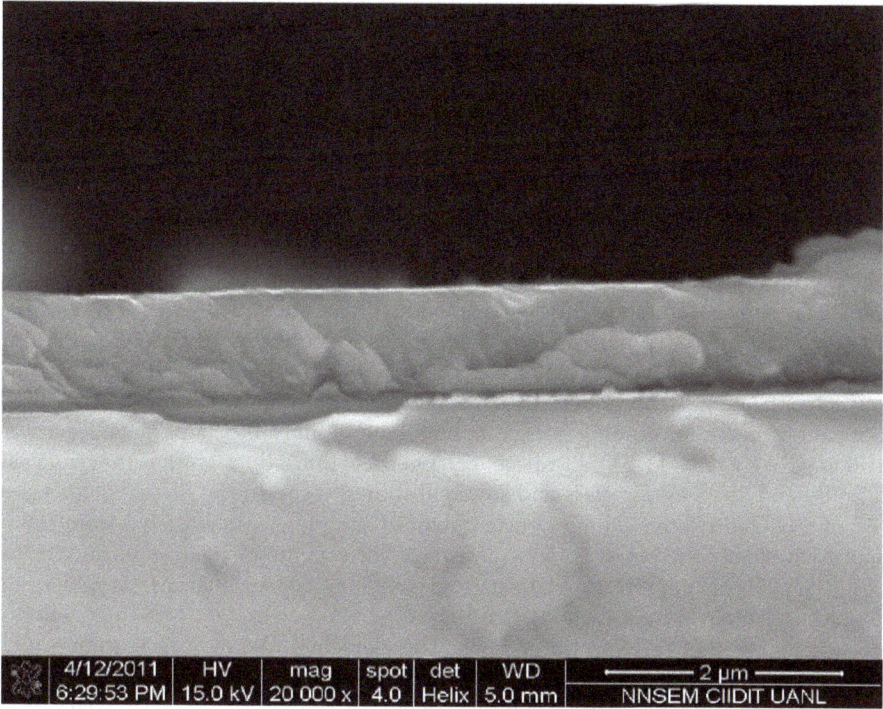

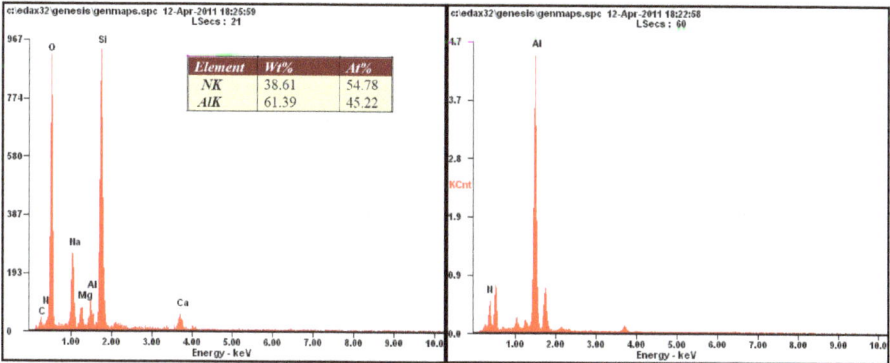

Fig. 5. Cross section FESEM micrograph of AlN film (S2). An homogeneous film deposition can be observed. In the right column an EDAX analysis of (a) film zone and (b) substrate zone is included.

4. Chemical characterization

The process of oxidation is a micro chemical event that was not completely detected by XRD. Because of that, XPS analyses were performed in order to detect and identify oxidized phases.

XPS measurements were obtained with a Perkin-Elmer PHI 560/ESCA-SAM system, equipped with a double-pass cylindrical mirror analyzer, and a base pressure of 1×10^{-9} Torr. To clean the surface, Ar^+ sputtering was performed with 4 keV energy ions and 0.36 $\mu A/cm^2$ current beam, yielding to about 3 nm/min sputtering rate. All XPS spectra were obtained after Ar^+ sputtering for 15 min. The use of relatively low current density in the ion beam and low sputtering rate reduces modifications in the stoichiometry of the AlN surface. For the XPS analyses, samples were excited with 1486.6 eV energy $Al_{K\alpha}$ X-rays. XPS spectra were obtained under two different conditions: (i) a survey spectrum mode of 0-600 eV, and (ii) a multiplex repetitive scan mode. No signal smoothing was attempted and a scanning step of 1 eV/step and 0.2 eV/step with an interval of 50 ms was utilized for survey and multiplex modes, respectively. The spectrometer was calibrated using the $Cu\ 2p_{3/2}$ (932.4 eV) and $Cu\ 3p_{3/2}$ (74.9 eV) lines. Al films deposited on the glass and silica substrates were used as additional references for Binding energy. In both kind of films, the BE of metallic (Al^0) $Al2p$-transition gave a value of 72.4 eV respectively. On these films, the $C1s$-transition gave values of 285.6 eV and 285.8 eV for glass and silica substrates, respectively. These values were set for BE of $C1s$. The relative atomic concentration of samples was calculated from the peak area of each element ($Al2p$, $O1s$, $N1s$) and their corresponding relative sensitivity factor values (RSF). These RSF were obtained from software system analysis (Moulder, 1992). Gaussian curve types were used for data fitting.

Figure 6 displays the XPS spectra of films. The elemental atomic concentration (atomic percent) calculated from the $O1s$, $N1s$ and $Al2p$ transitions is also included in the figure. Figure 6a shows the $Al2p$ high-resolution photoelectron spectrum of $S1$. The binding energies (BE) from the acquired $Al2p$ photoelectron transition are presented in table 3.

The survey spectra show the presence of oxygen in all films, regardless of the fact that some samples were grown without oxygen during deposition. From the XPS analysis, S2 and S5, our films with the best crystalline properties, a concentration of oxygen of 26.3% and 21.6% atomic percent respectively, was measured. The highest measured concentration of oxygen was of about 36.6%, corresponding to S8. This occurrence of oxidation was not directly detected by the XRD analysis, since these oxidized phases can be spread in a low amount throughout the film.

The nature of these phases can be inferred from the deconvoluted components of the $Al2p$ transition. In Figure 6a, the $Al2p$ core level spectrum is presented. This spectrum is composed of contributions of metallic Al ($BE=72.4\ eV$), nitridic Al in AlN ($BE=74.7\ eV$) and oxidic Al in Al_2O_3 ($BE=75.6\ eV$).

Despite the differences in experimental conditions, aluminium reacted with the nitrogen and the oxygen in different proportions. Even in S2, the thin film with the best crystalline properties, a proportion of about 30.6 % of aluminum reacted with oxygen to form an aluminium oxide compound. In S7, the relative contribution of Al in nitridic and oxidic state is almost similar, of 42.2% and 49.5%, respectively. A tendency, not absolute but in general, indicates that the higher the proportion of Al in oxidic state, the more amorphous the film.

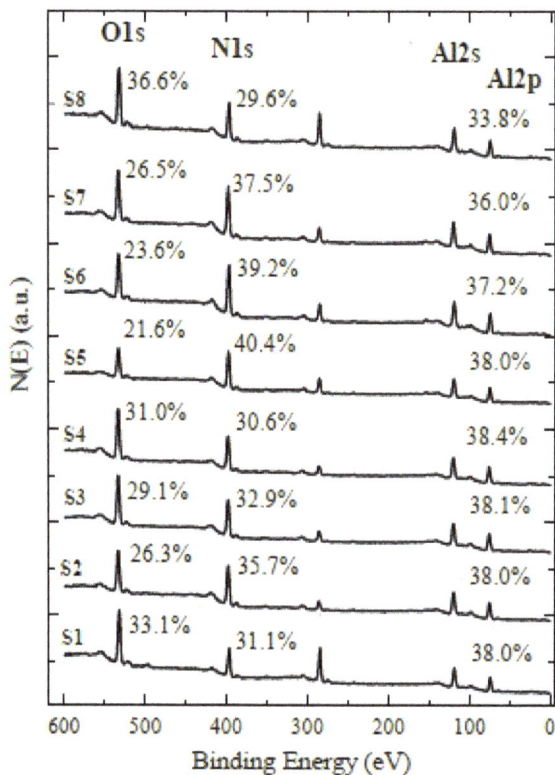

Fig. 6. XPS survey spectra of *dc* sputtered films. In this figure, the *O1s*, *N1s* and *Al2p* core-level principal peaks can be observed.

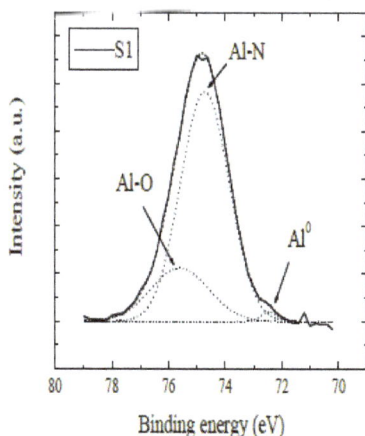

Fig. 6a. *Al2p* XPS spectrum of *S1*. The *Al2p* peak is composed of contributions of metallic aluminium (Al^0), aluminium in nitride (*Al-N*) and oxidic (*Al-O*) state.

	Binding energy (eV)			Contribution (%)		
	Al^0	Al-N	Al-O	Al-N	Al-O	ΔE (eV)
S1	72.4	74.7	75.6	77.3	21.6	0.9
S2		74.7	75.7	69.3	30.6	1.0
S3		74.7	75.2	74.7	22.9	0.5
S4	72.1	74.4	75.1	58.4	39.9	0.7
S5		74.7	75.5	64.4	35.1	0.8
S6	72.6	74.7	75.5	60.2	37.3	0.8
S7		74.7	75.4	42.2	49.5	0.7
S8		74.4	75.5	58.7	36.3	1.1

Table 3. Binding energy (eV) of metallic aluminium (Al^0), aluminium in nitridic (Al-N) and oxidic (Al-O) state obtained from deconvoluted components of Al2p transition. Percentage (relative %) of Al bond to N and O is also displayed.

For comparison purposes, some relevant literature concerning the binding energies of metallic-Al, AlN and Al_2O_3 has been reviewed and included in table 4. Aluminium in metallic state lies in the range of 72.5-72.8 eV. Aluminium in nitridic state lies in the range of 73.1-74.6 eV, while aluminium in oxidic state lies in the range of 74.0-75.5 eV. Also, there is an Al-N-O spinel-like bonding, very similar in nature to oxidic aluminium with a BE value of 75.4 eV. Another criteria used by various authors for phase identification, is to take the difference (ΔE) in BE of the Al2p transition corresponding to Al-N and Al-O bonds. This difference can take values of about 0.6 eV up to 1.1 eV (see Table 4).

Binding energy (eV)				
Al^0	Al-N	Al-O	ΔE (eV)	References
72.8	74.1	74.7	0.6	(Stanca, 2004)
	74.3	75.2	0.9	
	74.6	75.4 (spinel)	0.8	(Sohal et al., 2006)
72.8	74.1	74.7	0.7	(Jose et al., 2010)
	73.6	74.6	1.0	
72.8	74.4	75.2	0.8	(Wang et al., 1997)
72.7	74.5	75.5	1.0	(Gredelj et al., 2001)
72.8	74.6	75.6	1.0	
72.5				
	73.1	75.1	2.0	(Richthofen et al., 1996)
		74.2 (spinel)	1.1	

Table 4. Binding energy of (eV) of metallic aluminium (Al^0), aluminium in nitridic (Al-N) and oxidic (Al-O) state obtained from literature

In films, only small traces of metallic aluminium were detected in $S1$ at $72.4\ eV$. For $S4$ and $S8$, BE of Al in nitride gave a value of $74.4\ eV$, just below the BE of $74.7\ eV$, detected for the rest of the samples. This value of $74.4\ eV$ can be attributed to a substoichiometric AlN_x phase (Robinson et al., 1984; Stanca, 2004). On the other hand, the BE for aluminium in oxydic state varies from $75.1\ eV$ to $75.7\ eV$. The lowest values of BE of about $75.1\ eV$ and $75.2\ eV$, corresponding to $S3$ and $S4$, respectively, could be attributed to a substoichiometric Al_xO_y phase, although in our own experience, the reaction of aluminium with oxygen tends to form the stable γ-Al_2O_3 phase, which possesses somewhat higher value in BE. These finding agree with those reported in other works, where low oxidation states such as Al^{+1}, Al^{+2} can be found at a BE lower than the one of Al^{+3} (Huttel et al., 1993; Stanca, 2004). Oxidation states lower than $+3$ confer an amorphous character to the aluminium oxide (Gutierrez et al., 1997).

5. Theoretical calculations

Experimental results provided evidence that oxygen can induce important modifications in the structural properties of sputtered-deposited AlN films. In this way, theoretical calculations were performed to get a better understanding of how the position of the oxygen into the AlN matrix can modify the electronic properties of the film system.

The bulk structure of hexagonal AlN was illustrated in Figure 1. Additionally, hexagonal AlN can be visualized as a matrix of distorted tetrahedrons. In a tetrahedron, each Al atom is surrounded by four N atoms. The four bonds can be categorized into two types. The first type is formed by three equivalent Al-Nx, (x=1,2,3) bonds, on which the N atoms are located in the same plane normal to the [0001] direction. The second type is the Al-N_0 bond, on which the Al and N atoms are aligned parallel to the [0001] direction (see figure 7). This last bond is the most ionic and has a lower binding energy than the other three (Chaudhuri et al., 2007; Chiu et al., 2007; Zhang et al, 2005). When an AlN film is oxidized, the oxygen atom can substitute the nitrogen atom in the weakest Al-N0 bond while the displaced nitrogen atom can occupy an interstitial site in the lattice (Chaudhuri et al., 2007). For würzite AlN, there are four atoms per hexagonal unit cell where the positions of the atoms for Al and N are: Al(0,0,0), (2/3,1/3,1/2); N(0,0,u), (2/3,1/3, u+1/2), where "u" is a dimensionless internal parameter that represents the distance between the Al-plane and its nearest neighbor N-plane, in the unit of "c", according to the JCPDS database (Powder diffraction file, 1998).

The calculations were perfomed using the tight-binding method (Whangbo & Hoffmann, 1978) within the extended Hückel framework (Hoffmann, 1963) using the computer package YAeHMOP (Landrum, 1900). The extended Hückel method is a semiempirical approach that solves the Schrödinger equation for a system of electrons based on the variational theorem (Galván, 1998). In this approach, explicit correlation is not considered except for the intrinsic contributions included in the parameter set. For a best match with the available experimental information, experimental lattice parameters were used instead of optimized values. Calculations considered a total of 16 valence electrons corresponding to 4 atoms within the unit cell for AlN.

Band structures were calculated using 51 k-points sampling the first Brillouin zone (FBZ). Reciprocal space integration was performed by k-point sampling (see figure 8). From band structure, the electronic band gap (E_g) was obtained.

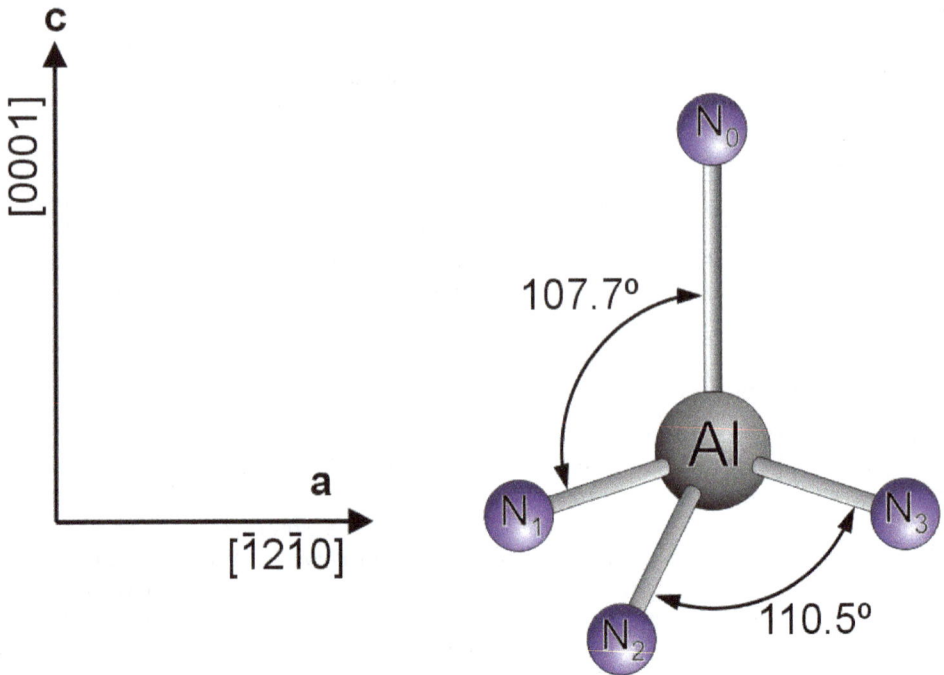

Fig. 7. Individual tetrahedral arrangement of hexagonal *AlN*.

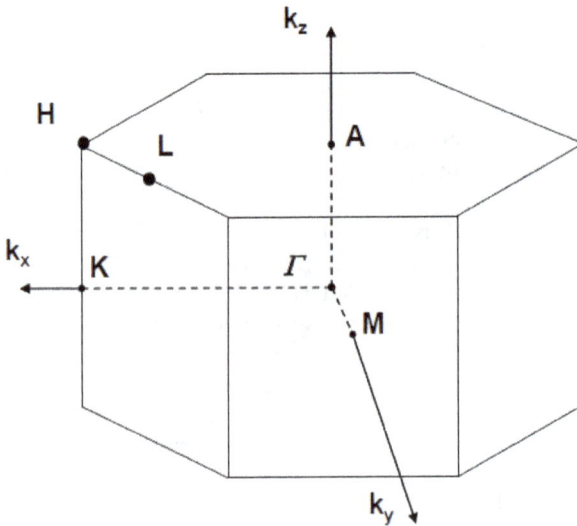

Fig. 8. Hexagonal lattice in k-space.

Calculations were performed considering four scenarios:
1.   A wurzite-like AlN structure with no oxygen in the lattice
2.   An oxygen atom inside the interstitial site of the tetrahedral arrangement (interstitial)
3.   An oxygen atom in place of the N atom in the weakest Al-N_0 bond (substitution)
4.   An oxygen atom on top of the AlN surface (at the surface).

Theoretical band-gap calculations are summarized in Table 5. Values are given in electron volts (eV).

	Energy gap (eV)
(1) AlN Hexagonal	7.2 – Direct M
(2) Interstitial O	1.3 – Direct M
(3) Substitution of N by O	0.82 – Direct H
(4) Surface O	6.2 – Indirect Γ-M

Table 5. Calculated energy gaps for pure AlN (würzite) and with oxygen in different atomic site positions.

For AlN hexagonal, a direct band gap of $\approx$ 7.2 eV at M was calculated (see Figure 9). When oxygen was taken into account in the calculations, the band gap value undergoes a remarkable change: *1.3 eV* for AlN with intercalated oxygen (2) and *0.8 eV* for AlN with oxygen substitution (3). In terms of electronic behavior, the system transformed from insulating (7.2 eV) to semiconductor (1.3 eV), and then from semiconductor (1.3 eV) to semimetal (0.82 eV).

This change in the electronic properties is explained by the differences between the ionic radius of Nitrogen (r_N) and Oxygen (r_O). The ionic radius of the materials involved was: r_N=0.01-0.02 nm, r_O=0.140 nm (Callister, 2006). Comparing these values, it can be noted that r_O is almost ten times higher than r_N. This fact would imply that when the oxygen atom takes the place of the nitrogen atom (by substitution o intercalation of O), the crystalline lattice expands because of the larger size of oxygen. Any change in the distance among atoms and the extra valence electron of the oxygen will alter the electronic interaction and in consequence, the band gap value

In calculation (4), the atoms of Al and N are kept in their würzite atomic positions while the oxygen atom is placed on top of the AlN lattice. In this case, the calculated band gap (6.31 eV) is closer in value to pure AlN (7.2 eV) than the calculated ones for interstitial (1.3 eV) and substitution (0.82 eV). In this case, theoretical results predicts that when the oxygen is not inside the Bravais lattice, the band gap will be close in value to the one of hexagonal AlN; conversely, the more the oxygen interacts with the AlN lattice, the more changes in electronical properties are expected; However, in energetic terms, competition between N and O atoms to get attached to the Al to form separated phases of AlN and Al_xO_y is the most probable configuration, as far as experimental results suggests.

Theoretical calculations of band structure for würzite AlN have been performed using several approaches; For comparison purposes, some of them are briefly described in Table 6.

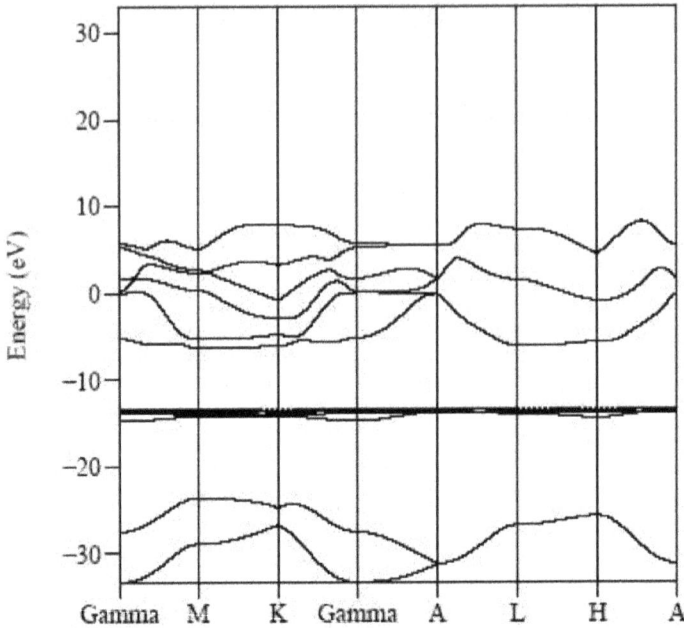

Fig. 9. Band structure for 2H-AlN hexagonal, sampling the first Brillouin zone (FBZ).

Energy band gap (eV)	Method/Procedure	Reference
6.05	Local density approximation (LDA) within the density functional theory (DFT) with a correction Δg, using a quasi-particle method: LDA+Δg	(Ferreira et al., 2005)
6.2	Empirical pseudopotential method (EPM). An analytical function using a fitting procedure for both symmetric and antisymmetric parts, and a potential is constructed	(Rezaei et al., 2006)
4.24	Full potential linear muffin-tin orbital (FPLMTO)	(Persson et al., 2005)
6.15	FPLMTO with a corrected band gap Δg	(Persson et al., 2005)

Table 6. AlN energy band gap values obtained from theoretical calculations.

From our results, the calculated band gap for AlN was 7.2 eV: slightly different to the reported experimental-value of ≈ 6.2 eV. About this issue, is important to take into account that in our calculations spin-orbit effects were not considered. Therefore, some differences arise, especially when an energy-band analysis is performed. Some bands could be shifted up or down in energy due to these contributions. However, it must be stressed out that our proposed method is simple, computationally efficient and the electronic structures obtained can be optimized to closely match the experimental and/or *ab-initio* results. More specific details about DOS graphs and PDOS calculations can be found in reference (García-Méndez et al., 2009), of our authorship.

6. Conclusions

In this chapter, the basis of DC reactive magnetron sputtering as well as experimental results concerning the growth of AlN thin films has been reviewed.

For instance, films under investigation were polycrystalline and exhibit an oriented growth along the [0002] direction. XRD measurements showed that films are composed mainly by crystals of hexagonal AlN. From XPS measurements, traces of aluminium oxides phases were detected. Films deposited without flux of oxygen presented the best crystalline properties, although phases of aluminum oxide were detected on them too. In this case, even in high vacuum, ppm levels of residual oxygen can subside and react with the growing film. Oxygen induces on films structural disorder that tends to disturb the preferential growth at the *c-axis*.

In other works of reactive magnetron sputtering, authors of ref. (Brien & Pigeat, 2008) found that for contamination of oxygen atoms (from 5% to 30 % atomic), AlN films can still grow in würzite structure at room temperature, with no formation of crystalline $AlNO$ or Al_2O_3 phases, just only traces of amorphous AlO_x phases, that leave no signature in diffraction recordings, which is consistent with our results, where a dominant AlN phase in the whole film was detected. On the other hand, authors of ref. (Jose et al., 2010) reports that even in high vacuum, ppm levels of oxygen can stand and promote formation of alumina-like phases at the surface of AlN films, where these phases of alumina could only be detected and quantified by XPS and conversely, X-ray technique was unable to detect. In other report, authors of ref. (Borges et al., 2010), stablished three regions: Metallic (zone M), transition (zone T) and compound (zone C), where chemical composition of $AlNO$ films varies depending the reactive gas mixture in partial pressure of N_2+O_2 at a fixed Ar gas partial pressure. Then, they found that when film pass from zone M to zone C, films grow from crystalline-like to amorphous-type ones, and the lattice parameters increase as more oxygen and nitrogen is incorporated into the films, which also represents the tendency we report in our results.

Thus, the versatility of the reactive DC magnetron sputtering that enables the growth of functional and homogeneous coatings in this case AlN films has been highlighted. To produce suitable films, however, it is necessary to identify the most favourable deposition parameters that maximize the sputtering yield, in order to get the optimal deposition rate: the sputter current that determines the rate of deposition process, the applied voltage that determines the maximum energy at which sputtered particles escape from target, the pressure into the chamber that determines the mean free path for the sputtered material, together with the target-substrate distance that both determines the number of collisions of particles on their way to the substrate, the gas mixture that determines the stoichiometry, the substrate temperature, all together influence the crystallinity, homogeneity and porosity of deposited films. As the physics behind the sputtering process and plasma formation is not simple, and many basic and technological aspects of the sputtering process and *AlN* film growth must be explored (anisotropic films, preferential growth, band gap changes), further investigation in this area is being conducted.

7. Acknowledgment

This work was sponsored by PAICyT-UANL, 2010.

8. References

Moreira, M.A.; Doi, I; Souza, J.F.; Diniz, J.A. (2011). Electrical characterization and morphological properties of AlN films prepared by dc reactive magnetron sputtering. Microelectronic Engineering, Vol. 88, No. 5, (May 2011), pp. 802-806, ISSN 0167-9317

Morkoç, H. (2008) *Handbook of Nitride Semiconductors and Devices*, Vol. 1: Materials Properties, Physics and Growth (2008) Weinheim Wiley-VCH. ISBN 978-3-527-40837-5, Weinheim, Germany.

Chaudhuri, J.; Nyakiti, L.; Lee, R.G.; Gu, Z.; Edgar, J.H.; Wen, J.G. (2007). Thermal oxidation of single crystalline aluminium nitride. *Materials Characterization*, Vol. 58, No. 8-9, (August 2007), pp. 672-679, ISSN 1044-5803

Chiu, K.H.; Chen, J.H.; Chen, H.R.; Huang, R.S. (2007). Deposition and characterization of reactive magnetron sputtered aluminum nitride thin films for film bulk acoustic wave resonator. *Thin Solid Films*, Vol. 515, No. 11, (April 2007), pp. 4819-4825, ISSN 0040-6090

Jang, K.; Lee, K.; Kim, J.; Hwang, S.; Lee, J.; Dhungel, S.K.; Jung, S.; Yi, J. (2006). Effect of rapid thermal annealing of sputtered aluminium nitride film in an oxygen ambient. *Mat Sci Semicon Proc*, Vol. 9, No. 6, (December 2006), pp. 1137-1141, ISSN 1369-8001

Kar, J.P.; Bose, G.; Tuli, S. (2006). A study on the interface and bulk charge density of AlN films with sputtering pressure. *Vacuum*, Vol. 81, No. 4, (November 2006), pp. 494-498, ISSN 0042-207X

Olivares, J.; González-Castilla, S.; Clement, M.; Sanz-Hervás, A.; Vergara, L.; Sangrador, J.; Iborra, E. (2007). Combined assessment of piezoelectric AlN films using X-ray diffraction, infrared absorption and atomic force microscopy. *Diamond & Related Materials*, Vol. 16, No. 4-7, (April 2007), pp. 1421-1424, ISSN 0925-9635

Prinz, G.M.; Ladenburger, A; Feneberg, M.; Schirra, M.; Thapa, S.B.; Bickermann, M.; Epelbaum, B.M.; Scholz, F.; Thonke, K.; Sauer, R. (2006). Photoluminescence, cathodoluminescence, and reflectance study of AlN layers and AlN single crystals. *Superlattices & Microstructures*, Vol. 40, No. 4-6, (December 2006), pp. 513-518, ISSN 0749-6036

Jose, F.; Ramaseshan, R.; Dash, S.; Tyagi, A.K.; Raj, B. (2010). Response of magnetron sputtered AlN films to controlled atmosphere annealing. Journal of Physics D: Applied Physics, Vol. 43, No. 7, (February 2010), pp. 075304-10, ISSN 0022-3727

Brien, V.; Pigeat, P. (2008). Correlation between the oxygen content and the morphology of AlN films grown by r.f. magnetron sputtering. *Journal of Crystal Growth*, Vol. 310, No. 16, (August 2008), pp. 3890-3895, ISSN 0022-0248

Jang, K.; Jung, S.; Lee, J.; Lee, K.; Kim, J.; Son, H.; Yi, J. (2008). Optical and electrical properties of negatively charged aluminium oxynitride films. *Thin Solid Films*, Vol. 517, No. 1, (November 2008), pp. 444-446, ISSN 0040-6090

Borges, J.; Maz, F.; Marques, L. (2010). AlNxOy thin films deposited by DC reactive magnetron sputtering. *Applied Surface Science*, Vol. 252, No. 257, No. 5, (December 2010), pp. 1478-1483, ISSN 0169-4332

Ianno, N.J.; Enshashy, H.; Dillon, R.O. (2002). Aluminum oxynitride coatings for oxidation resistance of epoxy films. *Surface and Coatings Technology*, Vol. 155, No. 2-3, (June 2002), pp. 130-135, ISSN 0257-8972

Erlat, A.G.; Henry, B.M.; Ingram, J.J.; Mountain, D.B.; McGuigan, A.; Howson, R.P.; Grovenor, C.R.M.; Briggs, G.A.D.; Tsukahara, T. (2001). Characterisation of aluminium oxynitride gas barrier films. *Thin Solid Films*, Vol. 388, No. 1-2, (June 2001), pp. 78-86, ISSN 0040-6090

Xiao, W.; Jiang X. (2004). Optical and mechanical properties of nanocrystalline aluminum oxynitride films prepared by electron cyclotron resonance plasma enhanced chemical vapor deposition. *Journal of Crystal Growth*, Vol. 263, No. 1-3, (March 2004), pp. 165-171, ISSN 0022-0248

Uchida, H.; Yamashita, M.; Hanaki, S.; Fujimoto, T. (2006). Structural and chemical characteristics of (Ti,Al)N films prepared by ion mixing and vapor deposition. *Vacuum*, Vol. 80, No. 11-12, (September 2006), pp. 1356-1361, ISSN 0042-207X

Sato, A.; Azumada, K.; Atsumori, T.; Hara, K. (2007). Characterization of AlN:Mn thin film phosphors prepared by metalorganic chemical vapor deposition. *Journal of Crystal Growth*, Vol. 298, (January 2007), pp. 379-382, ISSN 0022-0248

Takahashi, N.; Matsumoto, Y.; Nakamura, T. (2006). Investigations of structure and morphology of the AlN nano-pillar crystal films prepared by halide chemical vapor deposition under atmospheric pressure. *Journal of Physical and Chemistry of Solids*, Vol. 67, No. 4, (April 2006), pp. 665-668, ISSN 0022-3697

Brown, P.D.; Fay, M.; Bock, N., Marlafeka, M.; Cheng, T.S.; Novikov, S.V.; Davis, C.S.; Campion, R.P., Foxon, C.T. (2002). Structural characterisation of Al grown on group III-nitride layers and sapphire by molecular beam epitaxy. *Journal of Crystal Growth*, Vol. 234, No. 2-3, (January 2002), pp. 384-390, ISSN 0022-0248

Iwata, S.; Nanjo, Y.; Okuno, T.; Kurai, S.; Taguchi, T. (2007). Growth and optical properties of AlN homoepitaxial layers grown by ammonia-source molecular beam epitaxy. *Journal of Crystal Growth*, Vol. 301-302, (April 2007), pp. 461-464, ISSN 0022-0248

Lal, K.; Meikap, A.K. ; Chattopadhyay, S.K.; Chatterjee, S.K.; Ghosh, P.; Ghosh, M.; Baba, K.; Hatada, R. (2003). Frequency dependent conductivity of aluminium nitride films prepared by ion beam-assisted deposition. *Thin Solid Films*, Vol. 434, No. 1-2, (June 2003), pp. 264-270, ISSN 0040-6090

Matsumoto, T.; Kiuchi, M. (2006). Zinc-blende aluminum nitride formation using low-energy ion beam assisted deposition. *Nuclear Instruments and Methods in Physics Research*, Vol. 242, No. 1-2, (January 2006), pp. 424-426, ISSN 0168-583X

Ohring, M. (2002). *Materials Science of Thin Films: Deposition and Structure*, London Academic Press. ISBN 0-12-524975-6

Powder Diffraction File, JCPDS International Centre for Diffraction Data, PA, 1998 (www.icdd.com)

Brien, V.; Pigeat, P. (2007). Microstructures diagram of magnetron sputtered AlN deposits: Amorphous and nanostructured films. *Journal of Crystal Growth*, Vol. 299, No. 1, (February 2007), pp. 189-194, ISSN 0022-0248

Callister Jr, W.D. (2006). Materials Science & Engineering: an introduction. 6th edition (2006) Wiley & Sons. New York. ISBN 0471135763

Patterson, A.L. (1939). The Scherrer Formula for X-Ray Particle Size Determination. *Physical Review*, Vol. 56, No. 1, (July 1939), pp. 978-982, ISSN 1098-0121

Guo, Q.X.; Tanaka, T.; Nishio, M.; Ogawa, H. (2006). Growth properties of AlN films on sapphire substrates by reactive sputtering. *Vacuum*, Vol. 80, No. 7, (May 2006), pp. 716-718. ISSN 0042-207X

Swanepoel, R. (1983). Determination of the thickness and optical constants of amorphous silicon. *Journal of Physics E: Scientific Instruments*, Vol. 16, No. 12, (May 1983), pp. 1214-1222. ISSN 0022-3735

Zong, F.; Ma, H.; Du, W.; Ma, J.; Zhang, X.; Xiao, H.; Ji, F.; Xue, Ch. (2006). Optical band gap of zinc nitride films prepared on quartz substrates from a zinc nitride target by reactive rf magnetron sputtering. *Applied Surface Science*, Vol. 252, No. 22, (September 2006), pp. 7983-7986.ISSN 0169-4332

Guerra, J.A.; Montañez, L.; Erlenbach, O.; Galvez, G.; de Zela, F.; Winnacker, A.; Weingärtner, R. (2011). Determination of optical band gap of thin amorphous SiC and AlN films produced by RF magnetron sputtering. *Journal of Physics: Conference series*, Vol. 274, (September 2010), pp. 012113-012118, ISSN 1742-6588

Moulder, J.F.; Sticke, W.F.; Sobol, P.E.; Bomben, K.D. (1992) "Handbook of X-ray Photoelectron Spectroscopy". 2nd edition. Perkin-Elmer, Physical Electronics Division. Eden Prairie. ISBN 0962702625

Stanca, I. (2004). Chemical structure of films grown by AlN laser ablation: an X-ray photoelectron spectroscopy stydy. *Romanian Journal of Physics*, Vol. 49, No. 9-10, (May 2004), pp. 807-816, ISSN 1221-146X

Sohal, R.; Torche, M.; Henkel, K.; Hoffmann, P.; Tallarida, M.; Schmeiber, D. (2006). Al-oxynitrides as a buffer layer for Pr2O3/SiC interfaces. *Materials Science in Semiconductor Processing*, Vol. 9, No. 6, (December 2006), pp. 945-948, ISSN 1369-8001

Wang, P.W.; Sui, S.; Wang, W.; Durrer, W. (1997). Aluminum nitride and alumina composite film fabricated by DC plasma processes. *Thin Solid Films*, Vol. 295, No. 1-2, (February 1997), pp. 142-146, ISSN 0040-6090

Gredelj, S.; Gerson, A.R.; Kumar, S.; Cavallaro, G.P. (2001). Characterization of aluminium surfaces with and without plasma nitriding by X-ray photoelectron spectroscopy. *Applied Surface Science*, Vol. 174, No. 3-4, (April 2001), pp. 240-250, ISSN 0169-4332

Richthofen, A. von; Domnick, R. (1996). Metastable single-phase polycrystalline aluminium oxynitride films grown by MSIP: constitution and structure. *Thin Solid Films*, Vol. 283, No. 1-2, (September 1996), pp. 37-44, ISSN 0040-6090

Robinson, K.S.; Sherwood, P.M.A. (1984). X-Ray photoelectron spectroscopic studies of the surface of sputter ion plated films. *Surface and Interface Analysis*, Vol. 6, No. 6, (December 1984), pp. 261-266, ISSN: 0142-2421

Huttel, Y.; Bourdie, E.; Soukiassian, P.; Mangat, P.S.; Hurych, Z. (1993). Promoted oxidation of aluminum thin films using an alkali metal catalyst. *Journal of Vacuum Science and Technol A*, Vol. 11, No. 4, (July 1993), pp. 2186-2192, ISSN 0734-2101

Gutierrez, A.; Lopez, M.F.; Garcia-Alonso, C.; Escudero, M. In: I. Olefjord, L. Nyborg, D. Briggs, *7th European Conf. Applications on Surface & Interface Analysis*, Göteborg (1997) 1035-1038, John Wiley & Sons, Canada. ISBN 0471978272

Zhang, J.X.; Cheng, H.; Chen, Y.Z.; Uddin, A.; Yuan, S.; Geng, S.J.; Zhang, S. (2005). Growth of AlN films on Si (100) and Si (111) substrates by reactive magnetron sputtering. *Surface and Coatings Technology*, Vol. 198, No. 1-3, (August 2005), pp. 68-73, ISSN 0257-8972

Whangbo M.H.; Hoffmann, R. (1978). The band structure of the tetracyanoplatinate chain. *J. Am. Chem. Soc.*, Vol. 100, No. 19, (September 1978), pp. 6093-6098, ISSN 0002-7863

Hoffmann, R. (1963). An Extended Hückel Theory I Hydrocarbons. *J. Chem. Phys.*, Vol. 39, No. 6, (September 1963), pp. 1397-1413, ISSN 0021-9606

Landrum, G. A. YAeHMOP package:
http://overlap.chem.Cornell.edu:8080/yaehmop.html

Galván, D.H. (1998). Extended Hückel Calculations on Cubic Boron Nitride and Diamond. *Journal of Materials Science Letters*, Vol 17, No. 10 (May 2008), pp. 805-810, ISSN 1573-4811

Ferreira da Silva, A.; Souza Dantas, N.; de Almeida, J.S.; Ahuja, R.; Person C. (2005). Electronic and optical properties of würtzite and zinc-blende TlN and AlN. *Journal of Crystal Growth*, Vol. 281, No. 1, (July 2005), pp. 151-160, ISSN 0022-0248

Rezaei, B.; Asgari, A.; Kalafi, M. (2006). Electronic band structure pseudopotential calculation of wurtzite III-nitride materials. *Physica B*, Vol. 371, No. 1, (January 2006), pp. 107-111, ISSN 0921-4526

Persson, C.; Ahuja, R.; Ferreira da Silva, A.; Johansson, B. (2005). First-principle calculations of optical properties of wurtzite AlN and GaN. *Journal of Crystal Growth*, Vol. 231, No. 3, (October 2001), pp. 407-414, ISSN 0022-0248

García-Méndez, M.; Morales-Rodríguez, S.; Galván, D.H.; Machorro, R. (2009). Characterization of AlN thin films fabricated by reactive DC sputtering: experimental measurements and Hückel calculations. *International Journal of Modern Phisics B*, Vol. 33, No. 9, (April 2009), pp. 2233-2251, ISSN: 0217-9792

2

Crystal Growth Study of Nano-Zeolite by Atomic Force Microscopy

H. R. Aghabozorg, S. Sadegh Hassani and F. Salehirad
Research Institute of Petroleum Industry
Iran

1. Introduction

Mesoporous materials possess highly ordered periodic arrays of uniformly sized channels. Therefore, these compounds have attracted considerable attention for many researchers (Luo et al., 2000; Xu & Xue, 2006; Liu et al., 2009 a, Salehirad & Anderson, 1996; 1998a; 1998b; 1998c). Since the shape and texture of the materials strongly affect their properties, these materials due to their large surface areas, controllable pore size and easy functionalization have been used in separation science, drug delivery and various processes such as adsorption and catalysis (Liu et al., 2009a; Duan et al., 2008; Liu et al., 2009 b; Liu & Xue, 2008; Mohanty & Landskron, 2008; Salehirad, 2004; Alibouri, 2009a; 2009b).

Zeolites as crystalline aluminosilicates are microporous materials with well-defined channels and cavities have been used as ion-exchangers, adsorbents, heterogeneous catalysts and catalyst supports (Zhang et al., 2000; Zhang et al., 2000). Many parameters are found to be highly important in the physical appearance of final zeolite products (Aghabozorg et al., 2001). Morphology and crystal size of these compounds have an important role for their specific applications in industries (Zabala Ruiz et al., 2005). The ability to accommodate various organic and inorganic species in zeolites makes them ideal host materials for supramolecular organization (Bruhwiler & Calzaferri, 2004; Schulz-Ekloff et al., 2002; Sadegh Hassani et al., 2010a). In many cases, surface adsorption sites in zeolites lead to interesting photochemical properties for these compounds. (Zabala Ruiz et al., 2005; Hashimoto, 2003). Studying the surface structure of zeolites and dynamic phenomena occurring on the surface of these compounds under various conditions will accelerate the development of these compounds as catalyst or materials with versatile functions (Sugiyama Ono et al., 2001; Anderson, 2001; Sadegh Hassani et al., 2010a). Nucleation mechanisms and growth of zeolites are not well understood for many of the systems involved. In most cases, the problem is compounded with the presence of a gel phase. This gel also undergoes a continuous polymerization type reaction during nucleation and growth. Improved understanding of zeolite growth should enable a more targeted approach to zeolite synthesis in the future and may ultimately lead to the possibility of zeolite design to order. Numerous techniques have been used for study the structure of the zeolites. Among them, scanning probe microscopy (SPM) is an appropriate technique that can be applied in atmospheric condition. However, many SPM techniques can be used in essentially any environment, including ambient, UHV, organic solvent vapour and biological buffer, making it possible to observe the system in states that are simply inaccessible to other techniques with comparable resolution and they can be also used over a wide range of temperatures (Hobbs, et al., 2009).

Scanning probe microscopy, such as scanning tunneling microscopy (STM) and atomic force microscopy (AFM), has become a standard technique for obtaining topographical images of surface with atomic resolution (Hyon et al., 1999; Klinov & Magonov, 2004; Giessibl, 1995, Sadegh Hassani et al., 2088 a,b; 2010b). In addition, they may be used in many applications such as investigation of mechanical, chemical, electrical, magnetic and optical properties of surfaces, study of friction and adhesion forces, modifying a sample surface, crystal growth study and process controlling (Sundararajan & Bhushan, 2000; Burnham et al., 1991; Aime et al., 1994; Sadegh Hassani & Ebrahimpoor Ziaie, 2006; Ebrahimpoor Ziaie et al., 2008; Sadegh Hassani & Sobat, 2011; Sadegh hassani & Aghabozorg, 2011; Magonov, et al., 1997; Hobbs, et al., 2009; Leggett, et al., 2005; Williams, et al., 1999; Cadby, et al., 2005). There are some disadvantages for these techniques because they are limited to a surface view of a sample. Hence, extreme care has to be taken for interpretation of data for achieving the bulk properties (Franke, 2008; Hobbs, et al., 2009).These nondestructive methods do not require complex preparation of the desired sample and allows processes to be followed in-situ (Hobbs, et al., 2009).

This technique enables very high resolution imaging of nonconducting surfaces (Fonseca Filho et al., 2004) with the ability to measure the height of the surface very accurately and observe zeolite growth features not detectable by conventional methods, i.e. SEM, TEM, etc. (Sadegh Hassani et al., 2010). AFM have been used to image the surface of zeolites such as scolecite, stilbite, faujasite, heulandite and mordenite, since 1990.

In addition, by this method it is possible to follow in situ processes such as crystallization which need non-destructive methods. In addition, the effect of changes in temperature on structures can be monitored at the nanometer scale (Hoobs et al., 2009).

In this chapter, it is focused on using AFM for current progress toward the elucidation of zeolite growth. In this regard, after introducing AFM technique, an overview about controlling of crystal growth of materials, especially zeolites, by AFM is presented.

2. AFM technique

Atomic force microscope is a kind of microscope in which a sharp tip is mounted at the end of a spring cantilever of known spring constant. This microscope employs an optical detection system in which a laser beam is focused onto the backside of a reflective cantilever and is reflected from the cantilever onto a position sensitive photo detector. An image can be obtained based on the interaction between a desired sample and a tip. As the tip scans the surface of the sample, variation in the height of the surface is easily measured as flexing of the cantilever, then variation in the photodiode signal. This gives a 3-D profile map of surface topography.

There are feedback mechanisms that enable the piezo-electric scanners to maintain the tip at a constant force (to obtain height information), or height (to obtain force information) above the sample surface. AFM can be used not only for imaging the surfaces in atomic resolution but also for measuring the forces at nano-newton scale. The force between the tip and the sample surface is very small, usually less than 10^{-9} N. The detection system does not measure force directly. It senses the deflection of the micro cantilever (Ogletree et al., 1996; Sadegh Hassani et al., 2006; 2008c).

Imaging modes of operation for an AFM are dynamic and static modes. In the dynamic mode, including non contact and tapping modes, the cantilever is externally oscillated at or close to its fundamental resonance frequency. The oscillation amplitude, phase and resonance frequency are changed by tip-sample interaction forces. These changes in oscillation with respect to the external reference oscillation provide information about the sample's characteristics. Using dynamic mode, it is possible to monitor both the phase of the drive signal oscillating the cantilever, and the cantilever's response. This phase signal gives access to material properties, a combination of adhesive and viscoelastic properties (Tamayo& Garcia, 1998; Tamayo & Garcia, 1998; Hoobs et al,. 2009) with nanoscale resolution.

In non contact mode, the tip of the cantilever does not contact the sample surface. The cantilever is instead oscillated at a frequency slightly above its resonant frequency where the amplitude of oscillation is typically a few nanometers. The van der Waals forces, or any other long range force which extends above the surface acts to decrease the resonance frequency of the cantilever. This decrease in resonant frequency combined with the feedback loop system maintains a constant oscillation amplitude or frequency by adjusting the average distance between tip and sample surface. By measuring this distance at each point, a topographic image of the sample surface can be obtained.

In tapping mode, the cantilever is driven to oscillate up and down at near its resonance frequency similar to non contact mode. As the tip is approached to the surface the amplitude of oscillation is damped, and it is now the amplitude of the oscillation that is used as a feedback parameter.

In static mode, the cantilever is scanned across the surface of the sample and the topography of the surface are obtained directly using the deflection of the cantilever. In this operation mode, the static tip deflection is used as a feedback signal. Close to the surface of the sample, attractive forces can be quite strong, causing the tip to "snap-in" to the surface. Thus static mode AFM is almost always done in contact where the overall force is repulsive. Consequently, this technique is typically called "contact mode". In contact mode, the force between the tip and the surface is kept constant during scanning by maintaining a constant deflection.

3. Monitoring processes with AFM

The advent of atomic force microscopy (AFM) (Binnig et al. 1986) has provided the new possibilities to investigate the nanometer-sized events occurring at crystal surfaces during crystal growth and recrystallization. This is possible under ex situ and often in situ conditions as it can be suited to observe surfaces under solution. Under ex situ conditions a wide variety of synthetic parameters can be changed but some careful works such as quenching experiments must be performed before transferring the sample to AFM to prevent secondary processes caused by changing growth conditions.

Real-time images of growing crystals have provided the structural details revealing terrace growth, spiral growth, defect and intergrowth structure in a vast variety of growth studies (Mcpherson et al. 2000). The free energy for individual growth processes can be achieved by measuring real-time micrographs at a range of temperatures.

Over the last decade, the application of AFM to follow some processes in soft material has been reported. Hoobe et al. (2009) believed that scanning probe techniques have several capabilities that make them suitable for the investigation of soft materials, organic materials as they pass through a transformation, and directly observe processes at a near molecular scale. For example, in a macromolecular system, such as semicrystalline polymer, phase transitions of large molecule sometimes are in metastable states. To follow these non-equilibrium states which often control the final material properties, is important and can be performed by AFM (Strobl, 2007; Hoobs et al., 2009).

In the early study of soft material using AFM, contact mode was used (Magonov et al., 1997). The development of dynamic modes of operation, such as tapping mode and the subsequent development of phase imaging, allowed a substantial growth in the possibilities of the technique (Garcia & Perez, 2002; Hoobs et al., 2009). For example, atomic force microscope with tapping mode was applied to study the surface morphology of as-grown (111) silicon-face 3C-SiC mesaheterofilms (Neudeck et al., 2004). Their observation showed that wide variations in 3C surface step structure are as a function of film growth conditions and film defect content. The vast majority of as-grown 3C-SiC surfaces consisted of trains of single bilayer height (0.25 nm) steps. They reported that Macrostep formation (i.e., step-bunching) was rarely observed. As supersaturation is lowered by decreasing precursor concentration, terrace nucleation on the top (111) surface becomes suppressed.

AFM technique also has been used for study the structure of the zeolites. In this chapter, it is focused on crystal growth processes in open framework, inorganic materials (i.e. zeolites) studied by AFM. The zeolite crystal growth experiment is important to enhance the understanding of growth of zeolite crystals and nucleation, and controlling the defects in zeolites.

Many studies have been performed to investigate how zeolite crystals grow. Based on the results reported from these studies it is generally found that the growth linearly proceeds during the crystallization of most zeolites and this is applicable for both gel and clear solution syntheses (Subotic et al. 2003; Zhdanov et al. 1980; Bosnar et al. 1999; Iwasaki et al. 1995; Cundy et al. 1995; Bosnar et al. 2004; Cora et al. 1997; Kalipcilar et al. 2000; Schoeman et al. 1997; Caputo et al. 2000). Parameters such as gel composition, aging, stirring and temperature can affect on growth in zeolites (Subotic et al. 2003; Cundy et al. 2005). Imaging of zeolite surface by use of such a very powerful technique, has recently made possible the understanding the exact growth mechanism in the synthesis of these materials. The work performed on cleaved surface (0 0 1) of the natural zeolite scolecite (MacDougall et al. 1991) under ambient conditions has revealed the arrangement of 8-membered ring centers. Parameters such as lattice constants, angles and distances of the zeolite structure have been measured and compared favorably with its crystallographic structure. Feathered terraces with height ~9 Å, half the unit cell dimension of 17.94 Å, have been reported on the cleaved surface (010) of natural zeolite heulandites (Scandella et al. 1993). The outer surface of this zeolite investigated by AFM (Binder et al. 1996) has also revealed growth spiral at screw dislocation with the pitch of ~9 Å (sometimes double), which is consistent with the b-dimension of the zeolite. Further AFM works on heulandite (Yamamoto et al. 1998) crystals has revealed the presence of steps, suggesting a possible birth-and-spread mechanism.

The precipitated sodium aluminosilicate hydrogel has also been analyzed by AFM (Kosanovic et al. 2008). The obtained results have showed that predominantly true amorphous phase of the gel contained small proportions of partially crystalline (quasi-crystalline) or even fully crystalline phase. Some different methods such as FTIR, DTG, electron diffraction have confirmed this finding and also showed partially or even fully crystalline entities of the sample. AFM has also revealed that the particles of the partially or fully crystalline phase are nuclei for further crystallization of zeolite.

Anderson et al. (1996; 1998) have reported the first AFM study of synthetic zeolite Y and observed triangular (1 1 1) facet, revealing approximately triangular terraces. The height of the terraces is 15 Å which is consistent with one faujasite sheet and the step height observed by other techniques. By measuring the area of the terraces they showed that this area was growing at a constant rate which was consistent with a pseudo terrace-ledge-kink mechanism in comparison with the growth in dense phase structures.

Zeolite A due to water softening features is an industrially important zeolite. It has been studied to determine the mechanism of crystal growth, using AFM by some researchers (Agger et al. 1998; 2001; Sugiyama et al. 1999). This zeolite, similar to zeolie Y demonstrates a layer-type growth with terrace heights consistent with simple fractions of unit cell. The authors have reported terrace height of 12 Å, equivalent to half unit cell height consisting of a sodalite cage and double 4-ring (Agger et al. 1998; 2001). Sugiyama et al. (1999) have also reported values corresponding to the individual sodalite cage and double 4-ring with slight differences. Wakihara et al. (2005) have investigated the surface structure of zeolite A by AFM and compared their results with those obtained by the other techniques such as HRTEM and FE-SEM. They have found that the terminal structure of zeolite A is incomplete sodalite cages. These results support one of the terminal structures proposed by Sugiyama et al. (1999), although these findings may not be always applicable to all zeolites A synthesized by various methods.

Irregardless of these differences, the principal conclusions of both studies (Agger et al. 1998; Sugiyama et al. 1999) are the same and a layer-by-layer growth mechanism is operative for zeolite A (Agger et al.1998). The area of the terraces in this zeolite grows at a constant rate confirmed by the parabolic cross-section of the surface. Agger et al. (1998) have also reported a detailed simulation of the growth features in zeolite A and suggested that the rate of growth at kink site is the fastest growth process. They also report that since the terrace edges run parallel to the crystal edges so the rate of growth at edge or ledge sites is considerably less than that at kink sites. These findings of the relative rates of growth processes help to understand and then affect the relative rates to control defects and structure.

Umemura et al. (2008) have presented a computer program that simulates morphology as well as surface topology for zeolite A crystals. They compared favorably the simulation results with those obtained from AFM images on the {1 0 0}, {1 1 0} and {1 1 1} faces of synthetic crystals.

Slilicalite is a siliceous form of zeolite ZSM-5 which is industrially important for catalytic properties, so the control of the synthesis of this inorganic solid is of great interest. An extensive AFM study of crystal growth has been fulfilled for this material (Agger et al. 2001; 2003; Anderson et al. 2000). In the low temperature synthesis of this material (similar to zeolites A and Y) terraced, layer-by-layer growth has been observed on both the (1 0 1) and (1 0 0) facets. The height of the terraces is 10 Å corresponding to half the unit cell dimension in the [0 1 0] direction (or the height of one pentasil chain). A constant-area-deposition mechanism, not dominated by addition at kink site, has resulted in the approximately circular shape of the terraces, indicating no preferential growth direction. Terraces in the high temperature synthesis, which produces large crystal (not similar to the low temperature synthesis, producing small crystals), grow towards the crystal edges and have the height of several hundred angstroms (up to 110 nm high on the (0 1 0) face and up to 20 nm high on the (1 0 0) face). Such terraces have no relation to any structural element of the silicalite. Assuming of a layer growth mechanism the authors have concluded that an obstruction to terrace advance causes a build-up of the layers. They have suggested the defect inclusion mechanism to explain the relative terrace heights on the two faces.

4. Morphology study of Zeolite L by AFM

Zeolite L which was initially determined by Barrer et al. (1969) has hexagonal symmetry (Barrer et al., 1969; Baerlocher et al., 2001) with two-dimensional pores of about 0.71 nm aperture leading to cavities of about $0.48 \times 1.24 \times 1.07$ nm^3 and the Si/Al ratio is typically 3.0 (Pichat et al., 1975; Sig Ko & Seung Ahn, 1999). The zeolite crystals consist of cancrinite cages linked by double six-memberd rings, forming columns in the c direction. Connection of these columns gives rise to 12-membered rings with a free diameter varies from 0.71 nm (narrowest part) to 1.26 nm (widest part). The main channels are linked via non-planar eight membered rings forming an additional two-dimensional channel system with openings of about 0.15 nm. Studies have shown that the morphology of the crystals can be approximated by a cylinder, with the entrances of the main channels located at the base. The number of channels is equal to $0.265(dc)^2$, where dc is the diameter of the cylinder in nanometers (Zabala Ruiz et al., 2005; Bruhwiler & Calzaferri, 2005; Breck& Acara, 2005).

Brent and Anderson (2008) studied crystal growth mechanism in zeolite L and control the crystal habit by atomic force microscopy. They claimed that AFM was an excellent tool for determining crystal growth mechanisms in zeolites L and gave a snapshot in time as to how the shape of a crystal had developed, by imaging surface features. In their work, the surface features on both the (1 0 0) side walls, and the (0 0 1) hexagonal faces of zeolite L was investigated.

Sadegh Hassani et al. (2010a) reported synthesis and characterization of nano zeolite L. Nanosized zeolite L was synthesized from a gel mixture at 443 K with different aging times. The molar compositional ratio of the resulting gel was 7.6 Na$_2$O : 7.2 K$_2$O : 1.3 Al$_2$O$_3$: 40 SiO$_2$: 669 H$_2$O. This homogeneous gel mixture was transferred to a Teflon-lined autoclave and placed in an air-heated oven at 443K for different synthesis times (24, 45, 110, 160, and 200 h). The autoclave was removed from the oven at the scheduled times and quenched in cold water. The solid product was separated by centrifugation, washed thoroughly a few times with deionized water and oven dried at 353 K for 5 h.

The samples were characterized by XRD, XPS techniques and morphological changes investigated by TEM and AFM.

A commercial atomic force microscope (Solver P47 H, NT-MDT Company) operating in non-contact mode and equipped with a NSG11 cantilever was used to take images in nanometer scale. Samples were dispersed in ethanol by sonication and deposited on a suitable substrate to be applicable for AFM. Atomic force microscope was used to obtain detailed surface images, such as crystal dimensions, by zooming in on a fine particle.

The technique revealed the existence of a multitude of terraces with the height of either ≈ 1 or ≈ 2 nm. Figure 1.a shows a two dimensional image of the zeolite crystals after elapsing 24 h of the synthesis time. This image confirms again the hexagonal geometry of the zeolite crystal. Figure 1.b is a cross sectional profile of the image a. The terraces demonstrate growth direction; consequently, growth fronts develop a hexagonal profile. They are concentric, growing out to the crystal edge from a central nucleation point as shown in figure 1.c. It has been reported that the growth morphology is thermodynamically related to corresponding crystallographic structure, according to the chemical bonding theory of single crystal growth (Xu & Xue, 2006; Yan, 2007; Xue et al., 2009). Work carried out on synthetic zeolite crystals to date suggests crystal growth occurs through deposition and subsequent expansion of layers (Xu & Xue, 2006; Anderson & Agger, 1998). Their findings are in good agreement with this suggestion and show that zeolite L crystals grow via a layer mechanism. Further observation of terraces at atomic level, using a single crystal will help this suggestion.

AFM allowed the detailed observation of nanometer-size events at crystal surfaces. In addition, the images showed layer growth of the zeolite crystal and the height of terraces. Two-dimensional AFM images (Fig. 1.a) showed hexagonal structure which is in good agreement with the TEM results (Fig. 2). Furthermore, Three-dimensional structure of the zeolite crystal (Fig. 1.c) obtained by AFM (not possible by TEM) indicated hexagonal layers. In addition, Figure 3 exhibits the aggregation of zeolite L crystals.

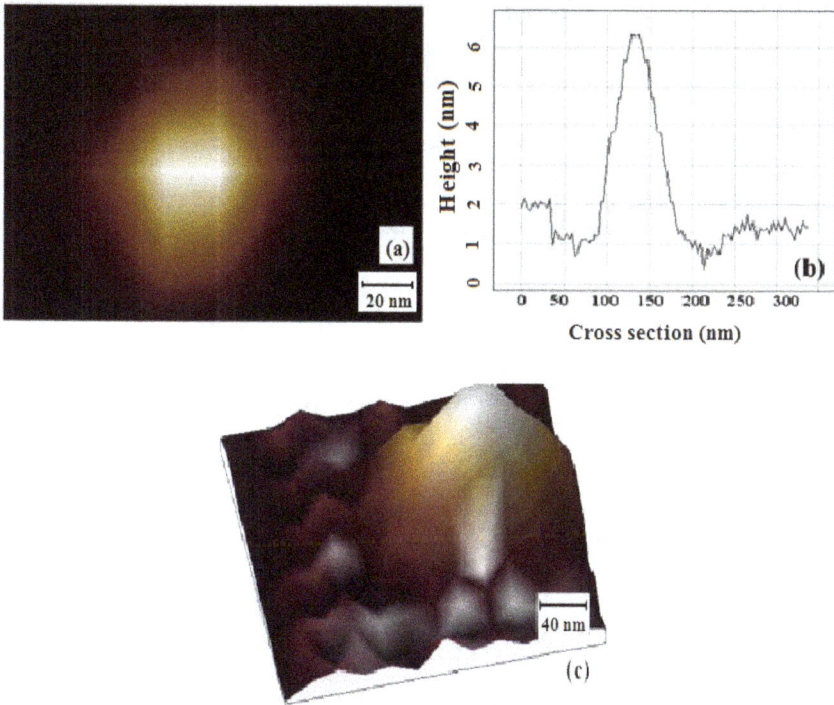

Fig. 1. AFM images of synthesized zeolite L. (a) two dimensional image of the crystal after 24 h and (b) its cross section, (c) three dimensional image of the crystal after 160 h (Sadegh Hassani et al., 2010a).

TEM images of the samples were recorded on CM260-FEG-Philips microscope and samples dispersed in acetone by sonication and deposited on a microgride. Figure 2 shows TEM images of the synthesized samples with various magnifications. These images indicate the average size of the sample crystallites (about 50 nm) possessing hexagonal geometry. A detailed surface image of the zeolite particles (Fig. 2c-d) indicated the parallel one dimensional channels arranged in a uniform pattern with hexagonal symmetry.

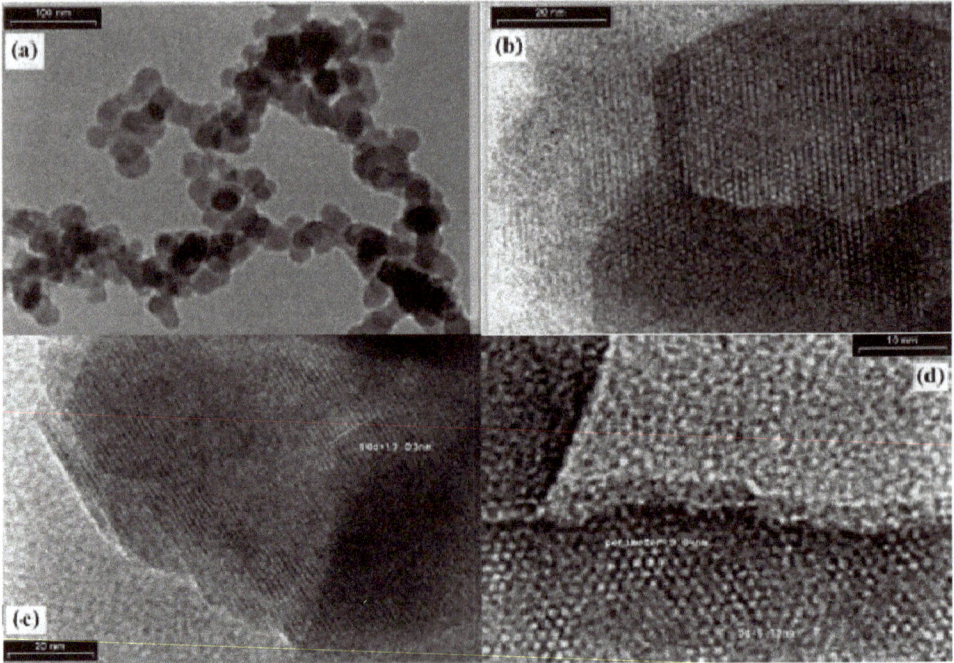

Fig. 2. TEM images of as-synthesized zeolite L with different resolutions for (a) the particle size and (b) morphology of the sample and (c) and (d) detailed surface images of the zeolite (Sadegh Hassani et al., 2010a).

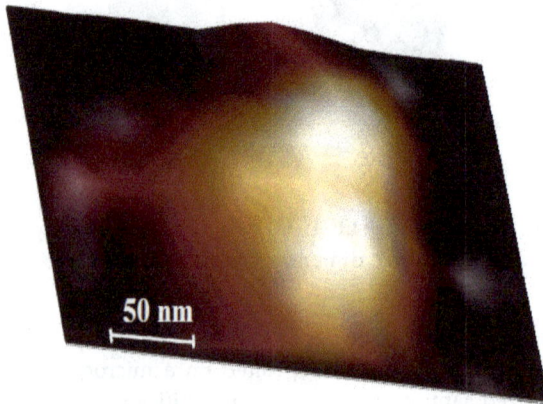

Fig. 3. Three dimensional AFM image of as synthesized zeolite L prepared in 160 h exhibiting aggregation of crystals.

In addition, AFM results exhibited bigger size of the crystal by increasing the synthesis time up to 160 h. Beyond this synthesis time, the size of the crystal decreased (Fig. 4). However, X-Ray diffraction patterns of the samples indicated that the synthesis times up to 110 h maintain almost the same crystallinity, whereas the synthesis times longer than that cause to decrease the crytallinity (89% for 160 h and 63% for 200 h). The apparatus used for XRD study was the powder X-ray diffractometer, Philips PW-1840, with a semi conductor detector and a Ni-filtered Kα (Cu) radiation source attachment.

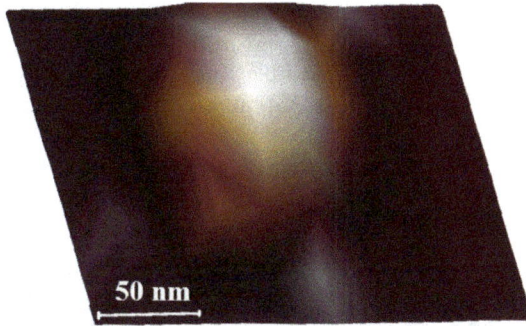

Fig. 4. Three dimensional AFM image of as synthesized zeolite L prepared in 200 h.

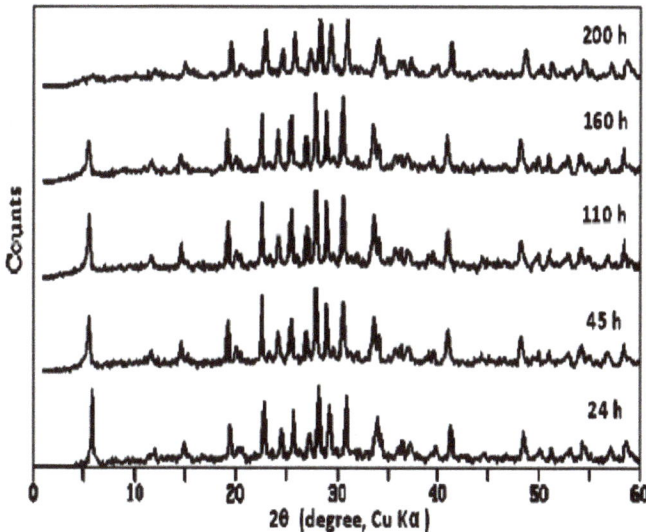

Fig. 5. X-ray diffraction patterns of the synthesized zeolite L obtained at different crystallization time at 443 K (Sadegh Hassani et al., 2010a).

Figure 5 shows the X-ray diffraction patterns of the as synthesized zeolite L samples obtained with different crystallization times. Characteristic XRD peaks showed that the fully crystalline phase (97% crystallinity) was obtained after 24 h. Reflections located at 2θ ≈ 5.5,

19.4, 22.7, 28.0, 29.1 and 30.7 were used to calculate crystallinity. Results show that the synthesis times up to 110h maintain almost the same crystallinity, whereas the synthesis times longer than that cause to decrease the crytallinity (89% for 160 h and 63% for 200 h).

It was reported that Si/Al molar ratio in zeolites structure could affect on morphology and crystal size of these compounds (Shirazi et al., 2008; Mintova et al., 2006; Celik et al., 2010). Therefore, Sadegh hassani et al. (2010a) using AAS and XPS performed the elemental analyses of the bulk and surfaces of the as synthesized nanozeolite L, respectivly. Elemental surface analysis of the zeolite sample was carried out on the X-ray Photoelectron Spectroscopy (XR3E2 Model-VG Microtech; Concentric Hemispherical Analyzer, EA 10 plus Model-Specsis).

Figure 6 shows the XPS spectrum of the sample. The spectrum depicts the surface analysis of the as synthesized nanozeolite L. The Si/Al ratios of the gel mixture, bulk and surface of the sample are shown in table 1.

The slight difference in Si/Al ratios (0.5) was observed between bulk and surface of the zeolite sample.

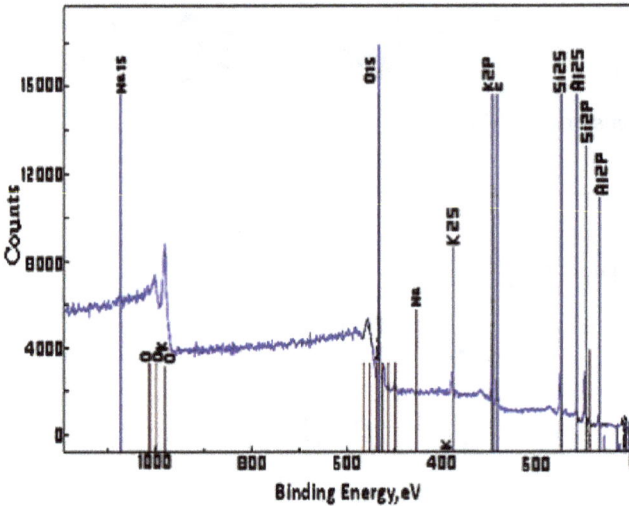

Fig. 6. Surface analysis results of synthesized zeolite L crystal by X-ray photoemission spectroscopy. (Sadegh Hassani et al., 2010a).

Sample	Si/Al
Gel mixture	15.4
Zeolite L (bulk)	3.5
Zeolite L (surface)	3.0

Table 1. Si/Al molar ratios of samples (Sadegh Hassani et al., 2010a).

5. Conclusion

This chapter is focused on the crystal growth study of various samples especially zeolites using AFM. Studies are revealed that atomic force microscopy is a powerful technique to follow in situ processes such as zeolite crystallization. In this regard, a study of crystal growth of nano-zeolite L is focused using atomic force microscopy (AFM). The results are compared with those of obtained from X-ray diffraction (XRD), X-ray photoelectron spectroscopy (XPS) and transmission electron microscopy (TEM) techniques.

TEM and two-dimensional AFM images indicate that the zeolite particles are in a nano-range and they have hexagonal structure. In addition, the AFM images show layer growth of the zeolite crystal and reveal the existence of a multitude of terraces with the height of either ≈1 or ≈2 nm.

The terraces demonstrate growth direction; consequently, growth fronts develop a hexagonal profile. They are concentric, growing out to the crystal edge from a central nucleation point. In addition, the AFM images exhibit layer growth, the height of terraces and the aggregation of zeolite L crystals. A detailed surface image of the zeolite particles indicate the parallel one dimensional channels arranged in a uniform pattern with hexagonal symmetry. AFM results also show bigger size of the crystal by increasing the synthesis time up to 160 h, beyond this synthesis time, the size of the crystal decrease.

6. References

Agger, J. R.; Hanif, N.; Cundy, C. S.; Wade, A. P.; Dennison, S.; Rawlinson, P. A. & Anderson, M. W. (2003). Silicalite crystal growth investigated by atomic force microscopy. *J. Am. Chem. Soc.* 125(3):830-9.

Agger, JR; Hanif, N; Anderson, MW. (2001). Fundamental zeolite crystal growth rate from simulation of atomic force microscopy. *Angew Chem Int* Ed, 40, 4065-4067.

Agger, J. R.; Pervaiz, N.; Cheetham, A. K. & Anderson, M. W. (1998). Crystallization in zeolite A studied by atomic force microscopy. *J. Am. Chem. Soc.* 120, 10754-410759.

Agger, JR; Hanif, N; Anderson, MW.; Terasaki, O. (2001). Growth Models in Microporous Materials *Microporouse Mesoporouse Mater.* 48, 1-9.

Agger, JR; Hanif, N; Anderson, MW; Terasaki, O. (2001). Crystal growth in framework materials. *Solid State Science.* 3. 809-819.

Aghabozorg, H. R.; Ghassemi, M. R.; Salehirad, F. et al. (2001). Investigation of parameters affecting zeolite NaA crystal size and morphology. *Iranian J. Cryst. Min.* 8, 2, 107-116.

Alibouri, M.; Ghoreishi, S. M. & Aghabozorg, H .R. (2009a). Hydrodesulfurization activity of using NiMo/Al-HMS nanocatalyst synthesized by supercritical impregnation. *Ind. Eng. Chem. Res.* 48, 4283-4292.

Alibouri, M.; Ghoreishi, S. M. & Aghabozorg, H. R. (2009b). Hydrodesulfurization of dibenzothiophene using CoMo/Al-HMS nanocatalyst synthesized by supercritical deposition. *The Journal of Supercritical Fluids* 49, 239-248.

Anderson, M. W.; Agger, J. R.; Thornton, J. T. & Forsyth, N. (1996). Crystal Growth in Zeolite Y Revealed by Atomic Force Microscopy. *Angewandte Chemie, International Edition in English.* 35, (11), 1210-1213.

Anderson, MW; Agger, JR; Pervaiz, N; Weigel, SJ; Cheetham, AK. (1998). Zeolite crystallization and transformation determined by atomic force microscopy. In: Treacy et al.; editor. Proceedings of the 12th International Zeolite Conference. Baltimore. MRS. 1487-94.

Anderson, M. W. (2001). Surface microscopy of porous materials. *Curr. Opinion Solid State Mat. Sci.* 5, 407 -415.

Anderson M. W. & Agger, J. A. (1998). Proc. Int. Zeolite 12th Conf., Vol. 3 (Materials Research Society, Warrendale, PA).

Anderson, MW; TErasaki, O. (2000). Understanding and Utilising novel microporous and mesoporous catalysts. In: Orafao, JM; Faria, JL; Figueiredo, JL. Editor. Actas do XVII Simposio Ibero-americano de Catalise. Porto. FEUP. 37-8.

Baerlocher, C.; Meier, W. M. & Olson, D. H. (2001) Atlas of Zeolite Framework Types (Elsevier, Amsterdam).

Barrer, R. M. & Villiger, H. Z. (1969). Crystal structure of synthetic zeolite L. *Z. Kristallogr.* 128, 352 -370.

Binder, G; Scandella, L; Schumacher, A; Kruse, N; Prins, R. (1996) Microtopographic and molecular scale observations of zeolite surface structures: Atomic force microscopy on natural heulandites. *Zeolites*, 16, 2-6.

Binnig, G.; Quate, CF.; Gerber, C. (1986). Atomic force microscopy. *Phys Rev Lett.* 56. 930-3.

Bosnar, S.; Subotic, B. (1999). Microporous Mesoporous Mater. 28. 483-493.

Bosnar, S.; Antonic, T.; Bronic, J.; Subotic, B. (2004). *Microporous Mesoporous Mater.* 76. 157-165.

Breck, D. W. & Acara, N. A. (1965). crystalline zeolite L. *U. S. Patent*: 3,216,789.

Brent, R. & Anderson, M. W. (2008). Fundamental Crystal Growth Mechanism in Zeolite L Revealed by Atomic Force Microscopy. *Angew. Chem.* 120, 5407 –5410.

Bruhwiler, D. & Calzaferri, G. (2004). Molecular sieves as host materials for supramolecular organization. *Micropor. Mesopor. Mat.* 72, 1-23.

Bruhwiler, D. & Calzaferri, G. (2005). Selective functionalization of the external surface of zeolite L. *Comptes Rendus Chimie.* 8, 391-398.

Cadby, A.; Dean, R.; Fox, A. M.; Jones, R. A. L. & Lidzey, D. G. (2005). Mapping the fluorescence decay lifetime of a conjugated polymer in a phase-separated blend using a scanning near-field optical microscope. *Nanoletters* . 5(11), 2232-7.

Caputo, D.; Gennaro, B. D.; Liguori, B.; Testa, F.; Carotenuto, I.; Piccolo, C. (2000). *Mater. Chem. Phys.* 66. 120-125.

Celik, F. E.; Kim, T. J. & Bell, A. T. (2010). Effect of zeolite framework type and Si/Al ratio on dimethoxymethane carbonylation. *Journal of Catalysis.* 270, 185–195.

Cleveland, J. P.; Anczykowski, B.; Schmid, A. E.; & Elings V. B. (1998). Energy dissipation in tapping-mode atomic force microscopy. *Appl. Phys. Letts.* 72(20), 2613-2615.

Cora, I.; Streletzky, K.; Thompson, R. W.; Phillies, G. D. J.(1997). *Zeolites.* 18. 119-131.

Cundy, C. S.; Henty, M. S.; Plaisted, R. J. (1995). Zeolite Synthesis Using a semi-continuous Reactor (part I). *Zeolites.* 15. 353-372.

Cundy, C. S.; Henty, M. S.; Plaisted, R. J. (1995). Zeolite Synthesis Using a Semi-continuous Reactor (part II). *Zeolites.* 15. 400-407.

Cundy, C. S.; Cox, P. A. (2005). The Hydrothermal Synthesis of Zeolites: Precursors, Intermediates and Reaction Mechanism. *Microporous Mesoporous Mater.* 82, 1-78.

Duan, G.; Zhang, Ch.; Li, A.; Yang, X.; Lu, L. & Wang, X. (2008). Preparation and Characterization of Mesoporous Zirconia Made by Using a Poly (methyl methacrylate). *Nanoscale Res. Lett.* 3, 118 -122.

Ebrahimpoor Ziaie, E.; Rachtchian, D. & Sadegh Hassani, S. (2008). Atomic force microscopy as a tool for comparing lubrication behavior of lubricants. *Materials Science: An Indian Journal.* 4, 2, 111-115.

Fonseca Filho, H. D.; Mauricio, M. H. P.; Ponciano, C. R. & Prioli, R. (2004). Metal layer mask patterning by force microscopy lithography. *Material Science and Engineering B.*, 112, 194-199.

Franke, M. & Rehse, N. (2008). Three-Dimensional Struc ture Formation of Polypropylene Revealed by in Situ Scanning Force Microscopy and Nanotomography. *Macromolecules* 41, 163-166.

Garcia, R. & Perez, R. (2002). Dynamic atomic force microscopy methods. *Surf. Sci. Rep.* (2002) 47, 197-301.

Giessibl, F. (1995). Atomic Resolution of the Silicon (111)-(7x7) Surface by Atomic Force Microscopy. *Science* 267, 68-71.

Hashimoto, S. (2003). Zeolite photochemistry: impact of zeolites on photochemistry and feedback from photochemistry to zeolite science. *J. Photochem. Photobio. C.* 4, 19 -49.

Hobbs, J. K. McMaster, T.J. Miles, M. J. & Barham, P.J. (1998). Direct observations of the growth of spherulites of poly(hydroxybutyrate-co-valerate) using atomic force microscopy. *Polymer.* 39(12), 2437-2446.

Hobbs, J. K. (1997). PhD Thesis, University of Bristol.

Hobbs, J. K. (2006). Progress in understanding polymer crystallization. Springer-Verlag, Berlin, Lecture Notes in Physics, Eds G Reiter and G Strobl.

Iwasaki, A; Hirata, M.; Kudo, I.; Sano, T.; Sugawara, S.; Ito, M.; Watanabe, M. (1995). In-situ measurement of crystal growth of rate of zeolite, *Zeolites.* 15, 308-314.

Hobbs, J. K.; Mullin, N.; Weber, Ch. H. M.; Farrance O. E. & Vasilev, C. (2009). Watching processes in soft matter with SPM. 12, 7-8, 26-33.

Kalipcilar, H.; Culfaz, A. (2000). Synthesis of Submicron Silicalite-1 Crystals from Clear Solutions. *Cryst. Res. Tecnol.* 35. 933-942.

Klinov, D. & Magonov, S. N. (2004). True molecular resolution in tapping-mode atomic force microscopy with high-resolution probes. *Appl. Phys. Lett.* 84, 2697-2699.

Kosanovic, C.; Bosnar, S.; Subotic, B.; Svetlicic, V.; Misic, T.; Drazic, G. & Havancsak, K. (2008). Study of the microstructure of amorphous aluminosilicate gel before and after its hydrothermal treatment. *Microporous and Mesoporous Materials.* 110, 2-3, 177-185.

Leggett, G. J.; Brewer N. J. & Chong K. S. L. (2005). Friction force microscopy: towards quantitative analysis of molecular organisation with nanometre spatial resolution. *Phys. Chem. Chem. Phys.* 7(6), 1107-1120.

Liu, J. & Xue, D. (2008). Thermal Oxidation Strategy towards Porous Metal Oxide Hollow Architectures. *Adv. Mater.* 20, 2622-2627.

Liu, J.; Liu, F.; Gao, K.; Wu, J. & Xue, D. (2009). Recent developments in the chemical synthesis of inorganic porous capsules. *J. Mat. Chem.* 19, 6073-6084.

Liu, J.; Xia, H.; Xue, D. & Lu, L.(2009b). Double-Shelled Nanocapsules of V2O5-Based Composites as High-Performance Anode and Cathode Materials for Li Ion Batteries. *J. Am. Chem. Soc.* 131, 12086 -12087.

Luo, Q. L.; Xue, Li, Z. & Zhao, D. (2000). Synthesis of nanometer-sized mesoporous oxides. *Studies Surf. Sci. Catal.* 129, 37-41.

MacDougall, J. E.; Cox, Sh. D.; Stucky, G. D.; Weisenhom, A. L.; Hansma, P. K.; Wise, W. S. (1991). Molecular resolution of zeolite surfaces as imaged by atomic force microscopy. *Zeolites.* 11, 426-427.

Magonov, S. N., & Reneker, D. H. (1997). Characterization of polymer surfaces with Atomic force microscopy. *Ann. Rev. Mater. Sci.* 27, 175-222.

Magonov, S.; Elings, N. V. & Papkov, V. S. (1997). AFM Study of Thermotropic Structural Transitions of Poly(diethylsiloxane). *Polymer.* 38, 297-307

Mcpherson, A.; Malkin, AJ.; Kuznetsov, YG. (2000). Atomic force microscopy in the study of macromolecular crystal growth. *Ann Rev Biophys Biomol Struct.* 29.361-410.

McMaster, T. J., Hobbs, J. K., Barham, P. J., Miles, M. J., (1997). AFM study of in situ real time polymer crystallization and spherulite structure, *Probe Microscopy.* 1, 43-56.

Mintova, S.; Valtchev, V.; Onfroy, T.; Marichal, C.; Knozinger, H. & Bein,T. (2006). Variation of the Si/Al ratio in nanosized zeolite Beta crystals. *Microporous and mesoporous materials.* 90, 1-3, 237-245.

Mohanty, P. & Landskron, K. (2008). Periodic Mesoporous Organosilica Nanorice. *Nanoscale Res. Lett.* 4, 169-172.

Neudeck, Ph. G.; Trunek, A. J. & Anthony Powell, J. (2004). Atomic Force Microscope Observation of Growth and Defects on As-Grown (111) 3C-SiC Mesa Surfaces. *Mat. Res. Soc. Symp. Proc.* Vol. 815, J5.32.1-J5.32.6.

Pichat, P.; Frannco Parra, C. & Barthomeuf, D. (1975). Infra-red structural study various type L zeolites. *J. Chem. Soc. Faraday Trans.* 71, 991-996.

Sadegh Hassani, S. & Ebrahimpoor Ziaie, E. (2006). Application of Atomic Force Microscopy for the study of friction properties of surfaces. *material science: An Indian journal,* 2(4-5), 134-141.

Sadegh Hassani, S.; Sobat, Z. & Aghabozorg, H. R. (2008a). Nanometer-Scale Patterning on PMMA Resist by Force Microscopy Lithography. *Iran. J. Chem. Chem. Eng.* Vol. 27, No. 4, 29-34.

Sadegh Hassani, S.; Sobat, Z. & Aghabozorg, H. R. (2008b). Scanning probe lithography as a tool for studying of various surfaces. *Nano Science and Nano Technology: An Indian journal,* Volume 2 Issue (2-3),94-98.

Sadegh Hassani, S.; Salehirad, F.; Aghabozorg, H. R. & Sobat, Z. (2010a). Synthesis and morphology of nanosized zeolite L. *Cryst. Res. Technol.* 45, No. 2, 183-187. (© 2010 WILEY-VCH Verlag GmbH & Co. KGaA, Weinheim)

Sadegh Hassani, S.; Sobat, Z. & Aghabozorg, H. R. (2010b). Force nanolithography on various surfaces by atomic force microscope. *Int. J. Nanomanufacturing,* Vol. 5, Nos. 3/4, 217-224.

Sadegh Hassani, S. & Sobat, Z. (2011). Studying of various nanolithography methods by using Scanning Probe Microscope. *Int .J. Nano .Dim.* 1(3), 159-175.

Sadegh Hassani, S. & Aghabozorg, H. R. (2011). Nanolithography study using scanning probe microscope, *Lithography* book , In TechWeb Publisher, in press.

Salehirad, F. & Anderson, M. W. (1996). Solid-State ^{13}C MAS NMR Study of Methanol-to-Hydrocarbon Chemistry over H-SAPO-34. *J. catal.*164, 301-314.

Salehirad, F. & Anderson, M. W. (1998a). Solid-State NMR studies of Adsorption Complexes and Surface Methoxy Groups on Methanol- Sorbed Microporous Materials. *J. Catal.* 177, 189-207.

Salehirad, F. & Anderson, M. W. (1998b). NMR studies of methanol-to-hydrocarbon chemistry-part1Primary products and mechanistic considerations using a wide-pore catalyst. *J. Chem. Soc., Faraday Trans.* 94, 2857-2866.

Salehirad, F. & Anderson, M. W. (1998c). Solid-state NMR study of methanol conversion over ZSM-23, SAPO-11 and SAPO-5 molecular sieves-part 2. *J. Chem. Soc., Faraday Trans.* 94, 2857-2866.

Salehirad, F.; Aghabozorg, H. R.; Manoochehri, M. & Aghabozorg, H. (2004). Synthesis of titanium silicalite-2 (TS-2) from methyamine-tetrabutylammonium hydroxide media. *Catalysis Communication.* 5 , 359-365.

Scandella, L; Kruse, N; Prins, R. (1993). Imaging of zeolite surface-structures by atomic force microscopy. *Surf Sci.* 281. L331-4.

Schoeman, B. J. (1997). Progress in Zeolite and Microporous Materials. Parts A-C. 647-654.

Schulz-Ekloff, G.; Wohrle, D.; Van, B. (2002). Chromophores in porous silicas and minerals: preparation and optical properties. *Micropor. Mesopor. Mat.* 51, 91-138.

Shirazi, L.; Jamshidi, E. & Ghasemi, M. R. (2008). The effect of Si/Al ratio of ZSM-5 zeolite on its morphology, acidity and crystal size. *Cryst. Res. Technol.* 43, 12, 1300-1306.

Sig Ko, Y. & Seung Ahn, W. (1999). Synthesis and characterization of zeolite. *Bull. Korean Chem. Soc.* 20, 2, 1-6.

Strobl, G. (2007). The Physics of Polymers: Concepts for understanding their structures and behaviour, Spinger.

Subotic, B; Bronic, J. (2003). In handbook of Zeolite Science and Technology (eds Auer bach, S. M.; Carrado, KA; Dutta, P. K.). Marcel Dekker New York. 1184.

Sugiyama Ono, S.; Matsuoka, O. & Yamamoto, S. (2001). Surface structures of zeolites studied by atomic force microscopy. *Micropor. Mesopor. Mat.* 48, 103-110.

Sugiyama, S.; Yamamoto, S.; Matsuoka, O.; Nozoye, H.; Yu, J.; Zhu, G.; Qiu,, S. & Terasaki, O. (1999). AFM observation of double 4-rings on zeolite LTA crystal surface. *Microporous Mesoporous Mater.* 28, 1-7.

Sugiyama, S; Yamamoto, S; Matsuoka, O; Nozoye, H; Yu, J; Zhu, G; et al. (1999). AFM observation of Double 4-rings on zeolite LTA crystal surface. *Microporous Mesoporous Mater.* 28. 1-7.

Tamayo, J. & Garcia, R. (1998). Relationship between phase shift and energy dissipation in tapping- mode atomic force microscopy. *Appl. Phys. Lett.* 73(20), 2926-2928.

Umemura, A.; Cubillas, P.; Anderson, M. W. & Agger, J. R. (2008). *Studies in Surface Science and Catalysis.* 705-708.

Wakihara, T.; Sasaki, Y.; Kato, H.; Ikuhara, Y. & Okubo, T. (2005). Investigation of the surface structure of zeolite A. *Phys. Chem. Chem. Phys.* 7, 3416–3418.

Williams, C. C. (1999). Two-Dimentional Dopant Profiling by Scanning Capacitance Microscopy*Ann Rev. Mater. Sci.* 29, 471-504.

Xu, J. & Xue, D. J. (2006). Fabrication of copper hydroxyphosphate with complex architectures. *Phys. Chem. B* 110, 7750 – 7756.

Xu, D. & Xue, D. (2006). Chemical bond analysis of the crystal growth of KDP and ADP. *J. Cryst. Growth* 286, 108-113.

Xue, D.; Yan, X. & Wang, L. (2009). Production of specific $Mg(OH)_2$ granules by modifying crystallization conditions. *Powder Technol.* 191, 98-106.

Yamamoto, S.; Sugiyama, S.; Matsuoka, O.; Honda, T.; Banno, Y.; Nozoye, H. (1998). AFM imaging of the surface of natural heulandite. *Microporous Mesoporous Mater.* 21. 1-6.

Yan, X.; Xu, D. & Xue, D. (2007). SO_4^{2-} ions direct the one- dimensional growth of $5Mg(OH)_2 \bullet MgSO_4 \bullet 2H_2O$ *Acta Materl.* 55, 5747 -5757.

Zabala Ruiz, A.; Bruhwiler, D.; Ban, T. and Calzaferri, G. (2005). Synthesis of zeolite L. Tuning size and morphology. *Monatshefte für Chemie.* 136, 77-89.

Zhang, X.; Zhang, Z.; Suo, J. & Li, Sh. (2000). Synthesis of mesoporous silica molecular sieves via a novel templating scheme Original Research Article; *Studies Surf. Sci. Catal.* 129, 23.

Zhdanov, S. P.; Samulevich, N. N. (1980). Proceedings of the 5th International Conference on zeolites. 75-84.

Three-Scale Structure Analysis Code and Thin Film Generation of a New Biocompatible Piezoelectric Material MgSiO$_3$

Hwisim Hwang, Yasutomo Uetsuji and Eiji Nakamachi
Doshisha University
Japan

1. Introduction

In this study, three subjects were investigated for a new biocompatible piezoelectric material generation:

1. Development of a numerical analysis scheme of a three-scale structure analysis and a process crystallographic simulation.
2. Design of new biocompatible piezoelectric materials.
3. Generation of MgSiO$_3$ thin film by using radio-frequency (RF) magnetron sputtering system.

Until now, lead zirconate titanate (Pb(Zr,Ti)O$_3$: PZT) has been used widely for sensors (Hindrichsena et al., 2010), actuators (Koh et al., 2010), memory devices (Zhang et al., 2009) and micro electro mechanical systems (MEMS) (Ma et al., 2010), because of its high piezoelectric and dielectric properties. The piezoelectric thin film with aligned crystallographic orientation shows the highest piezoelectric property than any polycrystalline materials with random orientations. Sputtering (Bose et al., 2010), chemical or physical vapor deposition (CVD or PVD) (Tohma et al., 2002), pulsed laser deposition (PLD) (Kim et al., 2006) and molecular beam epitaxy (MBE) (Avrutin et al., 2009) are commonly used to generate high performance piezoelectric thin films. Lattice parameters and crystallographic orientations of epitaxially grown thin films on various substrates can be controlled by these procedures. K. Nishida et al. (Nishida et al., 2005) generated [001] and [100]-orientated PZT thin films on MgO(001) substrate by using CVD method. They succeeded to obtain a huge strain caused by the two effect: the synergetic effect of [001] orientation with the piezoelectric strain; and the strain effect of [100] orientation caused by switching under conditions of the external electric field. Additionally, PZT-based piezoelectric materials, such as Pb(Zn$_{1/3}$Nb$_{2/3}$)O$_3$-PbTiO$_3$ (Geetika & Umarji, 2010) and PbMg$_{1/3}$Nb$_{2/3}$O$_3$-PbTiO$_3$ (Kim et al., 2010), have also been developed.

However, lead, which is a component of PZT-based piezoelectric material, is the toxic material. The usage of lead and toxic materials is prohibited by the waste electrical and electronic equipment (WEEE) and the restriction on hazardous substances (RoHS).

For alternative piezoelectric materials of the PZT, lead-free piezoelectric materials have been studied. J. Zhu et al. (Zhu et al., 2006) generated [111]-orientated BaTiO$_3$ on LaNiO$_3$(111) substrate, which had a crystallographic orientation with maximum piezoelectric strain constants. S. Zhang et al. (Zhang et al., 2009) doped Ca and Zr in $BaTiO_3$ and succeeded in generating the piezoelectric material with high piezoelectric properties. Further, P. Fu et al. (Fu et al., 2010) doped La_2O_3 in Bi-based $(Bi_{0.5}Na_{0.5})_{0.94}Ba_{0.06}TiO_3$ and succeeded in generating a high performance piezoelectric material. However, their goals were to develop an environmentally compatible piezoelectric material, and the biocompatibility of their piezoelectric materials has not been investigated. Therefore, their piezoelectric materials could not be applied for Bio-MEMS devices.

Recently, the Bio-MEMS, which can be applied to the health monitoring system and the drug delivery system, is one of most attractive research subject in the development of the nano- and bio-technology. Therefore, the biocompatible actuator for the micro fluidic pump in Bio-MEMS is strongly required. However, they remain many difficulties to design new biocompatible materials and find an optimum generation process. Especially, it is difficult to optimize the thin film generation process because there are so many process factors, such as the substrate material, the substrate temperature during the sputtering, the target material and the pressure in a chamber. Therefore, the numerical analysis scheme is necessary to design new materials and optimize the generation process.

The analysis scheme based on continuum theory is strongly required, due to time consuming experimental approach such as finding an optimum sputtering process and a substrate crystal structure through enormous experimental trials. The analysis scheme should predict the thin film deformation, strain and stress, which are affected by the imposed electric field and are constrained by the substrate.

Until now, the conventional analysis schemes, such as the molecular dynamics (MD) method (Rubio et al, 2003) and the first-principles calculation based on the density functional theory (DFT) (Lee & Chung, 2006), have been applied to the crystal growth process simulations. The MD method has been used mainly to analyze the crystal growth process of pure atoms. J. Xu et al. (Xu & Feng, 2002) calculated the Ge growth on Si(111). In the cases of the perovskite compounds, the MD method has been applied to analyze the phase transition, the polarization switching and properties of crystal depending on temperature and pressure. J. Paul et al. (Paul et al., 2007) analyzed the phase transition of $BaTiO_3$ caused by rising temperature and S. Costa et al. (Costa et al., 2006) analyzed the one of $PbTiO_3$ caused by rising temperature and pressure. However, the reliability of its numerical results is poor due to its uncertain inter-atomic potentials for the various combinations of atoms. The MD method could not predict the differences of poly-crystal structures and material properties caused by changing combinations of the crystals and the substrates. It can be concluded that the conventional MD method has many problems for the crystal growth prediction of perovskite compounds grown on the arbitrarily selected substrates.

On the other hand, the DFT can treat interactions between electrons and protons, therefore the reliable inter-atomic potentials can be obtained. The first-principles calculations based on the DFT were applied to the epitaxial growth of the ferroelectric material by O. Diegueaz et al. and I. Yakovkin et al.. O. Diegueaz et al. (Diegueaz et al., 2005) evaluated the stress increase and the polarization change caused by the lattice mismatch between a substrate and a thin film crystal, such as $BaTiO_3$ and $PbTiO_3$. Similarly, I. Yakovkin et al. (Yakovkin & Gutowski, 2004) has investigated in the case of $SrTiO_3$ thin film growth on Si substrate. However, these analyses adopted limited assumptions, such as fixing the conformations of thin film crystals and the growth orientations on the substrates. In this conventional algorithm, the grown orientation is determined by the purely geometrical lattice mismatch between thin films and substrates. This algorithm is not sufficient to predict accurately the preferred orientation of the thin film.

In order to generate the new piezoelectric thin film, a crystal growth process of the thin film should be predicted accurately. The stable crystal cluster of the thin film, which consists geometrically with substrate crystal, is grown on the substrate. Generally, the crystal cluster is an aggregate of thin film crystals. Their morphology and orientations were varied according to the combination of the thin film and the substrate crystals. Therefore, the numerical analysis scheme of the crystal growth process, which can find the best combination of the thin film and the substrate crystal, is strongly required, to optimize the new piezoelectric thin film.

In this chapter, following contents are discussed to develop the new biocompatible $MgSiO_3$ piezoelectric thin film.

1. The three-scale structure analysis algorithm, which can design new piezoelectric materials, is developed.

2. The best substrate of the $MgSiO_3$ piezoelectric thin film is found by using the three-scale structure analysis code.

3. The $MgSiO_3$ thin film is grown on the best substrate by using the RF magnetron sputtering system, and piezoelectric properties are measured.

4. An optimum generating condition of the $MgSiO_3$ piezoelectric thin film is found by using the response surface method.

Section 2 provides the description to the algorithm of the three-scale structure analysis code on basis of the first-principles calculation, the process crystallographic simulation and the crystallographic homogenization theory. Section 3 provides the best substrate of the new biocompatible $MgSiO_3$ piezoelectric thin film calculated by the three-scale structure analysis code. In section 4, the optimum generating condition of $MgSiO_3$ piezoelectric thin film is found. Finally, conclusions are given in section 5.

2. A three-scale structure analysis code

This section describes the physical and mathematical modelling of the three-scale structure and the numerical analysis scheme of three-scale structure analysis to characterize and design epitaxially grown piezoelectric thin films. The existing two-scale finite element analysis is the effective analysis tool for characterization of existing piezoelectric materials. This is because virtually or experimentally determined crystal orientations can be employed for calculation of piezoelectric properties of the macro continuum structure (Jayachandran et al., 2009). However, it can not be applied to a new piezoelectric material, due to unknown crystal structure and material properties.

Figure 1 shows the schematic description of the three-scale modelling of a new piezoelectric thin film, which is grown on a substrate. It shows the three-scale structures, such as a "crystal structure", a "micro polycrystalline structure" and a "macro continuum structure". In the crystal structure analysis, stable structures and crystal properties are evaluated by using the first-principles calculation. Preferred orientations and their fraction are calculated by using the process crystallographic simulation in the micro polycrystalline structure analysis. The macro continuum structure analysis provides the piezoelectric properties of the thin film by using the finite element analysis on basis of the crystallographic homogenization theory. Therefore, the three-scale structure analysis can predict the epitaxial growth process of not only the existent piezoelectric materials but also the new ones.

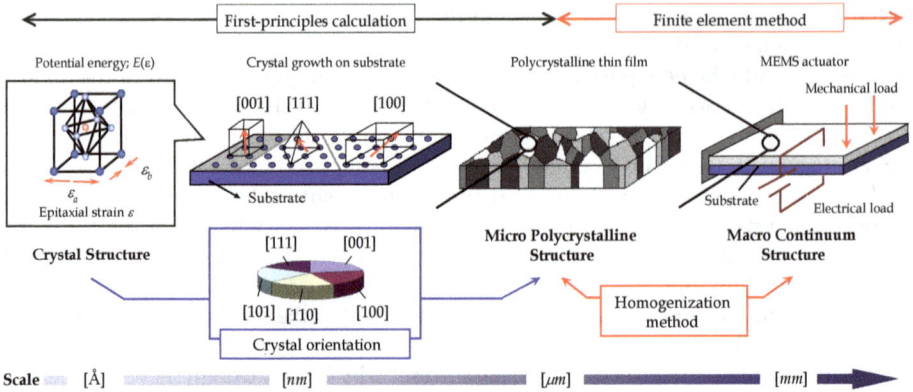

Fig. 1. Three-scale modelling of piezoelectric thin film as a process crystallography.

2.1 Crystal structure analysis by using the first-principles calculation
2.1.1 Stable crystal structure analysis
The stable structures of the perovskite cubic are calculated by the first-principles calculation based on the density functional theory (DFT) by using the CASTEP code (Segal et al., 2002). The stable structures are, then, computed using an ultra-soft pseudo potential method under the local density approximation (LDA) for exchange and correlation terms. A plane-wave basis set with 500eV cutoff energy is used and special k-points are generated by a 8x8x8 Monkhorst-pack mesh (Monkhorst et al., 1976).

The perovskite-type compounds ABX_3 provide well-known examples of displacive phase transitions. They are in a paraelectric non-polar phase at high temperature and have a cubic crystal structure (lattice constant $a = c$). The cubic crystal structure consists of A cations in the large eightfold coordinated site, B cations in the octahedrally coordinated site, and X anions at equipoint. The stability of cubic crystal structure can be estimated by an essential geometric condition, tolerance factor t. If ion radiuses of A, B and X are indicated with r_A, r_B, r_X, tolerance factor t can be described as

$$t = \frac{r_A + r_X}{\sqrt{2}\left(r_B + r_X\right)} \tag{1}$$

When tolerance factor t consists in the range from 0.75 to 1.10, the perovskite-type crystal structure has high stability. The cubic crystal structure often distorts to ferroelectric phase of lower symmetry at decreased temperature, which is a tetragonal crystal structure ($a > c$) with spontaneous polarization. These ferroelectric distortions are caused by a soft-mode of phonon vibration in cubic crystal structure, and it brings to good piezoelectricity. The soft-mode can be distinguished from other phonon vibration modes with negative eigenfrequency, and the transitional crystal structure depends strongly on the eigenvectors. Consequently, new biocompatible piezoelectric materials are searched according to the flowchart in Fig. 2. Firstly, biocompatible elements are inputted to A and B cations while halogens and chalcogens are set to X anion for the perovskite-type compounds. The combination of three elements is determined to satisfy the stable condition of the tolerance factor. The stable cubic structure of perovskite-type oxides is calculated to minimize the total energy. Next, the phonon vibration in the stable cubic structure is analyzed to catch the soft-mode which causes a phase transition from the paraelectric non-polar phase (cubic structure) to the ferroelectric phase (tetragonal structure). When the eigenfrequency of phonon vibration is positive, it is considered that the cubic structure is the most stable phase and does not change to other phase. On the other hand, in case that the eigenfrequency is negative, the cubic structure is guessed to be an unstable phase and change to other phase corresponding to the soft-mode. Additionally, if all eigenvectors of constituent atoms are parallel to c direction in crystallographic coordinate system, it is supposed to change from cubic to tetragonal structure. If not, it is supposed to transit to other structures except tetragonal one. On the base of phonon properties, the stable tetragonal structure with minimum total energy is searched using the eigenvector components for the initial atomic coordinates.

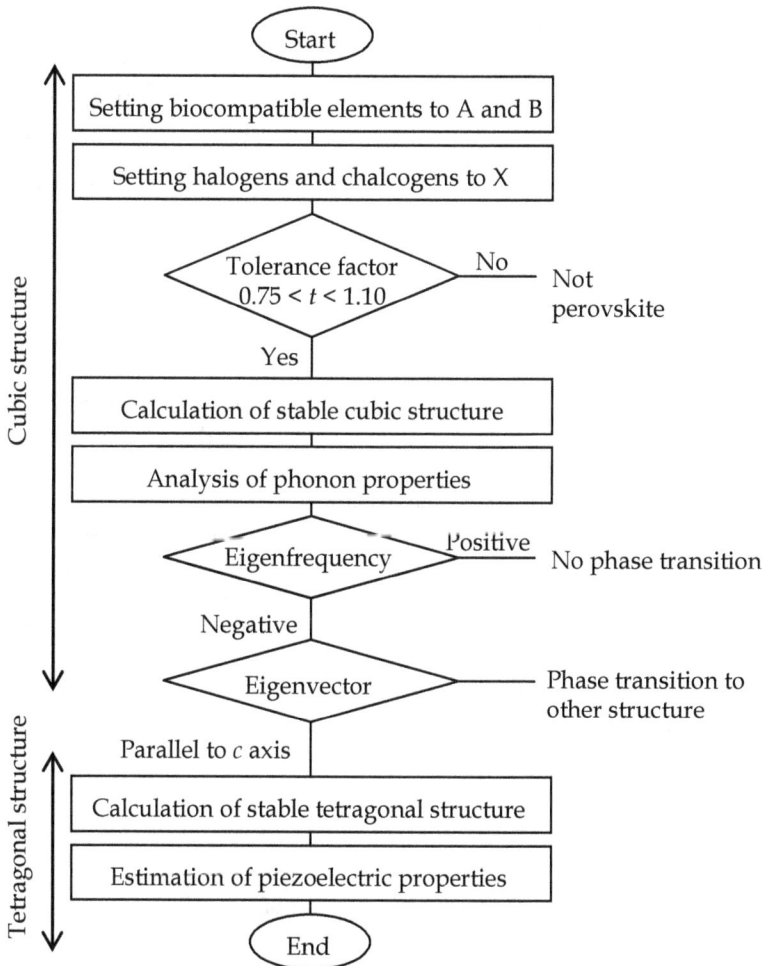

Fig. 2. The flowchart of searching new piezoelectric materials by the first-principles DFT.

Recently, many perovskite cubic crystals such as $SrTiO_3$ and $LaNiO_3$ have been reported. However, most of these materials could not be transformed into a tetragonal structure below Curie temperature, because most of perovskite cubic crystals are more stable than tetragonal crystals. Therefore, the tetragonal structure indicates a soft-mode of the phonon oscillation in cubic structure. Lattice parameters and piezoelectric constants of the tetragonal structure are calculated using the DFT.

2.1.2 Characterization of piezoelectric constants

The total closed circuit (zero field) macroscopic polarization of a strained crystal P_i^T can be described as,

$$P_i^T = P_i^S + e_{iv}\varepsilon_v \qquad (2)$$

where P_i^S is the spontaneous polarization of the unstrained crystal (Szabo et al., 1998, 1999). Under Curie temperature, ferroelectric crystal with tetragonal structure has a polarization along the c axis. The three independent piezoelectric stress tensor components are $e_{31} = e_{32}$, e_{33} and $e_{15} = e_{24}$. $e_{31} = e_{32}$ and e_{33} describe the zero field polarization induced along the c axis, when the crystal is uniformly strained in the basal a-b plane or along the c axis, respectively. $e_{15} = e_{24}$ measures the change of polarization perpendicular to the c axis induced by the shear strain. This latter component is related to induced polarization by $P_1 = e_{15}\varepsilon_5$ and $P_2 = e_{15}\varepsilon_4$. The total induced polarization along c axis can be described by a sum of two contributions.

$$P_3 = e_{33}\varepsilon_3 + e_{31}\left(\varepsilon_1 + \varepsilon_2\right) \qquad (3)$$

where $\varepsilon_1 = (a - a_0)/a_0$, $\varepsilon_2 = (b - b_0)/b_0$ and $\varepsilon_3 = (c - c_0)/c_0$ are strains along the a, b and c axes, respectively, and a_0, b_0 and c_0 are lattice parameters of the unstrained structure.

The electronic part of the polarization is determined using the Berry's phase approach (Smith & Vandelbilt, 1993), a quantum mechanical theorem dealing with a system coupled under the condition of slowly changing environment. One can calculate the polarization difference between two states of the same solid, under the necessary condition that the crystal remains an insulator along the path, which transforms the two states into each other through an adiabatic variation of a crystal Hamiltonian parameter λ_H. The magnitude of the electronic polarization of a system in state λ_H is defined only modulo eR/Ω, where R is a real-space lattice vector, Ω the volume of the unit cell, and e the charge of electron. In practice, the eR/Ω factor can be eliminated by careful inspection, in the condition where the changes in polarization are described as $|\Delta P| << |eR/\Omega|$. The electronic polarization can be described as,

$$P^{el}\left(\lambda_H\right) = -\frac{2e}{(2\pi)^3} \int_{BZ} d\mathbf{k}\, \frac{\partial}{\partial \mathbf{k}'}\phi^{(\lambda_H)}\left(\mathbf{k}, \mathbf{k}'\right)\Bigg|_{\mathbf{k}'=\mathbf{k}} \qquad (4)$$

where the integration domain is the reciprocal unit cell of the solid in state λ_H and $\phi^{(\lambda_H)}$ is quantum phase defined as phases of overlap-matrix determinants constructed from periodic parts of occupied valence Bloch states $v_n^{(\lambda_H)}(\mathbf{k})$ evaluated on a dense mesh of $\mathbf{k}$ points from $\mathbf{k}_0$ to $\mathbf{k}_0+\mathbf{b}$, where $\mathbf{b}$ is the reciprocal lattice vector.

$$\phi^{(\lambda_H)}(\mathbf{k},\mathbf{k}') = \text{Im}\left\{\ln[\det\langle v_m^{(\lambda_H)}(\mathbf{k})\left|v_m^{(\lambda_H)}(\mathbf{k}')\rangle\right]\right\} \tag{5}$$

The electronic polarization difference between two crystal states can be described as,

$$\Delta P^{\text{el}} = P^{\text{el}}(\lambda_{H2}) - P^{\text{el}}(\lambda_{H1}) \tag{6}$$

Common origins to determine electronic and core parts are arbitrarily assigned along the crystallographic axes. The individual terms in the sum depend on the choice, however, the final results are independent of the origins.

The elements of the piezoelectric stress tensor can be separated into two parts, which are a clamped-ion or homogeneous strain u, and a term that is due to an internal strain such as relative displacements of differently charged sublattices

$$e_{iv} = \left.\frac{\partial P_i^T}{\partial \varepsilon_v}\right|_u + \sum_k \left.\frac{\partial P_i^T}{\partial u_{k,i}}\right|_\varepsilon \frac{\partial u_{k,i}}{\partial \varepsilon_v} \tag{7}$$

where P_i^T is the total induced polarization along the ith axis of the unit cell.

Equation (7) can be rewritten in terms of the clamped-ion part and the diagonal elements of Born effective charge tensor.

$$e_{iv} = e_{iv}^{(0)} + \sum_k \frac{ea_i}{\Omega} Z_{k,ii}^* \frac{\partial u_{k,i}}{\partial \varepsilon_v} \tag{8}$$

where a_i is the lattice parameter, the clamped-ion term $e^{(0)}$ is the first term of Eq. (8). $e^{(0)}$ is equal to the sum of rigid core $e^{(0)}$, core and valence electronic $e^{(0),\text{el}}$ contributions. Subscript k corresponds to the atomic sublattices. Z^* is the Born effective charge described as,

$$Z_{k,iv}^* = Z_k^{core} + Z_{k,iv}^{*,\text{el}} = \left.\frac{\Omega}{ea_i}\frac{\partial P_i}{\partial u_{k,v}}\right|_\varepsilon \tag{9}$$

Piezoelectric response includes two contributions, that appear in linear response for finite distortional wave vectors $\mathbf{q}$, and contributions which appear at $\mathbf{q}= 0$. Improper polarization changes arise from the rotation or dilation of the spontaneous polarization P_i^s. The proper polarization of a ferroelectric or pyroelectric material is given by

$$P_i^P = P_i^T - \sum_j \left(\varepsilon_{ij}P_j^s - \varepsilon_{jj}P_i^s\right) \tag{10}$$

Proper piezoelectric constants e_{iv}^P can be described as,

$$e_{31}^P = \frac{\partial P_3^T}{\partial \varepsilon_1} + P_3^s \tag{11}$$

$$e_{15}^P = \frac{\partial P_1^T}{\partial \varepsilon_5} - P_3^s \tag{12}$$

and $e_{33}^P = e_{33}^T$, because the improper part of e_{33}^T is zero. The difference between proper polarization and total one is due to only homogeneous part, which can be described in the following equation for e_{31} ($e_{31}^{P,\text{hom}}$).

$$e_{31}^{P,\text{hom}} = e_{31}^{\text{hom}} + P_3^s = \frac{\partial P_3^{\text{el},T}}{\partial \varepsilon_1} + P_3^{\text{el},s} \tag{13}$$

This equation can use the similar expression for $e_{15}^{P,\text{hom}}$. The homogeneous part appears as a pure electronic term in the expression for the proper piezoelectric constants, which differ in crystal with nonzero polarization in the unstrained state.

The first term in Eq. (8) can be evaluated by polarization differences as a function of strain, with the internal parameters kept fixed at their values corresponding to zero strain. The second term, which arises from internal microscopic relaxation, can be calculated after determining the elements of the dynamical transverse charge tensors and variations of internal coordinate u_i as a function of strain. Generally, transverse charges are mixed second derivatives of a suitable thermodynamic potential with respect to atomic displacements and electric field. They evaluate the change in polarization induced by unit displacement of a given atom at the zero electric field to linear order. In a polar insulator, transverse charges indicate polarization increase induced by relative sublattice displacement. While many ionic oxides have Born effective charges close to their static value, ferroelectric materials with perovskite structure display anomalously large dynamical charges.

2.2 Micro polycrystalline structure analysis by using the process crystallographic simulation
2.2.1 Evaluation method of the total energy
The tetragonal crystal structure of perovskite compound and its five typical orientations [001], [100], [110], [101] and [111] are shown in Fig. 3. Considering a epitaxial growth of the crystal on a substrate, the lattice constants including a, b, c, θ_{ab}, θ_{bc} and θ_{ca} are changed because of the lattice mismatch with the substrate. These crystal structure changes can be determined by considering six components of mechanical strain in crystallographic coordinate system such as ε_a, ε_b, ε_c, γ_{ab}, γ_{bc} and γ_{ca}. In a general analysis procedure, the lattice mismatch in the specific direction was calculated and the crystal growth potential was derived. However, the epitaxially grown thin film crystal is in a multi-axial state. Therefore, the numerical results of the crystal energy of thin films are not corrected when considering only uni-axis strain.

In this study, the total energy of a crystal thin film with multi-axial crystal strain states is calculated by using the first-principles calculation, and is applied to the case of the epitaxial growth process. An ultra-soft pseudo-potentials method is employed in the DFT with the condition of the LDA for exchange and correlation terms. Total energies of the thin film crystal as the function of six components of crystal strain are calculated to find a minimum value. Total energies are calculated discretely and a continuous function approximation is introduced. A sampling area is selected by considering the symmetry between a and b axes in a tetragonal crystal structure. Sampling points are generated by using a latin hypercube sampling (LHS) method (Olsson & Sandberg, 2002), which is the efficient tool to get nonoverlap sapling points. The following global function model is generated by using a kriging polynomial hybrid approximation (KPHA) method (Sakata et al., 2007).

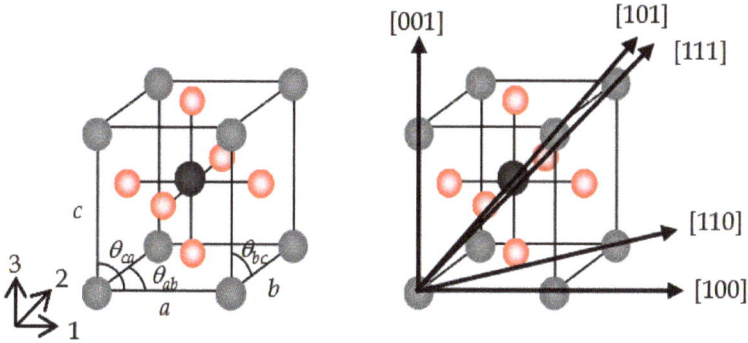

Fig. 3. Crystal structure and orientations of perovskite compounds.

$$E = A_h \varepsilon_h^2 + B_{ij} \varepsilon_i \varepsilon_j + C_l \varepsilon_l + E_{T0} \ (h, i, j = a, b, c, ab, bc, ca) \qquad (14)$$

where E_{T0} is the total energy of the stable crystal, ε_h, ε_i and ε_j epitaxial strains and A_h, B_{ij} and C_l coefficients generated by KPHA method. A gradient of total energy at each sampling point is calculated to generate an approximate quadratic function. The minimum point of a total energy can be found by using this function.

2.2.2 Algorithm of the process crystallographic simulation

In the process crystallographic simulation, it is assumed that several crystal unit cells of crystal clusters, which have certain conformations, can grow on a substrate as shown Fig. 4. The left-hand side diagram in Fig. 4 shows an example of conformation in cases of [001], [100], [110] and [101] orientations, and the right-hand side shows [111] orientation. O, A and B are points of substrate atoms corresponding to thin films ones within the allowable range of distance. l_{OA} and l_{OB} indicate distances of A and B from O, respectively. θ_{AOB} indicates the angle between lines OA and OB.

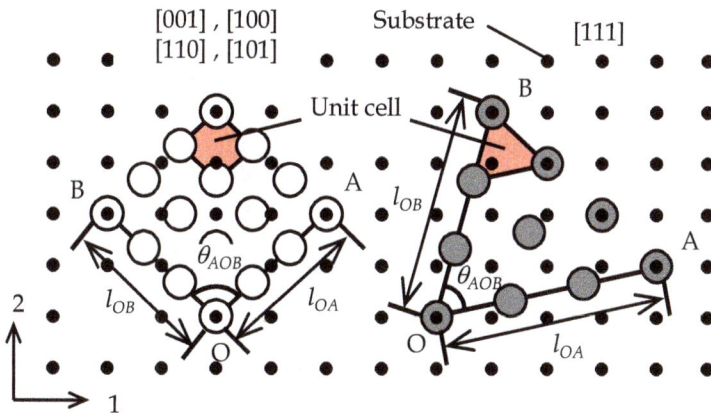

Fig. 4. Schematic of crystal conformations on a substrate.

	[001]	[100]	[110]	[101]	[111]
l_{OA}	a	c	c	$\sqrt{a^2+c^2}$	$\sqrt{a^2+c^2}$
l_{OB}	b	b	$\sqrt{a^2+b^2}$	b	$\sqrt{b^2+c^2}$
Epitaxial strain					
ε_a	$\dfrac{l_{OA}}{ka}-1$	ε_a^*	ε_a^*	ε_a^*	$\dfrac{\sqrt{(l_{OA}/k)^2-c^2(1+\varepsilon_c)^2}}{a}-1$
ε_b	$\dfrac{l_{OB}}{kb}-1$	$\dfrac{l_{OB}}{kb}-1$	$\dfrac{\sqrt{(l_{OB}/k)^2-a^2(1+\varepsilon_a^*)^2}}{b}-1$	$\dfrac{l_{OB}}{kb}-1$	$\dfrac{\sqrt{(l_{OB}/k)^2-c^2(1+\varepsilon_c)^2}}{b}-1$
ε_c	ε_c^*	$\dfrac{l_{OA}}{kc}-1$	$\dfrac{l_{OA}}{kc}-1$	$\dfrac{\sqrt{(l_{OA}/k)^2-a^2(1+\varepsilon_a^*)^2}}{c}-1$	$\dfrac{k-1}{kc}\sqrt{l_{OA}l_{OB}\cos\theta_{AOB}}-1$
γ_{ab}		γ_{ab}^*	γ_{ab}^*	$\theta_{AOB}-\theta_{ab}$	γ_{ab}^*
γ_{bc}	γ_{bc}^*	γ_{bc}^*	$\theta_{AOB}-\theta_{bc}$	γ_{bc}^*	$\dfrac{b}{c}\gamma_{ab}^*$
γ_{ca}	γ_{ca}^*	$\theta_{AOB}-\theta_{ca}$	$\theta_{AOB}-\theta_{ca}$	$\dfrac{a}{c}\gamma_{ab}$	$\dfrac{a}{c}\gamma_{ab}^*$

ε_i^* and γ_{ij}^* can be given from first-principles calculation to the minimize total energy

Table 1. Relationship of lattice constants and epitaxial strain with crystal orientations.

Table 1 summarizes the relationship between the lattice constants of the thin film and l_{OA} and l_{OB} according to crystal orientations. Additionally, Table 1 shows crystal strains, which can be determined in the corresponded crystal orientations. However, particular crystal strains, such as ε_i^* and γ_{ij}^* , cannot be determined by employing the lattice constants of the thin film and the geometric constants of the substrate. In this numerical analysis scheme, their unknown components are determined by employing the condition of minimum total energy of the crystal unit cell.

Figure 5 shows the flowchart of the crystal growth prediction algorithm. First, lattice constants of the thin film and the substrate are inputted. The following procedure is demonstrated. Substrate coordinates of A and B points, which are indicated as (m_A, n_A) and (m_B, n_B), are updated according to the numerical result under the condition of fixing O point in order to generate candidate crystal clusters with assumed conformations and orientations. The search range of the crystal cluster is settled as $0 < m_A, m_B < m$ and $0 < n_A, n_B < n$ by considering the grain size of the piezoelectric thin film crystal. $\vec{e}_1$ and $\vec{e}_2$ as shown in Figure 5 are unit vectors of the substrate coordinate system. Lattice constants of the crystal cluster are compared with geometrical parameters of the substrate, and candidate crystal clusters, which have extreme lattice mismatches, are eliminated. Crystal strains caused by the epitaxial growth are calculated for every candidates of the grown crystal cluster as shown in Table 1. The total energy of grown crystal cluster is estimated by using the total energy as a function of crystal strains. Total energies of candidate crystal clusters are compared with one of the free-strained boundary condition, in order to calculate total energy increments of candidate crystal clusters.

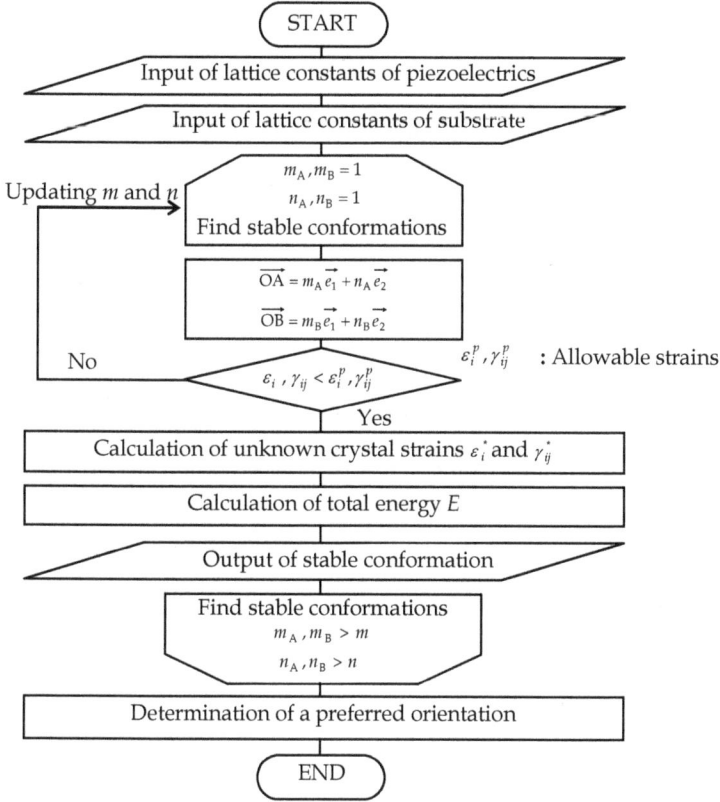

Fig. 5. Flowchart of the process crystallographic simulation.

The fraction of crystal cluster grown on the substrate is calculated by a canonical distribution (Nagaoka et al., 1994).

$$p_i = \frac{\exp\left[-\Delta E_i \big/ k_B T\right]}{\sum_n \exp\left[-\Delta E_n \big/ k_B T\right]} \tag{15}$$

Where ΔE is the total energy increment of the grown cluster, k_B the Boltzmann constant and T the temperature.

2.3 Macro continuum structure analysis by using the crystallographic homogenization method

The crystallographic homogenization method scales up micro heterogeneous structure, such as polycrystalline aggregation, to macro homogeneous structure, such as continuum body. The micro heterogeneous structure has the area Y and microscopic polycrystalline coordinate y, and the macro homogeneous structure has the area X and macroscopic sample coordinate x. Here, it relates to two scales by using the scale ratio λ_h.

$$\lambda_h = \frac{x}{y} \tag{16}$$

where λ_h is an extremely small value. Both coordinates of the micro polycrystalline and the macro continuum structures can be selected independently based on the Eq. (16). Coupling variables are affixed to the superscript λ_h, because the behaviour of the piezo-elastic materials is affected by the polycrystalline structure and λ_h.

The linear piezo-elastic constitutive equation is described as,

$$\sigma_{ij}^{\lambda_h} = C_{ijkl}^{E\lambda_h}\varepsilon_{kl}^{\lambda_h} - e_{kij}^{\lambda_h}E_k^{\lambda_h} \tag{17}$$

$$D_i^{\lambda_h} = e_{ikl}^{\lambda_h}\varepsilon_{kl}^{\lambda_h} + \epsilon_{ik}^{S\lambda_h} E_k^{\lambda_h} \tag{18}$$

The equation of the virtual work of piezoelectric material is written as,

$$\int_{\Omega^{\lambda_h}}\left(C_{ijkl}^{E\lambda_h}\varepsilon_{kl}^{\lambda_h} - e_{kij}^{\lambda_h}E_k^{\lambda_h}\right)\frac{\partial\delta u_i^{\lambda_h}}{\partial x_j}d\Omega + \int_{\Omega^\lambda}\left(e_{ikl}^{\lambda_h}\varepsilon_{kl}^{\lambda_h} + \epsilon_{ik}^{S\lambda_h} E_k^{\lambda_h}\right)\frac{\partial\delta\phi^{\lambda_h}}{\partial x_i}d\Omega$$
$$= \int_{\Gamma_d^\lambda}t_i\,\delta u_i^{\lambda_h}\,d\Gamma + \int_{\Gamma_e^\lambda}\rho\,\delta\phi^{\lambda_h}\,d\Gamma \tag{19}$$

where the strain tensor and the electric field tensor are

$$\varepsilon_{ij}^{\lambda_h} = \frac{1}{2}\left(\frac{\partial u_i^0(x)}{\partial x_j} + \frac{\partial u_j^0(x)}{\partial x_i}\right) + \frac{1}{2}\left(\frac{\partial u_i^1(x,y)}{\partial y_j} + \frac{\partial u_j^1(x,y)}{\partial y_i}\right) \tag{20}$$

$$E_i^{\lambda_h} = -\frac{\partial\phi^0(x)}{\partial x_i} - \frac{\partial\phi^1(x,y)}{\partial y_i} \tag{21}$$

It is assumed that the microscopic displacement and the electrostatic potential can be written as the separation of variables:

$$u_i^1(x,y) = \chi_i^{kl}(x,y)\frac{\partial u_k^0(x)}{\partial x_l} + \Phi_i^m(x,y)\frac{\partial\phi^0(x)}{\partial x_m} \tag{22}$$

$$\phi^1(x,y) = \varphi^{ij}(x,y)\frac{\partial u_i^0(x)}{\partial x_j} + R^k(x,y)\frac{\partial\phi^0(x)}{\partial x_k} \tag{23}$$

where $\chi_i^{kl}(x,y)$ is the characteristic displacement of a unit cell, $R^k(x,y)$ the characteristic electrical potential of a unit cell and $\phi^{ij}(x,y)$ and $\Phi_i^m(x,y)$ the characteristic coupling functions of a unit cell. The macroscopic dominant equations are described as,

$$\int_Y\left(C_{ijkl}^E\frac{\partial\chi_k^{mn}(x,y)}{\partial y_l} + e_{kij}\frac{\partial\varphi^{mn}(x,y)}{\partial y_k}\right)\frac{\partial\delta u_i^1(x,y)}{\partial y_j}dY$$
$$= -\int_Y C_{ijmn}^E\frac{\partial\delta u_i^1(x,y)}{\partial y_j}dY \tag{24}$$

$$\int_Y \left(e_{ikl}\frac{\partial \chi_k^{mn}(x,y)}{\partial y_l} - \epsilon_{ik}^S \frac{\partial \varphi^{mn}(x,y)}{\partial y_k}\right)\frac{\partial \delta\phi^1(x,y)}{\partial y_i}dY$$
$$= -\int_Y e_{imn}\frac{\partial \delta\phi^1(x,y)}{\partial y_i}dY \tag{25}$$

$$\int_Y \left(C_{ijkl}^E\frac{\partial \Phi_k^p(x,y)}{\partial y_l} + e_{kij}\frac{\partial R^P(x,y)}{\partial y_k}\right)\frac{\partial \delta u_i^1(x,y)}{\partial y_j}dY$$
$$= -\int_Y e_{pij}\frac{\partial \delta u_i^1(x,y)}{\partial y_j}dY \tag{26}$$

$$\int_Y \left(e_{ikl}\frac{\partial \Phi_k^p(x,y)}{\partial y_l} - \epsilon_{ik}^S \frac{\partial R^P(x,y)}{\partial y_k}\right)\frac{\partial \delta\phi^1(x,y)}{\partial y_i}dY$$
$$= \int_Y \epsilon_{ip}^S \frac{\partial \delta\phi^1(x,y)}{\partial y_i}dY \tag{27}$$

where, C_{ijkl}^E is the elastic stiffness tensor at constant electric field, ϵ_{ik}^S the dielectric constant tensor at constant strain and e_{kij} the piezoelectric stress constant tensor. They are calculated by experimentally measured crystal properties. Equations (24) - (27) have the solution under the condition of the periodic boundary. The homogenized elastic stiffness tensor, piezoelectric stress constant tensor and dielectric constant tensor are described by the following characteristic function tensor.

$$C_{ijmn}^{EH} = \frac{1}{|Y|}\int_Y \left(C_{ijmn}^E + C_{ijkl}^E\frac{\partial \chi_k^{mn}(x,y)}{\partial y_l} + e_{kij}\frac{\partial \varphi^{mn}(x,y)}{\partial y_k}\right)dY \tag{28}$$

$$e_{pij}^H = \frac{1}{|Y|}\int_Y \left(e_{pij} + e_{kij}\frac{\partial R^P(x,y)}{\partial y_k} + C_{ijkl}^E\frac{\partial \Phi_k^p(x,y)}{\partial y_l}\right)dY$$
$$= \frac{1}{|Y|}\int_Y \left(e_{pij} + e_{pkl}\frac{\partial \chi_k^{ij}(x,y)}{\partial y_l} - \epsilon_{pk}^S \frac{\partial \varphi^{ij}(x,y)}{\partial y_k}\right)dY \tag{29}$$

$$\epsilon_{ip}^{SH} = \frac{1}{|Y|}\int_Y \left(\epsilon_{ip}^S + \epsilon_{ik}^S\frac{\partial R^P(x,y)}{\partial y_k} - e_{ikl}\frac{\partial \Phi_k^p(x,y)}{\partial y_l}\right)dY \tag{30}$$

where superscript H means the homogenized value.
The conventional two-scale finite element analysis is based on the crystallographic homogenization method. In this conventional analysis, the virtually determined or experimentally measured orientations are employed for the micro crystalline structure to characterize the macro homogenized piezoelectric properties. However, this conventional analysis can not characterize a new piezoelectric thin film because of unknown crystal structure and material properties.

A newly proposed three-scale structure analysis can scale up and characterize the crystal structure to the micro polycrystalline and macro continuum structures. First, the stable structure and properties of the new piezoelectric crystal are evaluated by using the first-principles DFT. Second, the crystal growth process of the new piezoelectric thin film is analyzed by using the process crystallographic simulation. The preferred orientation and their fractions of the micro polycrystalline structure are predicted by this simulation. Finally, the homogenized piezoelectric properties of the macro continuum structure are characterized by using the crystallographic homogenization theory. Comparing the provability of crystal growth and the homogenized piezoelectric properties of the new piezoelectric thin film on several substrates, the best substrate is found by using the three-scale structure analysis. It is confirmed that the three-scale structure analysis can design not only existing thin films but also new piezoelectric thin films.

3. Three-scale structure analysis of a new biocompatible piezoelectric thin film

3.1 Crystal structure analysis by using the first-principles calculation

The biocompatible elements (Ca, Cr, Cu, Fe, Ge, Mg, Mn, Mo, Na, Ni, Sn, V, Zn, Si, Ta, Ti, Zr Li, Ba, K, Au, Rb, In) were assigned to A cation in the perovskite-type compound ABO_3. Silicon, which was one of well-known biocompatible elements, was employed on B cation. Values of tolerance factor were calculated by using Pauling's ionic radius. Five silicon oxides satisfied the geometrical compatibility condition, where $MgSiO_3$ = 0.88, $MnSiO_3$ = 0.93, $FeSiO_3$ = 0.91, $ZnSiO_3$ = 0.91 and $CaSiO_3$ = 1.01.

The stable cubic structure with minimum total energy was calculated for the five silicon oxides. As the cubic structure has a feature of high symmetry, the stable crystal structure was easy to estimate because of a little dependency on the initial atomic coordinates. Table 2 shows the lattice constants of the silicon oxide obtained by the first-principles DFT.

The phonon properties of cubic structure at paraelectric non-polar phase were calculated to consider phase transition to other crystal structures. Table 3 summarizes the eigenfrequency, the phonon vibration mode and the eigenvector components normalized to unity. $MgSiO_3$, $MnSiO_3$, $FeSiO_3$, $ZnSiO_3$ showed negative values of eigenfrequency. Cubic structures of these four silicon oxides became unstable owing to softening atomic vibration, and they had possibility of the phase transition to other crystal structure. On the other hand, the stable structure of $CaSiO_3$ was the cubic structure due to positive value of eigenfrequency.

The phonon vibration modes are also summarized in Table 3. All eigenvectors of $MgSiO_3$, $MnSiO_3$ and $FeSiO_3$ were almost parallel to c axis in crystallographic coordinate system. These three silicon oxides had a high possibility to change from the cubic structure to the tetragonal structure, which showed superior piezoelectricity. The eigenvector of O_I and O_{II} in $ZnSiO_3$, however, included a component perpendicular to c axis. It was expected that $ZnSiO_3$ changed from cubic structure to other structures consisting of a rotated SiO_6-octahedron, such as the orthorhombic structure with inferior piezoelectricity.

The stable tetragonal structure to minimize the total energy was calculated for the above three silicon oxides, $MgSiO_3$, $MnSiO_3$ and $FeSiO_3$. Total energies of these tetragonal structures were lower than those of the stable cubic structure. Table 4 shows lattice constants and internal coordinates of constituent atoms. In comparison of the aspect ratio among the three silicon oxides, the value of $MgSiO_3$ was larger than 1.0. On the other hand, the aspect ratio of $MnSiO_3$ and $FeSiO_3$ were smaller than 1.0. Generally, the tetragonal structure of typical perovskite-type oxides such as $BaTiO_3$ and $PbTiO_3$ had larger aspect ratio than 1.0. Consequently, the tetragonal structure of $MnSiO_3$ and $FeSiO_3$ could not be existed. The above results have indicated that $MgSiO_3$ was a remarkable candidate for the new biocompatible piezoelectric material.

Material	Lattice constant (nm)
$MgSiO_3$	0.3459
$MnSiO_3$	0.3431
$FeSiO_3$	0.3421
$ZnSiO_3$	0.3454
$CaSiO_3$	0.3520

Table 2. The lattice constants of cubic structure for candidates of the piezoelectric material.

Material	Eigenfrequency (cm⁻¹)	Mode	Phonon eigenvector components			
			Atom	ξ_1	ξ_2	ξ_3
$MgSiO_3$	-112		O_I	0.00	0.00	-0.37
			O_{II}	0.00	0.00	-0.37
			O_{III}	0.00	0.00	-0.22
			Si	0.00	0.00	-0.13
			Mg	0.00	0.00	0.88
$MnSiO_3$	-41		O_I	-0.09	0.00	-0.53
			O_{II}	-0.07	0.00	-0.53
			O_{III}	-0.09	0.00	-0.41
			Si	-0.07	0.00	-0.45
			Mn	0.04	0.00	0.23
$FeSiO_3$	-83		O_I	0.08	0.01	-0.32
			O_{II}	0.04	0.00	-0.32
			O_{III}	0.08	0.00	-0.17
			Si	0.03	0.00	-0.13
			Fe	-0.22	-0.02	0.83
$ZnSiO_3$	-267		O_I	0.24	0.00	-0.66
			O_{II}	0.00	-0.05	0.66
			O_{III}	-0.24	0.05	0.00
			Si	0.00	0.00	0.00
			Zn	0.00	0.00	0.00
$CaSiO_3$	238		O_I	0.00	0.00	-0.35
			O_{II}	0.01	0.00	-0.35
			O_{III}	0.01	0.00	0.29
			Si	-0.01	0.00	0.79
			Ca	0.00	0.00	-0.23

Table 3. Comparison of phonon properties of cubic structure among $MgSiO_3$, $MnSiO_3$, $FeSiO_3$, $ZnSiO_3$ and $CaSiO_3$.

	MgSiO₃	MnSiO₃	FeSiO₃
Lattice constant	$a = b = 0.3449$	$a = b = 0.3547$	$a = b = 0.3602$
(nm)	$c = 0.3538$	$c = 0.3440$	$c = 0.3349$
Aspect ratio	1.026	0.970	0.930
Internal coordinate			

Table 4. Lattice constants and internal coordinates of constituent atoms for tetragonal structure of $MgSiO_3$, $MnSiO_3$ and $FeSiO_3$.

	Mg	Si	O$_I$	O$_{II}$	O$_{III}$
Z_{11}^*	2.331	3.995	-3.024	-1.620	-1.682
Z_{22}^*	2.331	3.995	-1.620	-3.024	-1.682
Z_{33}^*	2.254	4.054	-1.637	-1.637	-3.035

Table 5. Born effective charge in tetragonal structure of $MgSiO_3$ perovskite.

		MgSiO₃	BaTiO₃	
			DFT	Experiment
Spontaneous polarization (C/m²)	P_3^S	0.471	0.226	0.260
Piezoelectric stress constant	e_{33}	4.57	6.11	3.66
(C/m²)	e_{31}	-2.20	-3.49	-2.69
	e_{15}	12.77	21.34	21.30

Table 6. Comparison of the spontaneous polarization and piezoelectric stress constant between $MgSiO_3$ and $BaTiO_3$.

Table 5 shows Born effective charges of each atoms of $MgSiO_3$. Piezoelectric properties, including the spontaneous polarization and piezoelectric stress constants e_{31}, e_{33} and e_{15}, were calculated by these Born effective charges. Table 6 summarizes piezoelectric properties of $MgSiO_3$, those of $BaTiO_3$ calculated by the DFT and observed by the experiment. It could be concluded that $MgSiO_3$ had larger spontaneous polarization than one of $BaTiO_3$. $MgSiO_3$ showed good piezoelectric properties, which were $e_{33} = 4.57$ C/m², $e_{31} = -2.20$ C/m² and $e_{15} = 12.77$ C/m².

3.2 Investigation of the best substrate of the biocompatible piezoelectric material MgSiO₃

Three biocompatible atoms, which include Au, Mo and Fe, were selected for the substrate candidate. This is because;

1. these atoms can be used for the under electrode.
2. chemical elements of these atoms have the cubic crystal structure.

Lattice constants of Au with FCC cubic structure are $a = b = c = 0.4080$ nm, and ones of Mo with BCC cubic structure are $a = b = c = 0.3147$ nm. Ones of Fe with BCC cubic structure are $a = b = c = 0.2690$ nm.

Crystal growth process of MgSiO₃ thin film on (100) and (111) facets of candidate substrates were demonstrated by using the process crystallographic simulation. Tables 7 - 9 show numerical results of MgSiO₃ thin film grown on (100) facets of these four substrates, and Tables 10 - 12 show results of one on (111) facets of the substrates.

Table 13 shows summary of the orientation fractions of MgSiO₃ thin film on substrates calculated by canonical distribution. In the case of Mo(100) substrate as shown in Table 8, MgSiO₃[100] and [001] were grown, and their orientation fraction were 61.5 % and 38.5 %, respectively. MgSiO₃[001] was grown on Au(100) and Fe(100) at 100 % probability. Comparing total energy increments of crystal clusters of MgSiO₃ grown on these substrates, MgSiO₃[001] on Fe(100) substrate was more stable due to the lowest total energy increment as shown in Table 9. MgSiO₃[111] was grown on Au(111) and Mo(111) substrates at 100 % provability.

Orientation	[001]	[001]	[001]
ε_a (%)	0.21	0.48	0.64
ε_b (%)	0.21	0.48	0.64
ε_c (%)	0.00	0.00	0.00
γ_{ab} (%)	0.00	0.00	0.00
γ_{bc} (%)	0.00	0.00	0.00
γ_{ca} (%)	0.00	0.00	0.00
Total energy of the unit cell (eV)	-2398.4025	-2398.3984	-2398.3946
Total energy increment (eV)	0.0749	0.1257	0.3158

Table 7. Analytical results for stable conformations and preferred orientations of MgSiO₃ thin film grown on Au(100) substrate.

Orientation	[100]	[001]	[001]
ε_a (%)	0.00	1.13	-0.50
ε_b (%)	1.13	1.13	-0.50
ε_c (%)	-1.41	-0.50	0.00
γ_{ab} (%)	0.00	0.00	0.00
γ_{bc} (%)	0.00	-0.50	0.00
γ_{ca} (%)	0.00	0.00	0.00
Total energy of the unit cell (eV)	-2398.3803	-2398.3772	-2398.3979
Total energy increment (eV)	0.0925	0.1046	0.4447

Table 8. Analytical results for stable conformations and preferred orientations of MgSiO₃ thin film grown on Mo(100) substrate.

Comparing total energy increments of crystal clusters of MgSiO$_3$ grown on these two substrates, Au(111) was better substrate than Mo(111) due to low total energy increment. MgSiO$_3$[111] and [001] on Fe(111) were grown at 97.8 % and 2.2 % probability, respectively. Consequently, it could be concluded that four substrates, which included Mo(100), Fe(100) and (111), Au(111), were candidates of the best substrate for MgSiO$_3$ thin film.

Orientation	[001]	[001]	[001]
ε_a (%)	-0.11	0.29	-0.56
ε_b (%)	-0.11	0.29	-0.56
ε_c (%)	0.00	0.00	0.00
γ_{ab} (%)	0.00	0.00	0.00
γ_{bc} (%)	0.00	0.00	0.00
γ_{ca} (%)	0.00	0.00	0.00
Total energy of the unit cell (eV)	-2398.4032	-2398.4016	-2398.3966
Total energy increment (eV)	0.0060	0.0906	0.1092

Table 9. Analytical results for stable conformations and preferred orientations of MgSiO$_3$ thin film grown on Fe(100) substrate.

Orientation	[111]	[001]	[111]
ε_a (%)	2.55	0.06	1.50
ε_b (%)	2.55	0.06	1.50
ε_c (%)	-0.03	0.00	-1.05
γ_{ab} (%)	-0.50	2.20	0.00
γ_{bc} (%)	-0.50	0.00	0.00
γ_{ca} (%)	-0.50	0.00	2.20
Total energy of the unit cell (eV)	-2398.2686	-2398.3450	-2398.3546
Total energy increment (eV)	0.5393	2.1038	2.3891

Table 10. Analytical results for stable conformations and preferred orientations of MgSiO$_3$ thin film grown on Au(111) substrate.

Orientation	[111]	[111]	[001]
ε_a (%)	0.73	0.93	-0.50
ε_b (%)	0.73	0.93	-1.53
ε_c (%)	-1.81	-1.61	0.50
γ_{ab} (%)	0.00	0.00	0.00
γ_{bc} (%)	0.00	0.00	0.50
γ_{ca} (%)	0.00	0.00	0.00
Total energy of the unit cell (eV)	-2398.3724	-2398.3717	-2398.3796
Total energy increment (eV)	0.7738	1.5518	1.9276

Table 11. Analytical results for stable conformations and preferred orientations of MgSiO$_3$ thin film grown on Mo(111) substrate.

Orientation	[111]	[001]	[111]
ε_a (%)	1.33	-0.56	3.19
ε_b (%)	1.33	-0.56	3.19
ε_c (%)	-1.22	0.50	0.60
γ_{ab} (%)	0.00	2.20	-0.50
γ_{bc} (%)	0.00	0.50	-0.50
γ_{ca} (%)	0.00	0.00	-0.50
Total energy of the unit cell (eV)	-2398.3618	-2398.3557	-2398.1793
Total energy increment (eV)	0.6661	0.7639	0.8962

Table 12. Analytical results for stable conformations and preferred orientations of MgSiO₃ thin film grown on Fe(111) substrate.

Substrate Atom	Facet	MgSiO₃ Orientation	Fraction (%)
Au	(100)	[001]	100.0
	(111)	[111]	100.0
Mo	(100)	[100]	61.5
		[001]	38.5
	(111)	[111]	100.0
Fe	(100)	[001]	100.0
	(111)	[111]	97.8
		[001]	2.2

Table 13. Analytical results of preferred orientations and their fractions for MgSiO₃ thin films grown on various substrates.

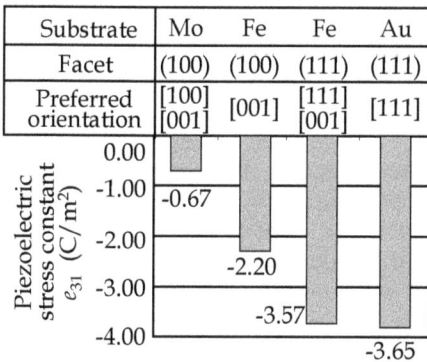

(a) Piezoelectric stress constant e_{31}

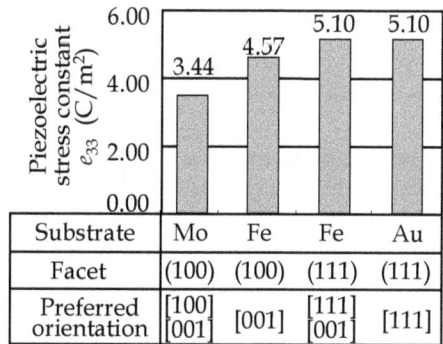

(b) Piezoelectric stress constant e_{33}

Fig. 6. Homogenized piezoelectric stress constant of MgSiO₃ thin film on various substrates.

Analytically determined piezoelectric stress constants and orientation fractions of MgSiO₃ were introduced into the macro continuum structure analysis. Homogenized piezoelectric strain constants of the MgSiO₃ thin film on four substrate candidates were calculated.

Figure 6(a) shows homogenized e_{31} constants and Fig. 6(b) e_{33}. Substrates, facets of substrates and orientation fractions of MgSiO$_3$ thin film were also shown in figures. MgSiO$_3$[111] on Au(111) substrate indicated the highest piezoelectric stress constants, e_{31} = -3.65 C/m^2 and e_{33} = 5.10 C/m^2. MgSiO$_3$[001] on Fe(100) showed e_{31} = -2.20 C/m^2 and e_{33} = 4.57 C/m^2. e_{31} of MgSiO$_3$[001] on Fe(100) was 39.7 % lower than one on Au(111) and e_{33} of MgSiO$_3$[001] on Fe(100) was 10.4 % lower than one on Au(111). In the case of Fe(111) substrate, e_{33} was equal to one on Au(111) substrate, however e_{31} was smaller than one on Au(111). Furthermore, MgSiO$_3$ on Au(111) was more stable than one on Fe(111) substrate. These results have concluded that Au(111) was the best substrate for MgSiO$_3$ thin film.

4. A new biocompatible piezoelectric MgSiO$_3$ thin film generation

4.1 Experimental method

MgSiO$_3$ tihn film is generated by radio-frequency magnetron sputtering. Three factors are selected for generating perovskite tetragonal structure and high piezoelectric property. These conditions are i) the substrate temperature T_s, ii) the post-annealing temperature T_a and iii) flow rate of oxygen f_{O2}. This is because that i) the substrate temperature contributes configuration and bonding of thin film crystals, and ii) the post-annealing temperature affects crystallization of amorphous crystal. iii) The flow rate of oxygen affects crystal morphology and composition of the MgSiO$_3$ crystal. These generation conditions are set as T_a = 300 °C, 350 °C, 400 °C, T_s = 600 °C, 650 °C, 700 °C, and f_{O2} = 1.0 sccm, 3.0 sccm, 5.0 sccm, respectively. The target material is used the mixed sinter of MgO and SiO$_2$, the substrate is Au(111)/SrTiO$_3$(110), which is determined by the three-scale structure analysis. The electric power is 100 W, flow rate of Ar gas is 10 sccm and the pressure in chamber during the sputtering is 0.5 Pa. Thin film is sputtered 4 hours and post-annealed an hour after sputtering.

The displacement–voltage curve of MgSiO$_3$ thin film is measured by ferroelectric character evaluation (FCE) system. Generally, displacement-voltage curve of the piezoelectric material shows butterfly-type hysteresis curve. The piezoelectric strain constant d_{33} can be calculated by gradient of the butterfly-type hysteresis curve. The response surface methodology (RSM) (Berger & Maurer, 2002) is employed to find the optimum combination of generation factor levels of the MgSiO$_3$ thin film.

4.2 Generation of the new biocompatible piezoelectric MgSiO$_3$ thin film

Displacement-voltage curves under the conditions of f_{O2} = 1.0, 3.0 and 5.0 sccm are shown in Fig. 7 - 9. All thin films showed the piezoelectric property due to butterfly-type hysteresis curves. The piezoelectric strain constant d_{33} could be calculated by the gradient at cross point of the butterfly-type hysteresis curve, and d_{33} was indicated in all graphs. d_{33} constants of all thin films were larger than the d_{33} constant (= 129.4 pm/V) of BaTiO$_3$, which was commonly used lead-free piezoelectric material generated in our previous study.

The optimum conditions for generating the MgSiO$_3$ thin film were found by using RSM. Figure 10 shows the response surface of d_{33} constant as a function of T_s and T_a under the condition of f_{O2}= 4.0 sccm. Figure 10(a) shows the aerial view and Fig. 10(b) top view. The black point indicates the highest point of d_{33} constant. The optimum condition, for T_s= 300 °C, T_a= 631 °C and f_{O2}= 4.0 sccm, was found.

MgSiO$_3$ thin film was generated at T_s = 250 °C, because the obtained best T_s was lowest temperature in the range of the substrate temperature which was set in this study. However, the displacement-voltage curves were not indicated the butterfly-type hysteresis curve. This is because the thin film was not crystallized to MgSiO$_3$, due to inactive adatoms and low collision rate of adatoms.

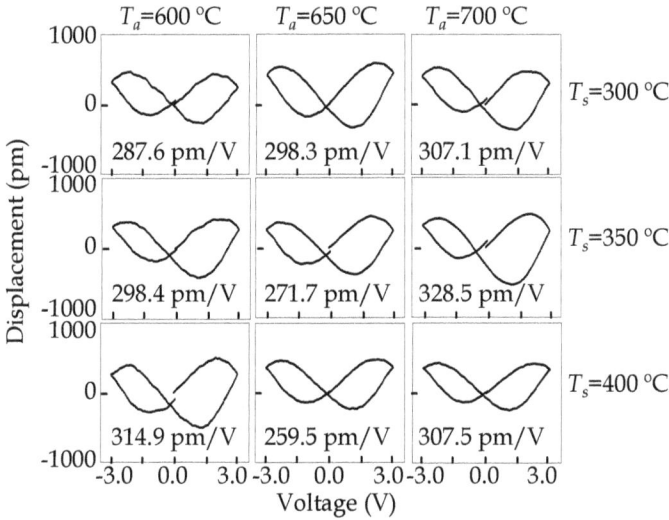

Fig. 7. Displacement-voltage curves of MgSiO$_3$ thin films in the case of f_{O2} = 1.0 sccm.

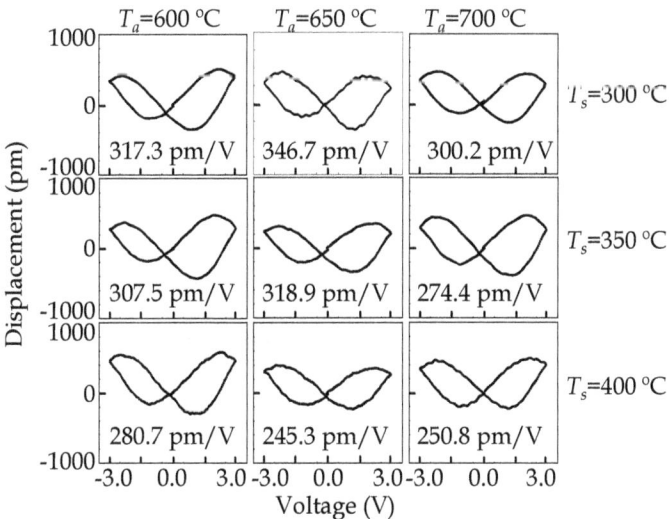

Fig. 8. Displacement-voltage curves of MgSiO$_3$ thin films in the case of f_{O2} = 3.0 sccm.

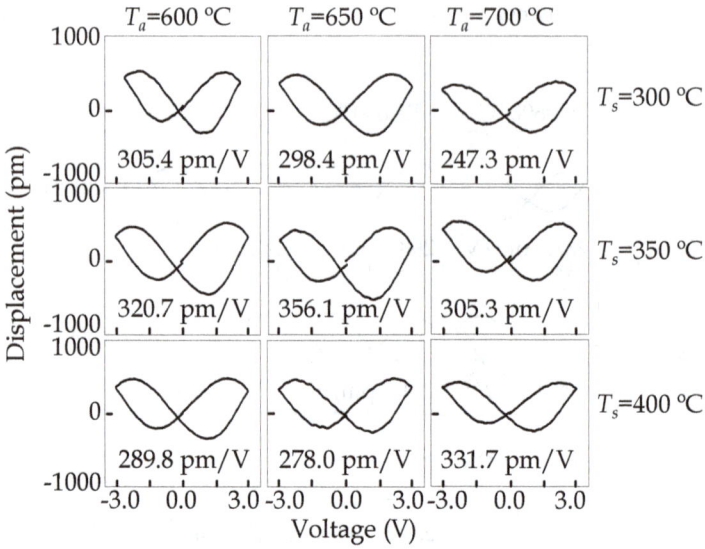

Fig. 9. Displacement-voltage curves of MgSiO$_3$ thin films in the case of f_{O2} = 5.0 sccm.

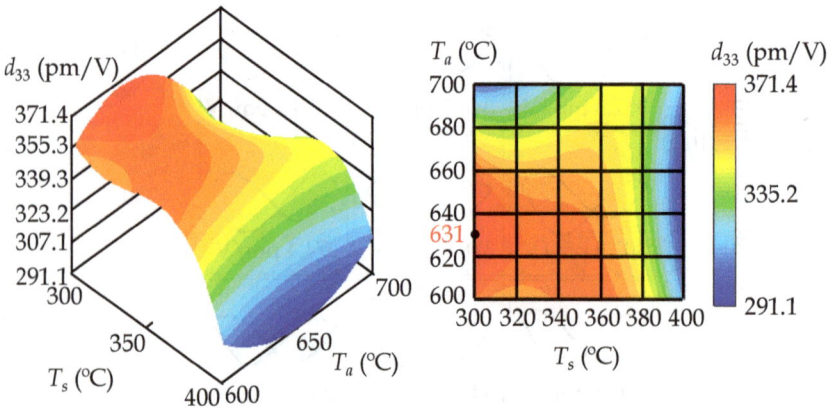

Fig. 10. Piezoelectric strain constant d_{33} as functions of T_s and T_a in the case of f_{O2} = 4.0sccm.

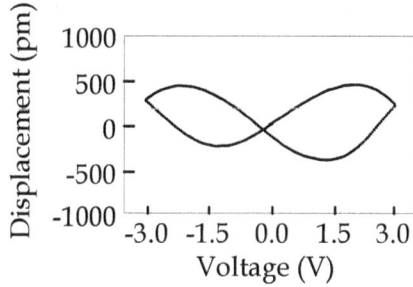

Fig. 11. Displacement-voltage curve of MgSiO$_3$thin film generated at the optimum condition, T_s = 300 °C, T_a = 631 °C and f_{O2} = 4.0 sccm.

Finally, MgSiO$_3$ thin film was generated at the optimum condition. Figure 11 shows its displacement-voltage curve. d_{33} constant was obtained as 359.2 pm/V and this value was higher than one of the pure PZT thin films, d_{33} = 307.0 pm/V, generated by Z. Zhu et al (Zhu et al, 2010).

Consequently, the piezoelectric MgSiO$_3$ thin film was generated successfully and it can be used for sensors and actuators for MEMS or NENS.

5. Conclusion

In this study, the three-scale structure analysis code, which is based on the first-principles density functional theory (DFT), the process crystallographic simulation and the crystallographic homogenization theory, was newly developed. Consequently, a new biocompatible MgSiO$_3$ piezoelectric material was generated by using the radio-frequency (RF) magnetron sputtering system, where its optimum generating condition has been found analytically and experimentally.

Section 2 discussed the algorithm of the three-scale structure analysis, which can design epitaxially grown piezoelectric thin films. This analysis was constructed in three-scale structures, such as a crystal structure, a micro polycrystalline structure and a macro continuum structure. The existing two-scale analysis could evaluate the property of the macro continuum structure by using experimentally observed information of crystal structure, such as crystallographic orientation and properties of the crystal. The three-scale structure analysis can design new biocompatible piezoelectric thin films through three steps, which were to calculate the crystal structure by using the first-principles DFT, to evaluate the epitaxial growth process by using crystallographic simulation, and to calculate the homogenized properties of thin film by using the crystallographic homogenization theory.

In section 3, in order to find a new biocompatible piezoelectric crystal and its best substrate, the three-scale structure analysis was applied to the silicon oxides. Consequently, MgSiO$_3$ had a large spontaneous polarization $P_3^S = 0.471$ C/m^2 and it could present good piezoelectric stress constants e_{33} = 4.57 C/m^2, e_{31} = -2.20 C/m^2 and e_{15} = 12.77 C/m^2. These results indicated that MgSiO$_3$ was one of the candidates of the new biocompatible piezoelectric thin film. Au(111) was the best substrate of MgSiO$_3$ thin film, because MgSiO$_3$[111] on Au(111) was most stable and showed highest piezoelectric stress constant e_{31}= -3.65 C/m^2 and e_{33}= 5.10 C/m^2.

In section 4, $MgSiO_3$ piezoelectric thin film was generated by using the RF magnetron sputtering system. The optimum condition was found by using the response surface methodology (RSM). Measuring the piezoelectric properties of the thin films by using the ferroelectric character evaluation (FCE) system, all $MgSiO_3$ thin films showed the piezoelectric property due to butterfly-type hysteresis curves. Finally, the optimum condition for T_s = 300 °C, T_a = 631 °C and f_{O2} = 4.0 sccm, was found and the best piezoelectric strain constant d_{33} = 359.2 pm/V was obtained. This value was higher than the one of the pure PZT thin films, d_{33}= 307.0 pm/V, generated by Z. Zhu et al.. Consequently, the piezoelectric $MgSiO_3$ thin film was generated successfully and it can be used for sensors and actuators for MEMS or NENS.

6. References

Hindrichsena, C., Møllerb, R., Hansenc, K. & Thomsena, E. (2010). Advantages of PZT Thick Film for MEMS Sensors, *Sensors and Actuators A: Physical*, (Oct. 2009) pp. 1-6, 0924-4247.

Koh, K., Kobayashi, T., Hsiao, F. & Lee, C. (2010). Characterization of Piezoelectric PZT Beam Actuators for Driving 2D Scanning Micromirrors, *Sensors and Actuators A: Physical*, (Sep, 2009), pp.1-12, 0924-4247.

Zhang, M., Jia, Z. & Ren, T. (2009). Effects of Electrodes on the Properties of Sol-Gel PZT Based Capacitors in FeRAM, *Solid-State Electronics*, Vol.53, (Aug. 2008), pp.473-477, 0038-1101.

Ma, Y., Kong, F., Pan, C., Zhang, Q. & Feng, Z. (2010). Miniature Tubular Centrifugal Piezoelectric Pump Utilizing Wobbling Motion, *Sensors and Actuators A: Physical*, Vol.157, (Jul. 2009), pp.322-327, 0924-4247.

Bose, A., Maity, T., Bysakh, S., Seal, A. and Sen, S. (2010). Influence of Plasma Pressure on the Growth Characteristics and Ferroelectric Properties of Sputter-Deposited PZT Thin Films, *Applied Surface Science*, Vol.256, (Feb. 2010), pp.6205-6212, 0169-4332.

Tohma, T., Masumoto, H. & Goto, T. (2002). Preparation of $BaTiO_3$ Films by Metal-Organic Chemical Vapor Deposition, *Japanese Journal of Applied. Physics Part1*, Vol.41, No.11B, (May 2002), pp.6643-6646, 0021-4922.

Kim, T., Kim, B., Lee, W., Moon, J., Lee, B. & Kim, J. (2006). Integration of Artificial $SrTiO_3/BaTiO_3$ Superlattices on Si Substrates using a TiN Buffer Layer by Pulsed Laser Deposition Method, *Journal of Crystal Growth*, Vol.289, No.2, (Aug. 2006), pp.540-546, 0022-0248.

Avrutin, V., Liu, H., Izyumskaya, N., Xiao, B., Ozgur, U. & Morkoc, H. (2009). Growth of $Pb(Ti,Zr)O_3$ Thin Films by Metal-Organic Molecular Beam Epitaxy, *Journal of Crystal Growth*, Vol.311, (Oct. 2008), pp.1333-1339, 0022-0248.

Nishida, K., Wada, S., Okamoto, S., Ueno, R., Funakubo, H. & Katoda, T. (2005). Domain Distributions in Tetragonal $Pb(Zr,Ti)O_3$ Thin Films Probed by Polarized Raman Spectroscopy, *Applied Physics Letters*, Vol.87, (2005), pp.232902.1-232902.3, 0003-6951.

Geetika & Umarji, A. (2010) The Influence of Zr/Ti Content on the Morphotropic Phase Boundary in the PZT–PZN System, *Materials Science and Engineering B*, Vol.167, (Sep. 2009), pp.171-176, 0921-5107.

Kim, K., Hsu, D., Ahn, B., Kim, Y. & Barnard, D. (2010). Fabrication and Comparison of PMN-PT Single Crystal, PZT and PZT-based 1-3 Composite Ultrasonic Transducers for NDE Applications, *Ultrasonics*, Vol.50, (Feb. 2009), pp.790-797, 0041-624X.

Zhu, J., Zheng, L., Luo, W., Li, Y., & Zhang, Y. (2006). Microstructural and Electrical Properties of $BaTiO_3$ Epitaxial Films on $SrTiO_3$ Substructures with a $LaNiO_3$ Conductive Layer as a Template, *Journal of Physic D*, Vol.39. (Feb. 2006), pp.2438-2443, 1361-6463.

Zhang, S., Zhang, H., Zhang, B. & Zhao, G. (2009). Dielectric and Piezoelectric Properties of $(Ba_{0.95}Ca_{0.05})(Ti_{0.08}Zr_{0.12})O_3$ Ceramics Sintered in A Protective Atmosphere, *Journal of European Ceramics Society*, Vol.29, (Apr. 2009), pp.3235-3242, 0955-2219.

Fu, P., Xu, Z., Chu, R., Li, W., Zhang, G. & Hao, J. (2010). Piezoelectric, Ferroelectric and Dielectric Properties of La_2O_3-doped $(Bi_{0.5}Na_{0.5})_{0.94}Ba_{0.06}TiO_3$ Lead-Free Ceramics, *Materials and Design*, Vol.31, (May 2009), pp.796-801, 0261-3069.

Rubio, J., Jaraiz, M., Bragado, I., Mangas, J., Barblla, J. & Gilmer, G. (2003). Atomistic Monte Carlo Simulations of Three-Dimensional Polycrystalline Thin Films, *Journal of Applied Physics*, Vol.94, (Aug. 2002), pp.163-168, 0021-8979.

Lee, S. & Chung, Y. (2006). Surface Characteristics of Epitaxially Grown Ni Layers on Al Surfaces: Molecular Dynamics Simulation, *Journal of Applied Physics*, Vol.100, No.7, (Feb. 2006), pp.074905.1-074905.4, 0021-8979.

Xu, J. & Feng, J. (2002). Study of Ge Epitaxial Growth on Si Substrates by Cluster Beam Deposition, *Journal of Crystal Growth*, Vol.240, No.3 (Jan. 2002), pp.407-404, 0022-0248.

Paul, J., Nishimatsu, T., Kawazoe, Y. & Waghmare, U. (2007). Ferroelectric Phase Transitions in Ultrathin Films of $BaTiO_3$, *Physical Review Letters*, Vol.99, No.7, (Dec. 2005), pp.077601.1-077601.4, 0031-9007.

Costa, S., Pizani, P., Rino, J. & Borges, D. (2006). Structural Phase Transition and Dynamical Properties of $PbTiO_3$ Simulated by Molecular Dynamics, *Journal of Condensed Matter*, Vol.75, No.6, (Sep. 2005), pp.064602.1-064602.5, 0953-8984.

Dieguez, O., Rabe, K. & Vanderbilt, D. (2005). First-Principles Study of Epitaxial Strain in Perovskites, *Physical Review B*, Vol.72, No.14, (Jun. 2005), pp.144101.1-144101.9, 1098-0121.

Yakovkin, I. & Gutowski, M. (2004). $SrTiO_3/Si(001)$ Epitaxial Interface: A Density Functional Theory Study, *Physical Review B*, Vol.70, No.16, (Nov. 2003), pp.165319.1-165319.7, 1098-0121.

Jayachandran, K., Guedes, J. & Rodrigues, H. (2009). Homogenization of Textured as well as Randomly Oriented Ferroelectric Polycrystals, *Computational Materials Science*, Vol.45, (Nov. 2007), pp.816-820, 0927-0256.

Segall, M., Lindan, P., Probert, M., Pickard, C., Hasinp, P., Clark, S. & Payne, M. (2002). First-Principles Simulation: Ideas, Illustrations and the CASTEP Code, *Journal of Physics: Condensed Matter*, Vol.14, (Jan. 2002), pp.2717-2744, 0953-8984.

Monkhorst, H. & Pack, J. (1976). Special Points for Brilloun-Zone Integrations, *Physical Review B*, Vol.13, No.12, (Jan. 1976), pp.5188-5192, 1098-0121.

Szabo, G., Choen, R. & Krakauer, H. (1998). First-Principles Study of Piezoelectricity in $PbTiO_3$, *Physical Review Letters*, Vol.80, No.19, (Oct. 1997), pp.4321-4324, 0031-9007.

Szabo, G., Choen, R. & Krakauer, H. (1999). First-Principles Study of Piezoelectricity in Tetragonal $PbTiO_3$ and $PbZr_{1/2}Ti_{1/2}O_3$, *Physical Review B*, Vol.59, No.20, (Sep. 1998), pp.12771-12776, 1098-0121.

King-Smith, R. & Vanderbilt, D. (1993). Theory of Polarization of Crystalline Solids, *Physical Review B*, Vol.47, No.3, (Jun. 1992), pp.1651-1654, 1098-0121.

Olsson, A. & Sandberg, G. (2002). Latin Hypercube Sampling for Stochastic Finite Element Analysis, *Journal of Engineering Mechanics*, Vol.128, No.1, (Sep. 1999), pp.121-125, 0733-9399.

Sakata, S., Ashida, F. & Zako, M. (2007). Hybrid Approximation Algorithm with Kriging and Quadratic Polynomial-based Approach for Approximate Optimization, *International Journal for Numerical Methods in Engineering*, Vol.70, No.6, (Jul. 2005), pp.631-654, 1097-0207.

Nagaoka, Y. (6th Jul. 1994). *Statistical Mechanics* (in Japanese), Iwanami Shoten, 4-000-07927-1, Japan, Tokyo.

Berger, P. & Maurer, R. (2002). *Experimental Design*, Duxbury Thomson Learning Inc, 0-534-35822-5, USA, Calfornia.

Zhu, Z., Li, J., Liu, Y. & Li, J. (2009). Shifting of the Morphotropic Phase Boundary and Superior Piezoelectric Response in Nb-doped $Pb(Zr, Ti)O_3$ Epitaxial Thin Films, *Acta Materialia*, Vol.57, (Feb. 2009), pp.4288-4295, 1359-6454.

The Influence of the Substrate Temperature on the Properties of Solar Cell Related Thin Films

Shadia J. Ikhmayies

Al Isra University, Faculty of Science and Information Technology, Amman, Jordan

1. Introduction

Polycrystalline films are generally considered to consist of crystallites joined together by the grain boundaries. The grain boundary regions are disordered regions, characterized by the presence of a large number of defect states due to incomplete atomic bonding or departure from stoichiometry for compound semiconductors. These states, known as trap states, act as effective carrier traps and become charged after trapping [1]. The density of defects and impurities in the grain boundaries is larger than that within the grains, so as the orientations of the grains change, the density of traps also changes [2]. The density of trap states depends critically on the deposition parameters [1] including the substrate temperature.

By increasing the substrate temperature the grain size increases, grain boundaries become narrower and their number decreases, the height of the potential barrier between grains decreases, and some impurities go out from the grain boundaries and become effectively incorporated in the lattice and other impurities migrate to the grain boundaries. Evaporation of some elements changes stoichiometry and may create other defects. These changes produce changes in the structure and phase of the films. As a result, the electrical, optical and electronic properties will change too. The presence of some of these changes in a film depends on the deposition technique followed in producing the film, raw materials used and deposition conditions.

There are different deposition techniques to prepare thin films in which the deposition temperature is one of the main parameters that should be controlled to get high quality films. These methods include thermal evaporation [3-7], spray pyrolysis (SP) [8-27], chemical bath deposition (CBD) [28-29], dc magnetron sputtering [30] etc.

In the following sections we will discuss the effect of the substrate temperature on the structural, morphological, electrical and optical properties of thin films deposited by different techniques. A review of experimental results obtained by different authors will be performed with discussions and comparisons between different results.

2. Structural properties

There is agreement between authors that the increase in the substrate temperature improves the crystallinity of the films and encourages the change from amorphous to polycrystalline structure and increases the grain size. X-ray diffraction (XRD) is the suitable tool to reveal these changes. For polycrystalline films, the variations of the intensity of Bragg peaks and their width at half maximum (FWHM) with substrate temperature are evidences on the changes in grain size. The narrowing of the lines of crystal growth at the higher substrate temperature (the decrease in FWHM) means that the grain size had increased. The shifts of the positions of the peaks refer to changes in lattice spacing and then lattice parameters. The appearance of some lines and disappearance of others with substrate temperature may mean a phase transition and/or the appearance or disappearance of other phases of the compounds under study or the presence of some elements. In this section different experimental results will be discussed to show the different effects of the substrate temperature on the structure of thin films through XRD diffractogramms.

A lot of experimental results are found about the change from amorphous to polycrystalline structure with substrate temperature. For films prepared by the spray pyrolysis (SP) technique, a lot of workers [8, 12, 14, 18] found that CdS films prepared at substrate temperatures more than or around 200 °C are polycrystalline. Our CdS:In thin films [8] were prepared at T_s = 350-490 °C and they are polycrystalline. Bilgin et al. [12] prepared CdS thin films by the SP technique at substrate temperatures 473-623 K and found them to be polycrystalline. But we reported that [15] SnO$_2$:F thin films were amorphous at temperatures lower than 360 °C. Gordillo et al. [22] found that SnO$_2$ films deposited at temperatures lower than 300 °C grow with an amorphous structure, but those deposited at T_s = 430 °C present a polycrystalline structure. Rozati [2] found that increasing the substrate temperature causes the SnO$_2$ thin films to exhibit a strong orientation along (200).

Films prepared by chemical path deposition (CBD) which is a low temperature technique are in most cases partially or totally amorphous [28]. Liu et al. [28] prepared CdS films by this technique at deposition temperatures in the range 55-85 °C and found that all of the produced films have some amorphous component and an improved crystallinity with the increase of deposition temperature was obtained.

Numerous experimental data showed that the orientations of crystal growth and preferential orientation are sensitive to the substrate temperature. For ZnO spray-deposited thin films of the hexagonal wurtzite-type, Hichou et al. [10] found that the intensity of the diffraction peaks is strongly dependent on the substrate temperature, where they got maximum intensity at T_s = 450 °C. They found that the [002] direction is the main orientation and it is normal to the substrate plane. For these films some orientations of crystal growth appeared and others disappeared with the variation in the substrate temperature. For CdS thin films prepared by SP technique Acosta et al. [14] found that the intensity of the (002) line increases with temperature, while the (101) peak tends to disappear, which is exactly opposite to what we have in our diffractograms for CdS:In thin films [8]. But we also showed that the preferential orientation of the crystal growth is very sensitive to the substrate temperature. At T_s = 350 °C the preferential orientation in our diffractogram [8] is the H(002)/C(111)- The peaks of these two lines are very close to each other, so it is difficult to distinguish them-, and at T_s = 460 °C it is the H(101), but at T_s = 490 °C it is the H(112)/C(311)- also it is difficult to distinguish the peaks of these two lines.

As we see the orientations of crystal growth and the preferential orientations for a certain compound are different for different authors [8, 14]. In some cases [10] the preferential orientation does not change with the substrate temperature. The preferential orientation in Ashour's [12] diffractograms for CdS thin films which showed just the hexagonal phase is the (101) which was not affected by the substrate temperature but all of the other lines are affected by the substrate temperature. Ashour [12], Pence et al. [13] and Bilgin et al. [18] did not find an influence of the substrate temperature on the preferential orientation for CdS films prepared by the SP technique. On the other hand, Abduev et al. [30] found that the position of the preferential orientation (002) of the hexagonal ZnO:Ga films of thickness 300 nm prepared by dc magnetron sputtering was shifted from 34.25° to 34.41° when the substrate temperature was increased from 50 to 300 °C.

A lot of authors observed a phase transition from cubic to hexagonal phase with the increase in substrate temperature [8, 14]. For spray-deposited CdS:In thin films our XRD diffractograms [8] showed a mixed (cubic and hexagonal phase) at T_s = 350 °C which was converted to only hexagonal phase at T_s = 490 °C. Also for CdS:In thin films prepared by the spray pyrolysis technique Acosta et al. [14] found that X-ray diffractograms of the samples prepared with In/Cd = 0.1 in the solution, the intensity of the (002) peak shows a noticeable increase while the (101) tends to disappear for higher T_s. They [14] say that these variations in peak intensity might be related with phase transition from a cubic to a hexagonal structure. For films prepared by CBD the phase change was observed too, where Liu et al. [28] found that the phase of CdS films was ambiguous, at low deposition temperatures. That is, it couldn't be distinguished (cubic or hexagonal) because the positions of the (002) and (110) lines of the hexagonal structure are similar to the (111) and (220) lines of cubic one, making it difficult to conclude whether the film is purely hexagonal or purely cubic or a mixture of the two phases. But the phase was predominantly hexagonal at higher temperatures, where the presence of the lines (102) and (203) of the hexagonal phase are evidences.

On the other hand, Ashour [12] observed spray-deposited CdS thin films with just one phase (wurtzite) in the temperature range 200-400°C. Their [12] XRD diffractograms showed a preferential orientation (002) along the c-axis direction perpendicular to the substrate plane. Also Bilgin et al. [18] observed just the hexagonal phase for CdS thin films prepared by ultrasonic spray pyrolysis (USP) technique onto glass substrates at different temperatures ranging from 473 to 623K in 50K steps.

We conclude that authors who got a preferential orientation that is independent on the substrate temperature, got just one phase (hexagonal), while those who got a change in the preferential orientation with substrate temperature have a phase transition (from cubic to hexagonal). From these results it is confirmed that increasing the deposition temperature promotes phase transformation from cubic to hexagonal and improves the crystallinity in CdS films. Fig.1 displays the XRD diffractograms of spray-deposited SnO$_2$:F thin films taken at different substrate temperatures by Yadav et al.[31].

The grain size of the polycrystalline films greatly depends on the substrate temperature during deposition [1]. The grain size was found to increase with the substrate temperature for thin films prepared by different deposition techniques [8, 18, 30]. Acosta et al. [14] found that grain size increases with the substrate temperature and presents a smaller dispersion as T_s is increased. This increase is evident from the decrease in FWHM that they observed in their XRD diffractograms. For spray deposited CdS:In thin films, we [8] got an increase in grain size from 10 to 33 nm for the change in the substrate temperatures from 350 to 490 °C, which was calculated by using XRD diffractograms and Sherrer's formula. Bilgin et al. [18] obtained an increase of the grain size of the CdS films from 126 to 336 °A with increasing substrate temperature from 473 to 623K, showing the improvement in the crystallinity of the films. Abduev et al. [30] got an increase from 32 to 36 nm for substrate temperature change from 50 to 300 °C and a decrease in FWHM from 0.27° to 0.24° for the same change in substrate temperature. This change was accompanied by a change in the lattice parameter c which decreased from c = 5.232 Å for the film deposited at 50 °C to c = 5.208 Å for the film deposited at the substrate temperature of 300 °C.

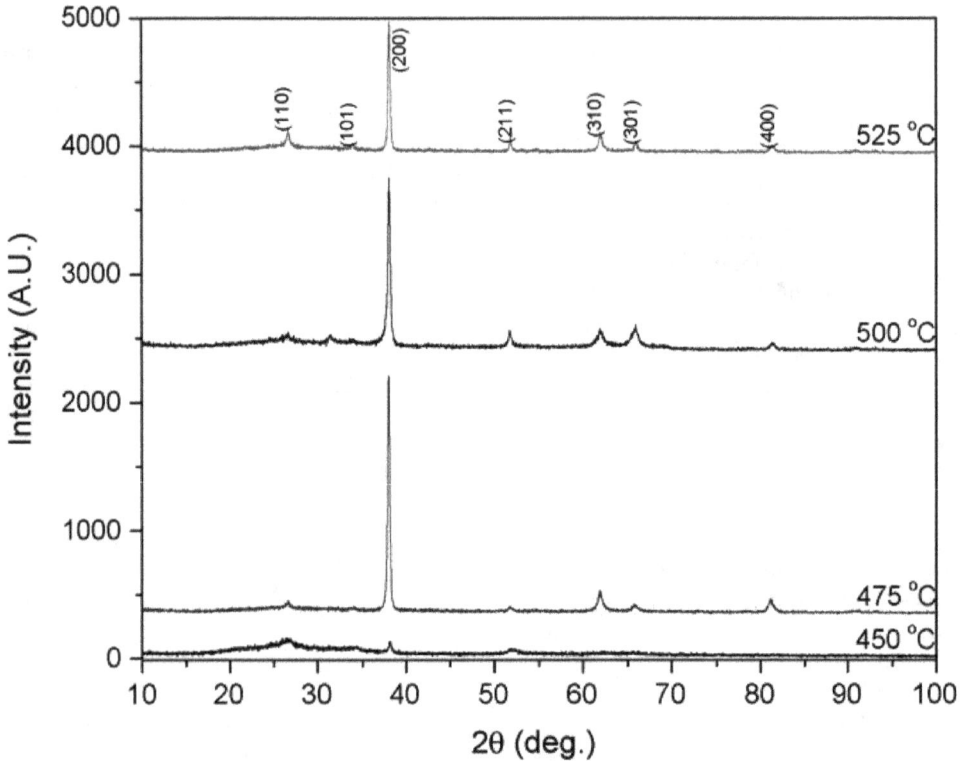

Fig. 1. XRD patterns of spray deposited SnO2:F thin films at various substrate temperatures. Reprinted with permission from Yadav et al. [31]; Copyright © 2009, Elsevier.

Different authors ploted the relation between grain size and the substrate temperature [12, 18]. Bilgin et al. [18] obtained a non-linear relation with the curve concaves down and the grain size increases then becomes constant after a certain value of T_s. Ashour [12] and Patil et al [32] got increasing in grain size with increasing the substrate temperature where the curve concaves up which means that the grain size did not reach a certain size after which there is no increase.

Stress is also varying with substrate temperature as seen by Abduev et al. [30] who found that for gallium doped ZnO films, the film stress had varied with increasing the substrate temperature from –1.915 GPa (the compression condition) at the room substrate temperature to 0.174 GPa (the tension condition) at the substrate temperature T = 300 °C.

In thin film solar cells it is found that the substrate temperature is also an effective parameter on the structure and the grain size. For CdS/CdTe thin film solar cells deposited on SnO$_2$-coated Corning 7059 borosilicate glass, or (100) Si wafers, the substrate temperature caused an increase in the grain size of the CdTe layer as found by Al-Jassim et al. [33]. Also for CdS/CdTe thin films Dhere et al. [34] used AFM measurements and showed that there was no CdTe grain growth, for samples deposited at different substrate temperatures after CdCl$_2$ heat-treatment, but samples deposited at lower temperatures have smaller grains and consequently higher grain boundary volume.

Substrate temperature enhances the interdiffusion in the interface region in CdS/CdTe polycrystalline thin films. Al-Jassim et al. [33] found that at deposition temperatures below 450 °C, only small amounts (~1%) of sulfur were detected in the CdTe films in the vicinity of the interface. On the other hand at deposition temperature of 625 °C, sulfur levels exceeding 10% in CdTe films were detected. This clearly indicates that CdTe devices deposited at high temperatures have an alloyed (CdS$_x$Te$_{1-x}$) active region. Dhere et al. [34] found that compositional analysis by small-area, energy dispersive X-ray analysis (EDS) revealed significant sulfur diffusion into the CdTe film. The amount of sulfur was below detection limit (<0.1 at.%) at the lowest deposition temperature, and increased with increasing deposition temperature.

3. Film morphology

Substrate temperature is an effective parameter in determining the shape and size of grains, surface roughness, porosity and density of voids as found by different authors.

For CdS:In thin films prepared by the SP technique at different substrate temperatures we [8] observed different surface morphologies. At 350 °C long rods or chains were observed, which are related to complex compounds. At T$_s$ = 460 °C we got open cubes and at T$_s$ = 490 °C spherical grains were observed. For SnO$_2$:F thin films prepared by the SP technique different morphologies were observed for films prepared at different substrate temperature [15] too. El Hichou et al. [10] observed a change in surface morphology for ZnO spray deposited thin films with the substrate temperature, where they have the larger grains in the film deposited at T$_s$ = 450 °C. The film deposited at the smallest substrate temperature T$_s$ = 350 °C had shown porous structure but films deposited at T$_s$ > 350 °C had a close-packed morphology.

For spray-deposited CdS:In films Acosta et al. [14] got AFM images which are shown in Figs. 2. Besides the grains size and topology details, it can be observed that grains present regular shape and surfaces for T$_s$ values ranging from 300 °C (Fig.2a) to 400 °C (Fig.2b). In samples obtained at T$_s$ = 425 °C (Fig.2c) and 450 °C (Fig.2d) respectively, aggregates of small grains covering grains with bigger sizes are found everywhere. Noting that these results are related with the XRD diffractograms in that reference. Since the substrate temperature is the only parameter that changes, Acosta et al. [14] say that the changes observed in surface morphology might have to do with particular specific thermodynamic parameters during the pyrolysis and nucleation processes.

Besides increasing the grain size, the increase in the substrate temperature decreases the density of voids. Fig.3 illustrates the SEM micrographs of the surfaces of the CdS films deposited by CBD at 55 °C, 65 °C, 75 °C and 85 °C taken by Liu et al. [28]. These micrographs show that increasing the deposition temperature results in an increase in grain size and consequently a decrease in voids. When the deposition temperature is 55 °C, CdS particles of 50 nm dot the surface of the glass substrate attributing to the controlled nucleation process associated with the low deposition rate. CdS films deposited at 65 °C have spherical particles of about 100 nm in size. The voids with different sizes ranging from 50 nm to 300 nm are still observed, indicating low packing density of the film. The surface of the CdS films deposited at 75 °C is compact and smooth, showing a granular structure with well-defined grain boundaries. It indicates that the increase of the bath temperature is an effective method to diminish voids on the CdS films. But it is noticed that CdS film deposited at higher temperature 85 °C displays a rather rough, inhomogeneous surface with overgrowth grains.

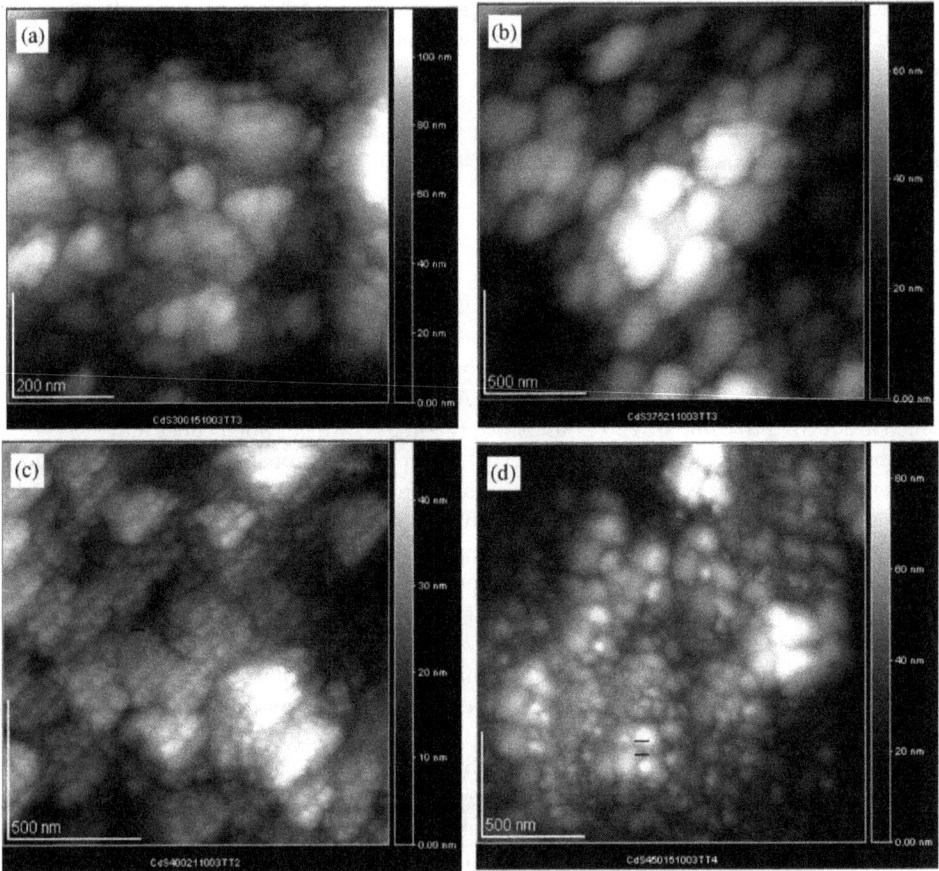

Fig. 2. AFM micrographs of CdS:In deposited by SP technique for different substrate temperatures. a) Ts = 300 °C: The grain size ranges from 50 to 75 nm. b) Ts = 375 °C: The grain size varies between 75 and 225 nm. c) Ts = 400 °C: The grain size between 45 and 60 nm and the size of grain agglomerates is between 170 and 350 nm. d) Ts = 450 °C: The grain size varies between 25 and 65 nm and the size of agglomerates is between 180 and 400 nm, respectively. Reprinted with permission from Acosta et al. [14]; Copyright © 2004, Elsevier.

Other authors showed that roughness increases with the substrate temperature such as Haug et al. [35] who found that the CdTe layers show a higher roughness with increasing substrate temperature, but they are less compact. Atomic Force Microscopy analysis showed that the root mean square (RMS) surface roughness ranges from 100 nm for 500 °C films to 550 nm for 600 °C films. On the other hand some authors found a decrease in roughness with the substrate temperature [14, 36]. For CdS:In thin films prepared by the SP technique Acosta et al. [14] found that as T_s begins to increase, the surface shows a decrease in roughness in zones surrounding pore-like configuration. Also surface roughness was found to decrease with substrate temperature by Abduev et al. [30] for ZnO thin films prepared by magnetron sputtering. Li Zhang et al. [36] also found that surface roughness decreases with the substrate temperature.

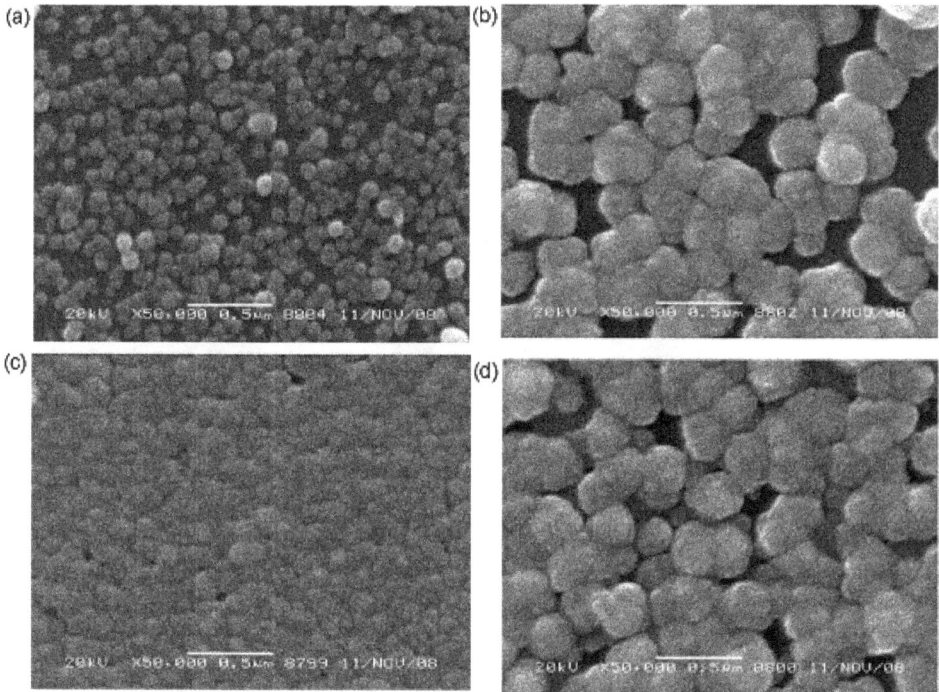

Fig. 3. SEM micrographs of CdS films grown at different temperatures by CBD: a) 55 °C. b) 65 °C. c) 75 °C. d) 85 °C. Reprinted with permission from Liu et al. [28]; Copyright © 2010, Elsevier.

4. Electrical properties

Changes in the structural and morphological properties of the films with the substrate temperature have their effect on the electrical properties of the films. These changes include the phase change, enlargement of grains, diminishing of grain boundaries, motion of impurities from or to the grain boundary region and evaporation of some elements during deposition process. For undoped compound semiconductors, it is found that stoichiometry increases with the substrate temperature due to the reduction in the density of defect states.

These variations will change the number of charge carriers which directly affect the resistivity of the films. They also affect the mechanism of carrier transport and then the linearity of current-voltage characteristics. We will discuss some of the experimental results which include some of these changes. The occurrence of all of these changes or some of them simultaneously has a net effect on the electrical properties as will be seen in the experimental results obtained by different authors.

I-V plots are usually used to investigate the electrical conduction mechanisms and to determine the resistivity of the films. Linear I-V plots mean that the ohmic conduction mechanism is predominant (i.e. electronic conduction through grains). Nonlinear I-V plots mean that other conduction mechanisms are found which are non- electronic and the conduction through the grain boundaries is predominant. It is known that the trap states, which act as effective carrier traps and become charged after trapping result in the appearance of a potential barrier which impedes the flow of majority carriers from one grain to another and affects the electrical conductivity of the films [1]. Three possible mechanisms may govern the grain-to-grain carrier transport through the potential barrier mentioned above:

i.   over-the-barrier thermionic emission of carriers having sufficient energy to surmount the barrier;

ii.   quantum mechanical tunneling from grain to grain through the barrier by carriers having energy less than the barrier height; and

iii. hopping through the localized states.

The relative magnitudes of the barrier height and the width of the barrier will depend critically on the crystallite size and carrier concentration. Depending on the above, with respect to the energy of carriers, one of the above processes will be operative in charge transport in polycrystalline semiconductor films. The thermionic process is limited by the height of the barrier. The thermionic emission will be temperature dependent, with activation of the order of the barrier height, while the tunneling process would be almost independent of temperature. For films with a barrier height larger than what could be surmounted by the carriers with the energy at lower temperatures, tunneling seems to be the dominant mechanism of charge transform in the films. The films grown at lower temperatures will have a smaller crystallite size, and as such the grain boundary region will be substantially larger than the grains. The grain boundary regions being disordered and highly resistive, the film will look like a conglomeration of crystallites embedded in the amorphous matrix [1].

Linear I-V characteristics were recorded by us at all deposition temperatures under study for CdS:In [8], SnO$_2$:F thin films prepared by the SP technique on glass substrates [15, 25-26], undoped ZnO thin films [37-38], Al-doped ZnO thin films [39] and CdTe thin films prepared by vacuum evaporation [40]. Bilgin et al. [18] have linear I-V curve for a CdS thin film prepared by ultrasonic spray pyrolysis (USP) technique at 523 K (250 °C) in the voltage range 0-100 V. This means that this film has not got trapped structure and so, the ohmic conduction mechanism is dominant for this film in the whole voltage range. For the sample obtained at 473 K they [18] found four regions with different slopes where the drawing was performed on log-log scale. The ohmic conduction is dominant in the 0.1–8V voltage range where the slope is 1.11. The slope is 2.24 in the second region which is called space charge limited (SCL) region. The existence of this region shows that CdS films have shallow trapped structure. Then, the trap filled limited (TFL) region comes as the third region. This implies that all traps are filled. The last region with a slope of 2.42 shows trap free region. The other two films prepared by Bilgin et al. [18] at 573 and 623K have deep trapped structure. The mechanism in the sample obtained at 623K is more complex. There are three deep trap levels with different energies for this film. The voltage ranges of these three regions are 16–24, 40–56 and 80–100V, respectively.

Numerous experimental results showed that the resistivity of semiconducting thin films decreases with the deposition temperature [8, 12, 15, 23-24]. For spray-deposited CdS:In thin films we [8] got a decrease of the room temperature resistivity in the dark from 1.5×10^8 Ω.cm at T_s = 380 °C to 1.2×10^6 Ω.cm at T_s = 490 °C and we explained this by the encouragement of crystal growth at higher substrate temperature as concluded from the XRD diffractograms. For CdS thin films prepared by SP technique Ashour [12] got a decrease of room temperature resistivity (10^5-10^3) with substrate temperature in the range 200-400 °C and related it to the growth of the grain size and the improvement in film stoichiometry. For SnO_2:F thin films prepared by SP technique [15] we found a rapid decrease of the resistivity with the substrate temperature. The same behavior was also observed by Shanthi et al. [23] for undoped SnO_2 films prepared by the spray pyrolysis, where they recorded a gradual decrease in the resistivity with the deposition temperature in the range 340-540 °C. Also the same behavior was observed by Zaouk et al. [24] for fluorine-doped tin oxide thin films prepared by electrostatic spray pyrolysis at substrate temperatures in the range 400-550 °C.

Other authors found a decrease in resistivity until a certain temperature and then it increases again [15, 18-20, 30]. Abduev et al. [30] found that for ZnO thin films prepared by dc magnetron sputtering, the growth temperature dependence of resistivity is nonmonotonic. They found that the lowest resistivity (3.8×10^{-4} Ω.cm) is attained at the substrate temperature of 250 °C; then, it increases insignificantly. On the other hand some authors found an increase in resistivity followed by a decrease [28, 36]. Liu et al. [28] measured electrical resistivity for CdS thin films prepared by CBD and found that it arises to 5×10^5 Ω.cm level for temperature 55-75 °C, then it decreases to 7.5×10^4 Ω.cm for 80 °C and 8.5×10^3 Ω.cm for 85 °C sharply. They explained this variation as can be due to the cubic-hexagonal transformation in agreement with structural and optical analysis. Li Zhang et al. [36] also observed an increase in the resistivity with substrate temperature followed by a decrease for Cu(In, Ga)Se$_2$ films prepared by the three-stage co-evaporation process.

From the results of Hall coefficients measurements taken by Liu et al. [28] for CdS thin films prepared by CBD it is found that the CdS films are of n-type conductivity. It is also found that mobility increases from 3.228×10^{-1} cm^2/(V.s) to 6.517 cm^2/(V.s) with the increase of deposition temperature from 55 °C to 75 °C tardily which can be understood by considering the increase of the grain sizes and decrease of the grain boundaries. However, the mobility increases to 6.513×10^1 cm^2/(V.s) at 80 °C and 1.183×10^2 cm^2/(V.s) at 85 °C promptly in contrast with the behavior of resistivity. This behavior can be attributed to the transition from the cubic to the hexagonal phase again, besides the improvement of crystallinity.

The investigation of Hall parameters by Abduev et al. [30] showed that the charge carrier mobility continuously grows with increasing the substrate temperature, and the free carrier concentration has the peak (1.27×10^{21} cm^{-3}) at the substrate temperature of 250 °C. Such a character of the temperature dependence of the free carrier concentration in doped ZnO films is caused by the fact that in zinc oxide there are always intrinsic donor defects in the bulk and at the surface grains in addition to the impurity donors introduced in the ZnO lattice (the substitutional impurity). The multiple experimental and theoretical data indicate that oxygen vacancies play an important role in the conductivity of transparent conducting films. It is observed that the behavior of the resistivity to a large extent is reflected by the carrier density and only little by the mobility; low resistivity corresponds to high carrier density and vice versa.

Abduev et al. [30] explains this behavior by: At low film growth temperatures ($T \leq 150$ °C), the main contribution to the charge carrier concentration is made by intrinsic defects and the efficiency of the Ga incorporation in the ZnO lattice is low, which is confirmed by the small values of Hall mobility in these films. With increasing substrate temperature, the efficiency of impurity atom incorporation into the crystal lattice increases and the concentration of intrinsic defects inside ZnO grains decreases, which is confirmed by the data of the X-ray diffraction analysis and by a substantial increase in the Hall mobility values at a deposition temperature of 200 °C. The increase in Hall mobility at $T \geq 200$ °C is caused also by the lowering of potential barriers for free carriers on the grain boundaries due to the intensification of the process of oxygen thermal desorption from the grain surface during the film growth in vacuum.

For thin film solar cells, the performance is dependent on the substrate temperature. That is, the short-circuit current density J_{sc}, open circuit voltage V_{OC}, Fill factor FF and efficiency η are all dependent on the substrate temperature.

Li Zhang et al. [36] showed that for CIGS solar cells the cell performance increases with the increase in the growth temperature. It is noticed that the cell efficiency increases with increase in the growth temperature. When the substrate temperature is 380 °C the efficiency is very low. The best efficiency at 550 °C is related to the better structural and electrical properties. It is noticed that the effects of the substrate temperature on fill factor (FF) and open circuit voltage (V_{OC}) shows similar trends with cell efficiency. That means the dependence of cell efficiency on the substrate temperature is dominated by the value of FF and V_{OC}. That can be explained by the improvement of carrier concentration and resistivity of CIGS films dominated by Na incorporation diffused from the glass substrate which is expected to be temperature dependent.

Julio et al. [27] investigated the electrical and photovoltaic properties of ZnO/CdTe heterojunctions where ZnO was prepared by the SP technique on CdTe single crystal under the effect of varying the substrate temperature and post deposition temperature for annealing in H_2. For substrate temperatures in the range $T_s = 430\text{-}490$ °C for the spray-pyrolysis deposition the optimum behavior was obtained for $T_s = 460$ °C. They [27] found that as the substrate temperature is increased from 430 to 460 °C the dark J-V characteristics improved considerably and shifted towards higher bias voltages, remaining almost parallel to one another and exhibiting a strong reduction in J_0 with increasing T_s. The reverse current characteristics show similar improvement. Under simulated illumination, large values of short-circuit current were observed: typically of the order of 20 mA/cm^2 for illumination of 87 mW/cm^2 for T_s less than 470 °C. The solar conversion efficiency increased markedly with increasing T_s up to 460 °C, primarily because of an increase in V_{OC} and a fill factor which can be correlated with the decrease in J_0. The improvement in junction characteristics observed with increasing substrate temperature up to 460 °C according to Jullio et al. [27] may have several explanations: The density of the interface states may depend on the orientation of the film; preferential orientation increases as a characteristic temperature is reached.

5. Optical properties

Since the substrate temperature affects the structural properties of the films including lattice parameters and phase, and the electrical properties including the density of charge carriers and density of traps, the optical properties will change.

The absorption coefficient is dependent on the conductivity which is a function of the density of charge carriers. The change in the absorption coefficient will change the transmittance of the films. Some authors found that the transmittance of thin films increases with the substrate temperature [12, 31]; other workers found a decrease in the transmittance with substrate temperature [14] and others found no change in the transmittance of thin films with the substrate temperature [8].

The increase of transmittance with substrate temperature was recorded by Ashour [12] who found an increase of the transmittance with substrate temperature in the range 200-400 °C for undoped spray pyrolyzed CdS thin films of thickness 500 nm. He attributed this improvement in transmittance with substrate temperature to either the decrease in thickness or the improvement in perfection and stoichiometry of the films. Yadav et al. [31] found an increase in transmission with the increase in the substrate temperature for SnO_2:F thin films prepared by the spray pyrolysis technique on glass substrates at substrate temperatures 450-525 °C (Fig.5a). At lower temperatures, i.e. at 450 °C, relatively lower transmission is due to the formation of whitish milky films due to incomplete decomposition of the sprayed droplets.

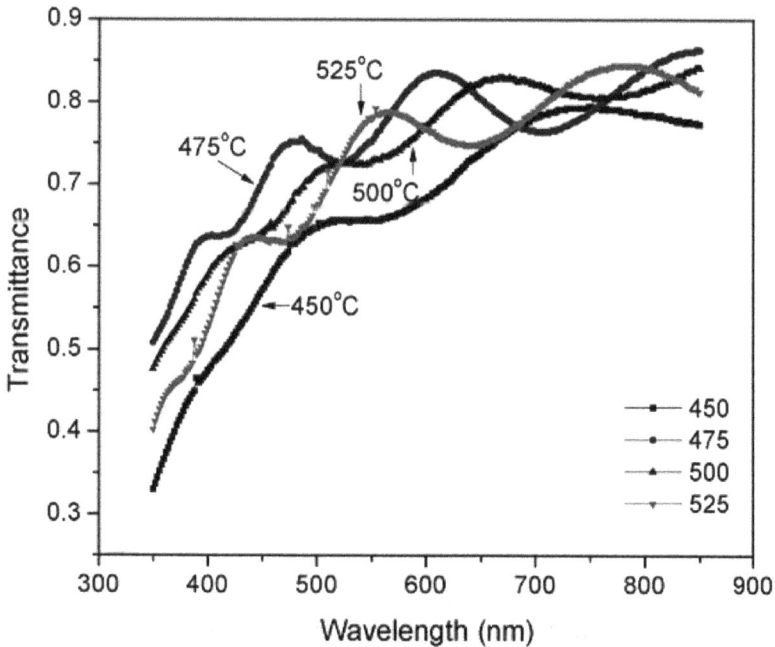

(a)

(b)

(c)

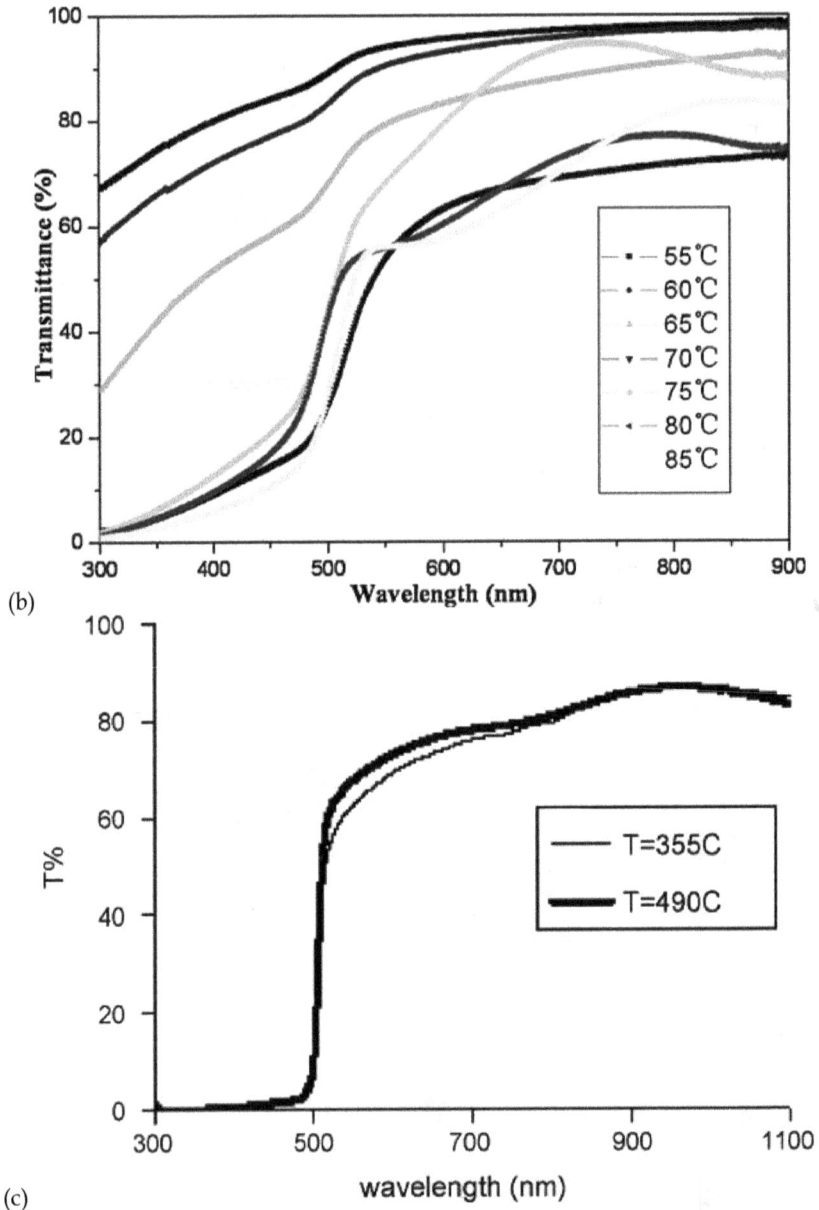

Fig. 4. The optical transmittance of thin films at different substrate temperatures against the wavelength of incident radiation. a) SnO2:F thin films. Reprinted with permission from Yadav et al. [31]; Copyright © 2009, Elsevier. b) CdS films. Reprinted with permission from Liu et al. [28]; Copyright © 2010, Elsevier. c) CdS:In thin films [8]. permission from [9], S. A. Studenikin et al. Journal of Applied physics, 84 (4) (1998), 2287-2294. Copyright [1998], American Institute of Physics.

The decrease in transmittance with substrate temperature was observed by Acosta et al. [14] for CdS:In thin films prepared by the spray pyrolysis technique, but at the same time they have a variable thickness with substrate temperature (decrease then increase) which may be the main reason of the decrease in transmittance. Also the decrease of transmittance with substrate temperature was recorded by Liu et al. [28] for CdS films prepared by CBD (Fig.5b). It can be observed that the transmittance of the film decreases rapidly with the increase of the deposition temperature from 55 °C up to 70 °C, which is caused by reducing voids and increasing film thickness mainly. For higher deposition temperatures, the transmittance initially increases to 84% for the film deposited at 75 °C due to less light scattering by its smoothest surface. It decreases to about 68% at deposition temperatures above 80 °C, which may be due to either more light scattering on their rough surfaces or the transition of the CdS phase from the cubic to hexagonal structure [28]. Another observation about these transmission spectra is that the absorption edge shifts towards higher wavelength side, suggesting a reduction in the bandgap value, and it becomes steeper with deposition temperature rising.

No dependence of transmittance on the substrate temperature was recorded by us [8] for CdS:In thin films prepared by the spray pyrolysis technique (Fig.5c). We think that the transmittance of our films was independent of the substrate temperature due to the way of spraying that we used. We sprayed for 10 s, waited 1–3 min and then sprayed again. The preparation of a set of films by this method takes a long period of time depending on the required thickness of the films (around 4 h for films of thickness around 1 µm). Ashour [12] did not mention the deposition time that he used or the way he followed in spraying, while Acosta et al. [14] produced their films with a deposition time of 5 min in all cases. From our trials we found that using longer deposition times results in less transparent films, and the short period of spraying results in highly transparent films. Also for transparent conducting gallium_doped ZnO films prepared by magnetron sputtering on glass substrates at T_s = 100-300 °C, Abduev et al. [30] have high transmittance which is approximately independent on the substrate temperature, but they observed a shift of the absorption edge in spectra to shorter wavelengths.

The dependence of the bandgap energy on substrate temperature was recorded by different workers [8, 12, 18, 30]. One reason of this dependence is that stress is greatest in films deposited at low temperatures, which results in wider bandgap than bulk. So the increase in substrate temperature reduces stress and then reduces the bandgap energy. Another reason is the increase in interplanner distances or equivalently the lattice parameter with the substrate temperature which appears as a shift in the XRD diffractogram towards smaller angles. It is well known that the lattice parameter and energy gap have opposite behavior [41]. Other reasons include the change in the density of charge carriers with the substrate temperature and the movement of dopants from grain boundaries to the grains to be effectively incorporated in the crystal lattice.

A slight increase in the optical bandgap energy with substrate temperature was observed by different authors [8, 12, 18, 42]. For spray-deposited CdS:In thin films we [8] found that E_g slightly increases with the substrate temperature. This increase can be related to the phase change from mixed (cubic and hexagonal) to hexagonal phase as seen in XRD diffractograms in reference [8]. We found that the E_g = 2.42 eV for a film deposited at 355 °C and E_g = 2.44 eV for a film deposited at 490 °C. Bilgin et al. [18] observed slight increase of

E_g for CdS films prepared by ultrasonic spray pyrolysis (USP) technique onto glass substrates at different temperatures ranging from 473 to 623K. Ashour [12] got E_g = 2.39 - 2.42 eV for CdS films prepared by chemical spray-pyrolysis technique on glass at substrate temperatures in the range 200-400 °C. Melsheimer and Ziegler [42] observed this increase of E_g with substrate temperature for tin dioxide thin films prepared by the spray pyrolysis technique. Values of E_g = 2.51-3.05 eV where obtained for amorphous and partially polycrystalline thin films prepared at T_s = 340-410 °C, and E_g = 3.35-3.43 eV for polycrystalline tin dioxide thin films produced at T_s = 420-500 °C.

An increase followed by a decrease in bandgap energy with substrate temperature was observed by Abduev et al. [30] for ZnO:Ga thin films prepared by dc magnetron sputtering (from 3.52 to 3.72 eV when T_s increases to 250 °C) then a decrease to 3.65 eV at 300 °C. This result was consistent with their electrical properties. Other authors got a decrease then an increase in the bandgap energy with substrate temperature. For spray-deposited indium doped CdS thin films on glass substrates, Acosta et al. [14] got a decrease in the bandgap energy with substrate temperature from 300-425 °C then it increased at T_s = 450 °C. They interpreted the increase observed in E_g by saying that it might be related with the variations in size and morphology of grains.

The decrease of bandgap energy with substrate temperature was observed by some authors such as Liu et al. [28] who observed this for CdS films prepared by CBD. But it is important to notice that the thickness of their films is not constant, which means that the decrease in bandgap is also related to the increase in film thickness not only to the increase in substrate temperature.

Urbuch tail width E_e which is known to be constant or weakly dependent on temperature and is often interpreted as the width of the tail of localized states in the band gap [43] was also found to be randomly affected by the substrate temperature as shown by Bilgin et al. [18] for CdS thin films prepared by USP technique, where it has values in the range 122-188 meV for substrate temperatures in the range 473-623 K. But Melsheimer and Ziegler [42] observed a decrease of E_e with substrate temperature for spray-deposited tin dioxide thin films. For amorphous and partially polycrystalline films prepared at T_s = 340-410 °C, it decreased from 530 to 350 meV. For polycrystalline films prepared at T_s = 420-500 °C it decreased from 240-200 meV.

Photoluminescence and cathodolumenescence always used to explore defects and traps. But the density of trap states depends critically on the deposition parameters [1] and hence on the substrate temperature. Changes in phase, bandgap and density of traps will be reflected on the photoluminescence and cathodoluminescence spectra. It is found that the luminescence intensity depends strongly on the deposition temperature [10].

Fig.6 displays the photoluminescence (PL) spectra for a set of ZnO films deposited by the SP technique by Studenikin et al. [9] at different substrate temperatures and annealed identically in forming gas at 750 °C for 40 min. As we said before, the photoluminescence intensity depends strongly on the substrate temperature. Fig.7 shows the relation between the PL intensity and the substrate temperature for the green peak in the same reference [9]. As the figure shows, the maximum PL intensity is at T_s = 200 °C. They attributed the green PL to oxygen deficiency. This means that much lower temperatures could be used to produce oxygen-deficient ZnO in a reducing atmosphere. Stoichiometry increases with temperature so the green peak becomes smaller with temperature due to the decrease of oxygen deficiency.

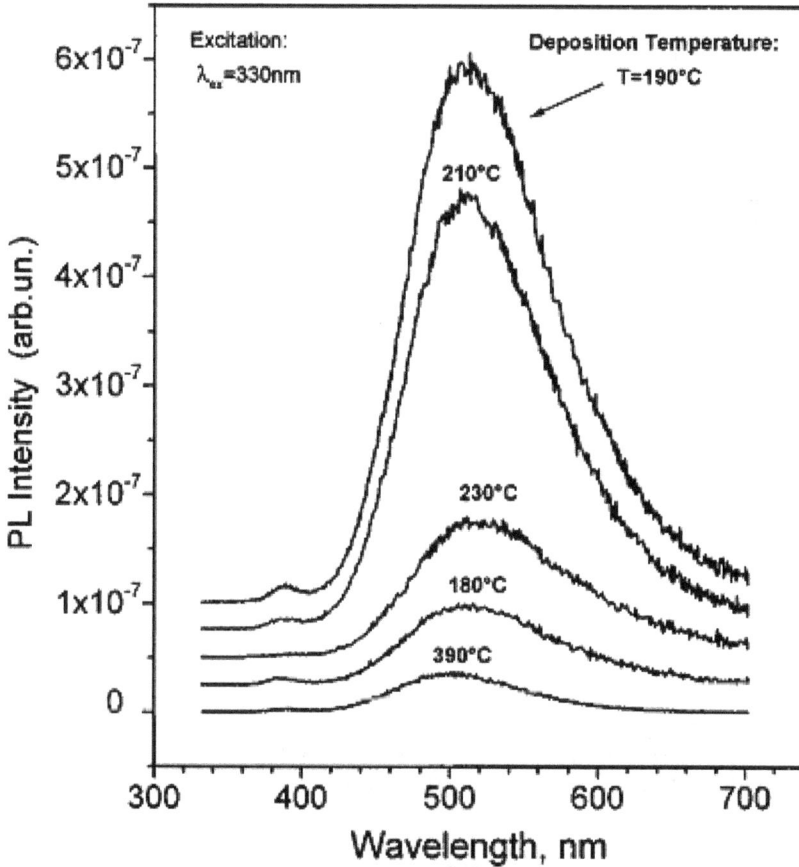

Fig. 5. Photoluminescence spectra of undoped ZnO films grown at different temperatures and annealed in one process in forming gas at 750 °C during 40 minutes. Reprinted with permission from [9], S. A. Studenikin et al. Journal of Applied physics, 84 (4), 2287-2294(1998). Copyright [1998], American Institute of Physics.

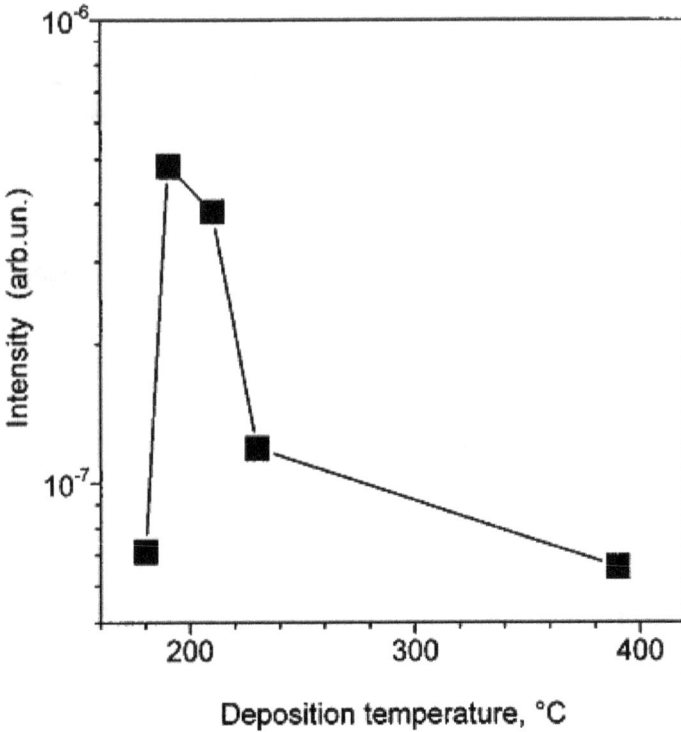

Fig. 6. Intensity of the green photoluminescence of undoped ZnO films as a function of deposition temperature. Reprinted with permission from [9], S. A. Studenikin et al. Journal of Applied physics, 84 (4), 2287-2294(1998). Copyright [1998], American Institute of Physics.

Fig.7 shows the cathodoluminescence spectra for ZnO films prepared by the SP technique at different substrate temperatures taken by El Hichou et. al. [10]. They found that when the substrate temperature increases, the surface of the films is entirely covered by grains and condensed. They observed that extinction of the blue-green emission (centred around 510 nm) is at substrate temperature of 350 and 400 °C, whereas the near UV emission at 382 nm becomes more dominant than other transitions (blue-green and red emissions) at 450 °C. The blue-green emission (510 nm) appears above substrate temperature 450 °C but the red emission (640 nm) appears at different substrate temperature. At T_s = 500 °C, the UV transition shifts to higher wavelength and becomes comparable in cathodoluminescence intensity with blue-green emission. The maximum value of cathodoluminescence intensity for three bands is obtained at T_s = 450 °C [10].

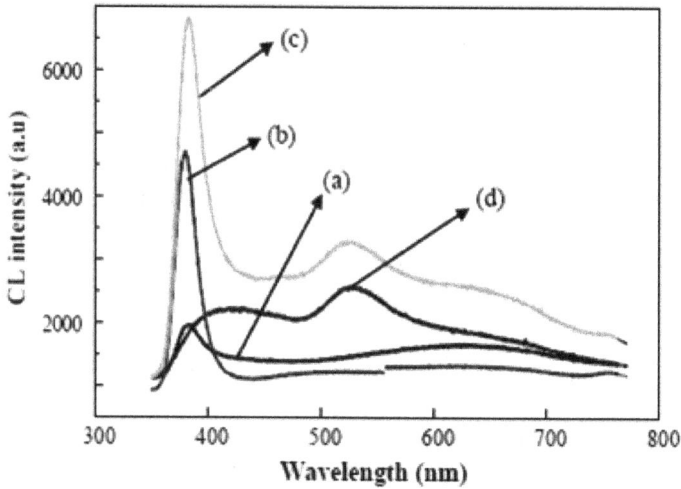

Fig. 7. Cathodoluminescence spectra of ZnO sprayed at flow rate f = 5 ml/min at different substrate temperatures: a) Ts = 350 °C, b) Ts = 400 °C, c) Ts = 450 °C, and d) Ts = 500 °C. Reprinted with permission from El Hichou et al. [10]; Copyright © 2005, Elsevier.

6. Conclusions

Experimental results show that there are influences of the substrate temperature on the properties of semiconducting thin films which are related to solar cells. Change of state from amorphous to polycrystalline, phase change from cubic to hexagonal, increase in grain size and decrease in the number and width of grain boundaries were observed with the increase of the substrate temperature. Morphological changes such as shape of grains, surface roughness, porosity and density of voids were also observed by different authors. The electrical properties were also found to change due to the changes in the density of charge carriers and density of traps with substrate temperature, beside changes in the structural and morphological properties. The optical properties are also sensitive to these changes, and so the transmittance, optical bandgap, width of Urbuch tail, photoluminescence and cathodolumenescence were found to change with changes in the substrate temperatures too.

7. References

[1] D. Bhattacharyya, S. Chaudhuri, and A. K. Pal, Electrical conduction at low temperatures in polycrystalline CdTe and ZnTe films, Materials Chemistry and Physics 40 (1995), 44-49.

[2] S.M. Rozati, The effect of substrate temperature on the structure of tin oxide thin films obtained by spray pyrolysis method, Materials Characterization 57 (2006), 150-153.

[3] O.A. Fouad , A.A. Ismail, Z.I. Zaki and R.M. Mohamed, Zinc oxide thin films prepared by thermal evaporation deposition and its photocatalytic activity, Applied Catalysis B: Environmental 62 (2006), 144-149.

[4] M. Lepek, B. Dogil and R. Ciecholewski, A Study of CdS thin film deposition, Thin Solid Films, 109 (1983), L103-L107.

[5] E. Bertran, J. L. Morenza, J. Esteve and J. M. Codina, Electrical properties of polycrystalline indium-doped cadmium sulfide thin films, J. Physics D, 17(8) (1984), 1679-1685.

[6] D. Patidar, R. Sharma, N. Jain, T. P. Sharma and N. S. Saxena, Optical properties of thermally evaporated CdS thin films, Cryst. Res. Technol., 42(3) (2007), 275-280.

[7] P.P. Sahay, R.K. Nath, S. Tewari, Optical properties of thermally evaporated CdS thin films, Cryst. Res. Technol. 42 (3) (2007), 275-280.

[8] Shadia. J. Ikhmayies, Riyad N. Ahmad-Bitar, The influence of the substrate temperature on the photovoltaic properties of spray-deposited CdS:In thin films, Applied Surface Science 256 (2010), 3541-3545.

[9] S. A. Studenikin, Nickolay Golego and Michael Cocivera, Fabrication of green and orange photoluminescent, undoped ZnO films using spray pyrolysis, Journal of Applied physics, 84 (4) (1998), 2287-2294.

[10] A. El Hichou, M. Addou, J. Ebothé , M. Troyon, Influence of deposition temperature (Ts), air flow rate (f) and precursors on cathodoluminescence properties of ZnO thin films prepared by spray pyrolysis, Journal of Luminescence 113 (2005), 183-190.

[11] Shadia J. Ikhmayies, Production and Characterization of CdS/CdTe Thin Film Photovoltaic Solar Cells of Potential Industrial Use, PhD Thesis, University of Jordan, 2002.

[12] A. Ashour, Physical properties of spray pyrolysed CdS thin films, Turk. J. Phys. 27 (2003), 551-558.

[13] Steve Pence, Elizabeth Varner, Clayton W. Bates Jr., Substrate temperature effects on the electrical properties of CdS films prepared by chemical spray pyrolysis, Mater. Lett. 23 (1995), 13-16.

[14] Dwight R. Acosta, Carlos R. Maganã, Arturo I. Martínez, Arturo Maldonado, Structural evolution and optical characterization of indium doped cadmium sulfide thin films obtained by spray pyrolysis for different substrate temperatures, Solar Energy Mater. Solar Cells 82 (2004), 11-20.

[15] Shadia. J. Ikhmayies, Riyad N. Ahmad-Bitar, Effect of the substrate temperature on the electrical and structural properties of spray-deposited SnO2:F thin films, Materials Science in Semiconductor Processing 12 (2009), 122-125.

[16] M. Krunks, E. Mellikov, E. Sork, Formation of CdS films by spray pyrolysis, Thin Solid Films 145 (1986), 105-109.

[17] L.W. Chow, Y.C. Lee, H.L. Kwok, Structure and electronic properties of chemically sprayed CdS films, Thin Solid Films 81 (1981), 307-318.

[18] V. Bilgin, S. Kose, F. Atay, I. Akyuz, The effect of substrate temperature on the structural and some physical properties of ultrasonically sprayed CdS films, Mater. Chem. Phys. 94 (2005), 103-108.

[19] Chitra Agashe, B. R. Marathe, M. G. Takwale, and V. G. Bhide, Structural properties of SnO2:F films deposited by spray pyrolysis technique. Thin Solid Films 164 (1988), 261- 264.

[20] E. Shanthi, A. Banerjee, V. Dutta, and K. L. Chopra, Electrical and optical properties of tin oxide films doped with F and (Sb+F). J Appl Phys; 53(3) (1982), 1615-1621.

[21] A. E. Rakhshani, Y. Makdisi, and H. A. Ramazaniyan , Electronic and optical properties of fluorine-doped tin oxide films. J Appl Phys 83(2) (1998), 1049–57.

[22] G. Gordillo, L. C. Moreno, W. de la. Cruz., and P. Teheran, Preparation and characterization of SnO2 thin films deposited by spray pyrolysis from SnCl2 and SnCl4 precursors. Thin Solid Films 252 (1994), 61–66.

[23] E. Shanthi, V. Dutta, A. Banerjee, K. L. Chopra, Electrical and optical properties of undoped and antimony-doped tin oxide films. J Appl Phys 51(12) (1980), 6243–6250.

[24] D. Zaouk, Y. Zaatar, A. Khoury, C. Llinares, J.-P. Charles, J. Bechara, Electrical and optical properties of pyrolytically electrostatic sprayed fluorine-doped tin-oxide: dependence on substrate-temperature and substrate-nozzle distance. J Appl Phys 10(87) (2000), 7539–7543.

[25] Shadia J. Ikhmayies, Riyad N. Ahmad-Bitar., Effect of processing on the electrical properties of spray-deposited SnO2:F thin films. Am J. Appl Sci. 5(6) (2008), 672–677.

[26] Shadia J. Ikhmayies, Riyad N. Ahmad-Bitar, The effects of post-treatments on the photovoltaic properties of spray-deposited SnO2:F thin films, Applied Surface Science 255 (2008), 2627–2631.

[27] Julio A. Aranovich, Dolores Golmayo, Alan L. Fahrenbruch, and Richard H. Bube, Photovoltaic properties of ZnO/CdTe heterojunctions prepared by spray pyrolysis. J. Appl. Phys. 51(8) (1980), 4260-4268.

[28] Fangyang Liu, Yanqing Lai , Jun Liu, Bo Wang, Sanshuang Kuang, Zhian Zhang, Jie Li, Yexiang Liu, Characterization of chemical bath deposited CdS thin films at different deposition temperature, Journal of Alloys and Compounds 493 (2010), 305–308.

[29] Hülya Metin, R. Esen., Potoconductivity studies on CdS films grown by chemical bath deposition technique, Erciyes Üniversitesi Fen Bilimleri Enstitüsü Dergisi, 19(1-2) (2003), 96-102.

[30] A. Kh. Abduev, A. K. Akhmedov, A. Sh. Asvarov, A. A. Abdullaev, and S. N. Sulyanov, Effect of growth temperature on properties of transparent conducting gallium_doped ZnO films, Semiconductors 44(1) (2010), 32–36.

[31] A.A. Yadav, E.U. Masumdar, A.V. Moholkar, M. Neumann-Spallart, K.Y. Rajpure, C.H. Bhosale, Electrical, structural and optical properties of SnO2:F thin films: Effect of the substrate temperature, Journal of Alloys and Compounds 488 (2009), 350–355.

[32] P.S. Patil, S.B. Sadale, S.H. Mujawar, P.S. Shinde, P.S. Chigare, Synthesis of electrochromic tin oxide thin films with faster response by spray pyrolysis, Applied Surface Science 253 (2007), 8560–8567.

[33] M.M. Al-Jassim, R.G. Dhere, K.M. Jones, F.S. Hasoon, and P. Sheldon, The morphology, microstructure, and luminescent properties of CdS/CdTe films, Presented at the 2nd World Conference and Exhibition on Photovoltaic Solar Energy Conversion; Vienna, Austria; 6-10 July (1998).

[34] R. Dhere, D. Rose, D. Albin, S. Asher, M. Al-Jassim, H. Cheong, A. Swartzlander, H. Moutinho, T. Coutts, and P. Sheldon, Influence of CdS/CdTe interface properties on the device properties, Presented at the 26th IEEE Photovoltaic Specialists Conference, Anaheim, California, September 29- October 3, (1997).

[35] F.-J. Haug, Zs. Geller, H. Zogg, A. N. Tiwari, and C. Vignali, Influence of deposition conditions on the thermal stability of ZnO:Al films grown by rf magnetron sputtering, J. Vac. Sci. Technol. A 19(1) (2001), 171-174.

[36] Li Zhang, Qing He, Wei-Long Jiang, Fang-Fang Liu, Chang-Jian Li, Yun Sun, Effects of substrate temperature on the structural and electrical properties of Cu(In,Ga)Se2 thin films , Solar Energy Materials & Solar Cells 93 (2009), 114–118.

[37] Shadia J. Ikhmayies, Naseem M. Abu El-Haija and Riyad N. Ahmad-Bitar., Characterization of undoped spray-deposited ZnO thin films of photovoltaic applications, FDMP: Fluid Dynamics & Materials Processing 6(2) (2010), 165-178.

[38] Shadia J. Ikhmayies, Naseem M. Abu El-Haija and Riyad N. Ahmad-Bitar., The influence of annealing in nitrogen atmosphere on the electrical, optical and structural properties of spray- deposited ZnO thin films, FDMP: Fluid Dynamics & Materials Processing 6(2) (2010), 219-232.

[39] Shadia J. Ikhmayies, Naseem M. Abu El-Haija and Riyad N. Ahmad- Bitar, Electrical and optical properties of ZnO:Al thin film prepared by the spray pyrolysis technique, Physica Scripta 81(1) (2010) art. no.015703

[40] Shadia J. Ikhmayies and Riyad N Ahmad-Bitar, Electrical, optical and structural properties of vacuum evaporated CdTe thin films, Collected Proceedings of TMS 2009 138th Annual Meeting & Exhibition, San Francisco, California, USA, February 15-19 (2009), 427-434.

[41] O. De. Melo, L. Hernández., O. Zelaya-Angel, R. Lozada-Morales, M. Becerril, and E. Vasco, Low resistivity cubic phase CdS films by chemical bath deposition technique, Appl. Phys. Lett. 65(10) (1994), 1278-1280.

[42] Melsheimer and D. Ziegler., Band gap energy and Urbach tail studies of amorphous, partially crystalline and polycrystalline tin dioxide. Thin Solid Films, 129 (1985), 35-47.

[43] Y. Natsume, H. Sakata, and T. Hirayama, Low temperature electrical conductivity and optical absorption edge of ZnO films prepared by chemical vapor deposition, phys. stat. Sol. (a) 148 (1995), 485-495.

5

Green Synthesis of Nanocrystals and Nanocomposites

Mallikarjuna N. Nadagouda

Water Supply and Water Resources Division,
National Risk Management Research Laboratory,
U. S. Environmental Protection Agency, Ohio,
USA

1. Introduction

Metal nanomaterials have attracted considerable attention because of their unique magnetic, optical, electrical, and catalytic properties and their potential applications in nanoelectronics (1–5) as well as in various wet chemical synthesis methods (6–14). There is also great interest in synthesizing metal and semiconductor nanoparticles due to their extraordinary properties—properties which are different than when they are in bulk. Green chemistry principles are also regaining popularity for this type of synthesis (8, 15–25). Green chemistry is the design, development, and implementation of chemical products and processes to reduce or eliminate the use and generation of substances that are hazardous to human health and to the environment (25). An example of a greener application of metal nanoparticles is the use of silver and gold nanoparticles, produced from vegetable oil, that are being used in antibacterial paints (26).

Polymer-inorganic nanocomposites have also attracted a lot of attention recently due to their unique, size-dependent chemical and physical properties (26–30). In response to this, different methods of preparing novel nanocomposites with desired properties and functions have been developed (31–35). Such methods should produce materials in which the unique properties of the nanoparticles are preserved (30). One of the main approaches is the dispersion of the previously prepared nanocrystals in polymers. Another is the generation of nanocrystals in polymers *in situ*. In the latter approach, various nanocables, nanowires and nanoparticulates, generated *in situ*, have been reported (36–45).

2. Production of nanomaterials using greener methods

Three areas of opportunity to engage in green chemistry when synthesizing metal nanoparticles by the reduction of the corresponding metal ion salt solutions are: (i) choice of solvent, (ii) the reducing agent employed, and (iii) the capping agent (or dispersing agent). There has also been growing attention in identifying environmentally friendly materials that are multifunctional in this area. For example, the vitamin B_2 can function as both a reducing and a capping agent for Au and Pt metals (15). In addition to its high water solubility, biodegradability and low toxicity when compared to other reducing agents, such as sodium borohydride ($NaBH_4$), sodium citrate and hydroxylamine hydrochloride, and B_2 is the most widely used, behaviorally-active drug in the world. By using natural, available resources like B$_2$, it is possible to prepare nanospheres, nanowires, and nanorods by using solvents of varying densities. It is possible to make multiple shape nanostructures by altering the density of the solvents. This green approach can also be extended to silver and palladium noble nanostructures.

Similarly, vitamin B$_1$ has also been used as a reducing and a capping agent (46). The method is a one-pot method and is greener in nature. By this method, bulk quantities of nanoballs of aligned nanobelts as well as nanoplates of the noble metal palladium in water can be synthesized without the need of any external capping, surfactant agents, and/or large amounts of insoluble templates that have been commonly deployed.

Vitamin C has also been used to fabricate novel core-shell (Fe and Cu), metal (noble metals) nanocrystals. Transition metal salts such as Cu and Fe were reduced using ascorbic acid in solution, a benign, naturally-available antioxidant, and then the simultaneous addition of noble metal salts. This process resulted in the formation of a core-shell structure, depending on the core and shell material used for the preparation (21). Pt yielded a tennis ball-shaped structure, with a Cu core; whereas Pt and Au formed regular spherical nanoparticles. Au, Pt, and Pd formed cube-shaped structures with Fe as the core.

Another interesting route to the synthesis of dendritic Ag structures without the use of any reducing chemical is the transmetallic reaction between copper and silver. The copper–carbon substrate of a transmission electron microscopy (TEM) grid reacted with the aqueous silver nitrate solution within minutes to yield spectacular tree-like silver dendrites. This occurred without using any added capping or reducing reagents (47). These results demonstrate a facile, aqueous, room-temperature synthesis of a range of noble metal nano- and meso-structures (see Figures 1 and 2) that have widespread technological potential in the design and development of next-generation fuel cells, catalysts, and antimicrobial coatings.

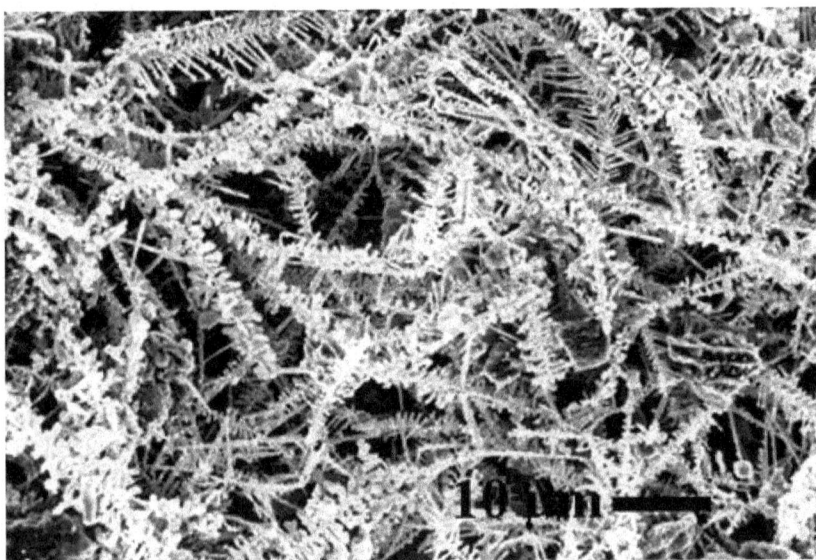

Fig. 1. Scanning electron microscopy image of silver dendrite, formed with copper shavings and activated carbon.

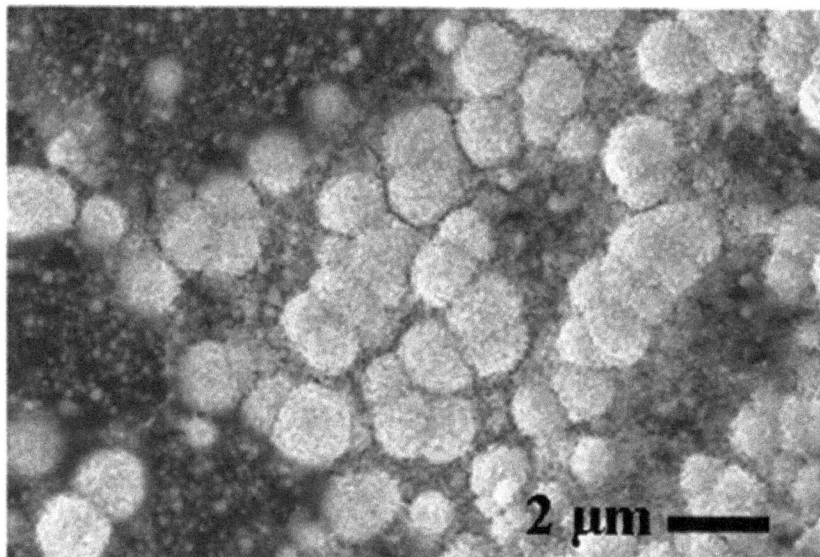

Fig. 2. Scanning electron microscopy image of spongy Pd, formed on a transmission electron microscopy copper grid.

Another material that was investigated in this study was green tea. Green tea has attracted significant attention recently, both in the scientific and consumer communities due to its health benefits for a variety of disorders, ranging from cancer to weight loss. This publicity has led to the increased consumption of green tea by both the general and the patient population, and the inclusion of green tea extract in several nutritional supplements, including multivitamin supplements. There are several polyphenolic catechins in green tea such as viz, (−) epicatechin (EC), (−) epicatechin-3-gallate (ECG), (−) epigallocatechin (EGC), (−) epigallocatechin-3-gallate (EGCG), (+) catechin, and (+) gallocatechin (GC). These compounds are strong antioxidants and hence, can reduce metals salts. One such example is the preparation of noble metals using tea/coffee extract (48). This one-pot method uses no surfactant, capping agent, and/or template. The size of the obtained nanoparticles ranges from 20–60 nm (see Figures 3 and 4) and are crystallized in face-centered cubic symmetry. This method is general and may be extended to other noble metals such as gold (Au) and platinum (Pt).

To prepare the coffee extract, 400 mg of coffee powder (Tata Bru coffee powder 99%) was dissolved in 50 mL of water. Then, 2 ml of 0.1NAgNO₃ (AgNO₃, Aldrich, 99%) was mixed with 10 ml of coffee extract and shaken to ensure thorough mixing. The 40 reaction mixture was allowed to settle at room temperature. For the tea extract, 1 g of tea powder (Red label from Tata, India Ltd. 99%) was boiled in 50 ml of water and filtered through a 25 µl Teflon filter. A similar procedure was repeated for Pd Q4 nanoparticles (using 0.1 N PdCl2, Aldrich, 99%). To evaluate 45, the source (tea and coffee extract) effect on morphology of the Ag and Pd nanoparticles was prepared and several experiments were performed using the above described procedure using the sources as shown in Table 1.

Sl No.	Item Brand Names
1	Sanka coffee
2	Bigelow tea
3	Luzianne tea
4	Starbucks coffee
5	Folgers coffee
6	Lipton tea

Table 1. Various brands of tea/coffee used to generate nanoparticles.

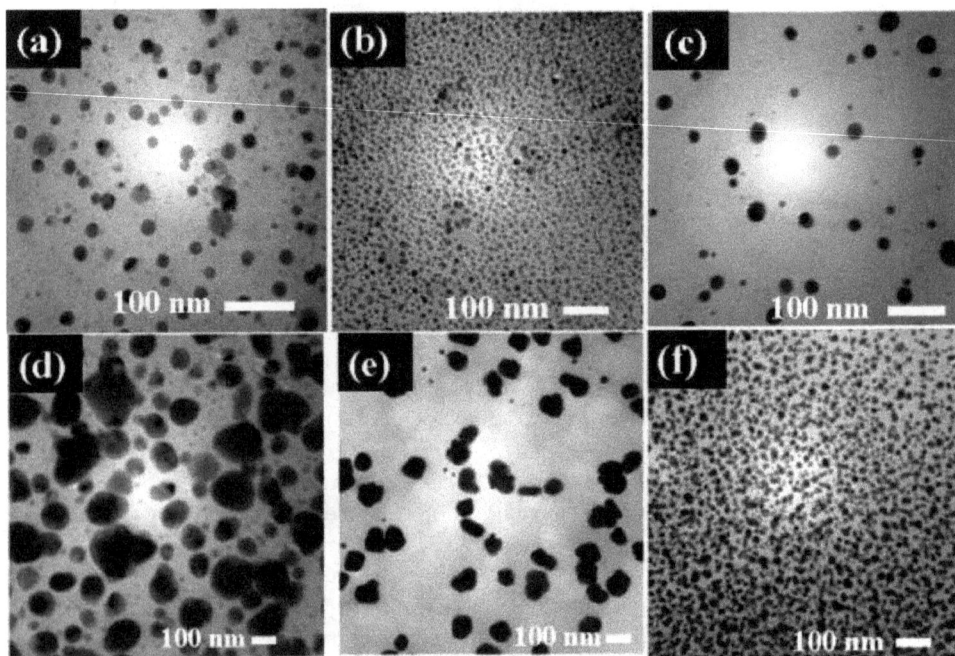

Fig. 3. TEM image of silver nanoparticles, synthesized using (a) Bigelow tea, (b) Folgers coffee, (c) Lipton tea, (d) Luzianne tea, (e) Sanka coffee, and (f) Starbucks coffee extract at room temperature. The process involved one step and did not use any hazardous reducing chemicals or non-degradable capping agents.

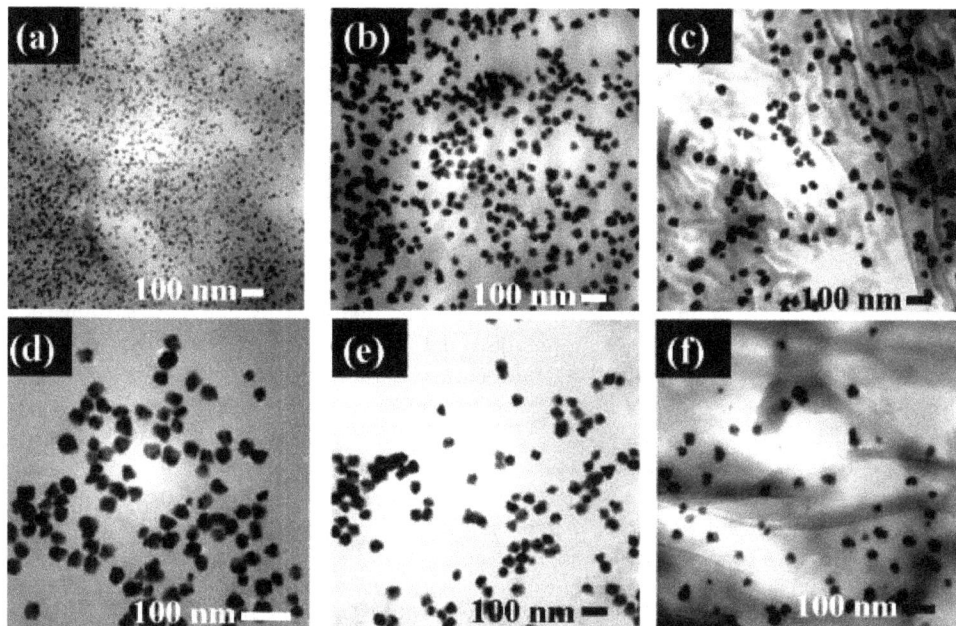

Fig. 4. TEM image of palladium nanoparticles, synthesized using (a) Sanka coffee, (b) Bigelow tea, (c) Luzianne tea, (d) Starbucks coffee, (e) Folgers coffee and (f) Lipton tea extract at room temperature. The process involved one step and did not use any hazardous reducing chemicals or non-degradable capping agents.

Apart from our work with noble nanometals, we have developed a greener, more straight forward, single-step approach for the synthesis of bulk quantities of nanofibers of the electronic polymer, fully-reduced polyaniline (leucoemarldine) without using any reducing agents, surfactants, and/or large amounts of insoluble templates. The nanofibers undergo a spontaneous redox reaction with noble metal ions under mild aqueous conditions, resulting in deposition of various shapes such as leaves, particulates, nanowires, and cauliflower for Ag, Pd, Au, and Pt, respectively. Thus, this approach affords a facile entry into this technologically important class of metal-polymer nanocomposites (49).

3. Microwave assisted synthesis of noble nanostructures and composites

Microwaves play an important role in green chemistry. The use of microwaves can reduce energy consumption and the time used to obtain desired materials. Over the past couple of years, microwave (MW) chemistry has moved from a laboratory curiosity to a well-established, synthetic technique used in many academic and industrial laboratories around the world. Even though the overwhelming number of MW-assisted applications used today are still performed on a laboratory scale, it expected that this technology may be used on a larger, perhaps even production-size, scale in conjunction with radio frequency or conventional heating. Microwave chemistry is based on two main principles: the dipolar mechanism and the electrical conductor mechanism.

The dipolar mechanism occurs when, under a very high frequency electric field, a polar molecule attempts to follow the field in the same alignment. When this happens, the molecules release enough heat to drive the reaction forward. In the later mechanism, the irradiated sample is an electrical conductor and the charge carriers, ions and electrons, move through the material under the influence of the electric field and lead to polarization within the sample. These induced currents and any electrical resistance will heat the sample.

Microwave heating has received considerable attention as a promising new method for the one-pot synthesis of metallic nanostructures in solutions. Because of this, the microwave-assisted synthetic approach for producing silver nanostructures has recently been reviewed. In the review process, researchers have successfully demonstrated the application of this method in the preparation of silver (Ag), gold (Au), platinum (Pt), and palladium (Pd) nanostructures. MW heating conditions allow not only for the preparation of spherical nanoparticles within a few minutes, but also for the formation of single crystalline polygonal plates, sheets, rods, wires, tubes, and dendrites. The morphologies and sizes of the nanostructures can be controlled by changing different experimental parameters, such as the concentration of metallic salt precursors, the surfactant polymers, the chain length of the surfactant polymers, the solvents, and the operation reaction temperature. In general, nanostructures with smaller sizes, narrower size distributions, and a higher degree of crystallization have been obtained more consistently via MW heating than by heating with a conventional oil-bath.

The use of microwaves to heat samples is a practical boulevard for the greener synthesis of nanomaterials (50) and provides many desirable features, such as shorter reaction times, reduced energy consumption, and better product yields. For example, Kundu et al. (51) have synthesized electrically conductive gold nanowires within 2-3 min using DNA as a reducing and nonspecific capping agent using a MW irradiation method. Similarly, uniform and stable polymer-stabilized colloidal clusters of Pt, Ir, Rh, Pd, Au, and Ru have been synthesized by MW irradiation with a modified domestic MW oven (52). The resulting colloidal clusters have small average diameters and narrow size distributions. Further, polychrome silver nanoparticles have been prepared using a soft solution approach under MW irradiation from a solution of silver nitrate ($AgNO_3$) in the presence of poly (N-vinyl-2-pyrrolidone) without any other reducing agent. Different morphologies of silver colloids with attractive colors could be obtained using different solvents as the reaction medium (53).

The MW method can find diversified applications; for example, bulk quantities of nanocarbons with pre-selected morphology can be synthesized in a simple and rapid MW heating approach directly from conducting polymers (54). On the same grounds, the successful preparation of highly active and dispersed metal nanoparticles on a mesoporous material has been accomplished in a conventional MW oven using an eco-friendly protocol in which ethanol and acetone–water were employed as both solvents and reducing agents. The materials exhibited different particle sizes, depending on the metal and the time of MW irradiation and the ensuing nanoparticles were found to be very active and selective in the oxidation of styrene (55).

Recently, Nadagouda et al. (23) have accomplished bulk syntheses of Ag and Fe nanorods using polyethylene glycol (PEG) under MW irradiation conditions. Due to tremendous increases in the biological applications of these nanostructures, there is a continued interest in using biodegradable polymers or surfactants to cap these nanoparticles in order to prevent their aggregation. Most of these biodegradable polymers or surfactants have the tendency to be soluble in water and it is of great interest to know that good dispersion or capping can be obtained using these biodegradable polymers or surfactants. The PEG was chosen as a reducing agent and stabilizing agent for several reasons. First, PEG is biodegradable (as well as non-toxic) and has high water solubility at room temperature, unlike other polymers. It can also form complexes with metal ions and, thereafter, reduce to metals. Finally, it contains alcoholic groups that were exploited for the reduction and the stabilization of the nanoparticles. Favorable conditions to make Ag nanorods were established and the process was expanded to make Fe nanorods with uniform size and shape. The nanorods' formation depended upon the concentration of PEG used in the reaction with Ag salt (see schematic diagram 1). Ag and Fe nanorods crystallized in face-centered cubic symmetry. In a typical procedure, aqueous silver nitrate (AgNO$_3$) solution (0.1 M) and different molar ratios of PEG (molecular weight 300) were mixed in a 10 mL test tube at room temperature to form a clear solution. The reaction mixture was irradiated in a CEM Discover focused MW synthesis system maintaining a temperature of 100 °C (monitored by a built-in infrared sensor) for 1 h with a maximum pressure of 280 psi. The resulting precipitated Ag nanorods (see Figures 5–7) were then washed several times with water to remove excess PEG.

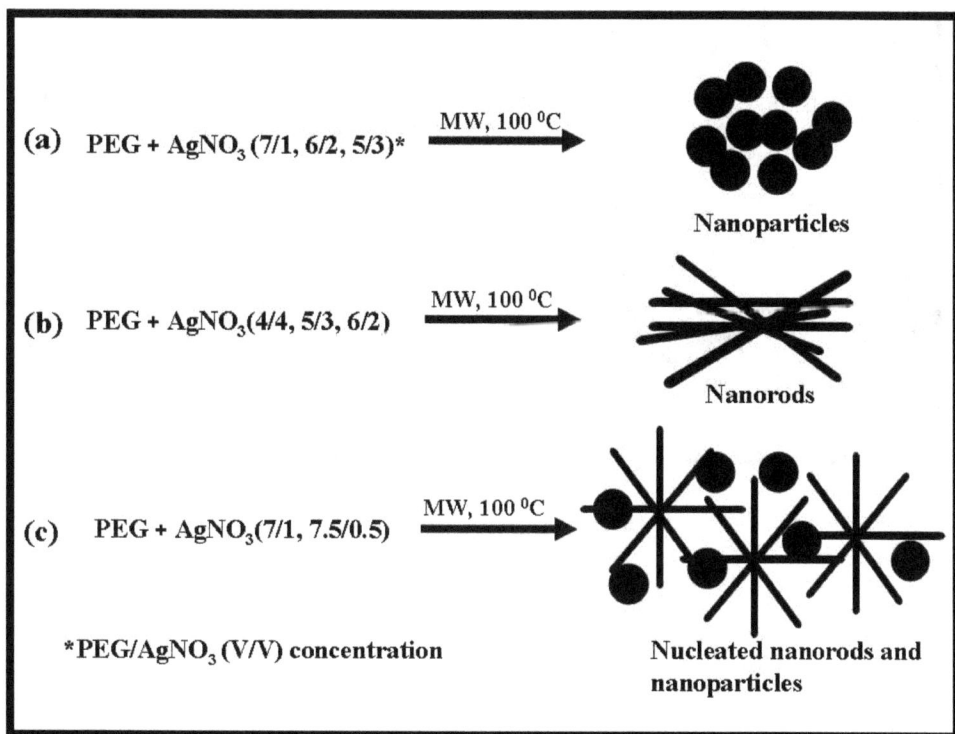

Scheme 1. Schematic illustrations of experimental mechanisms that generated Ag (a) nanoparticles, (b) nanorods, and (c) nucleated nanorods and nanoparticles.

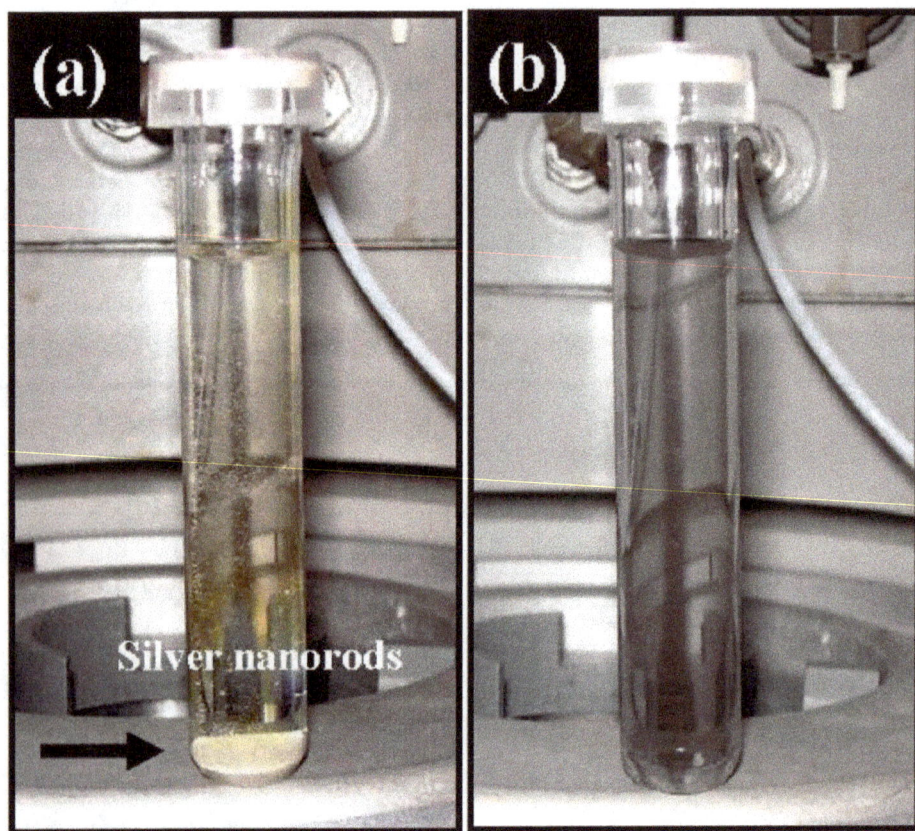

Fig. 5. Photographic image of (a) precipitated Ag nanorods after microwave irradiation for 2 min; and (b) control reaction of the same reaction composition carried out using an oil bath at 100 °C for 1 h.

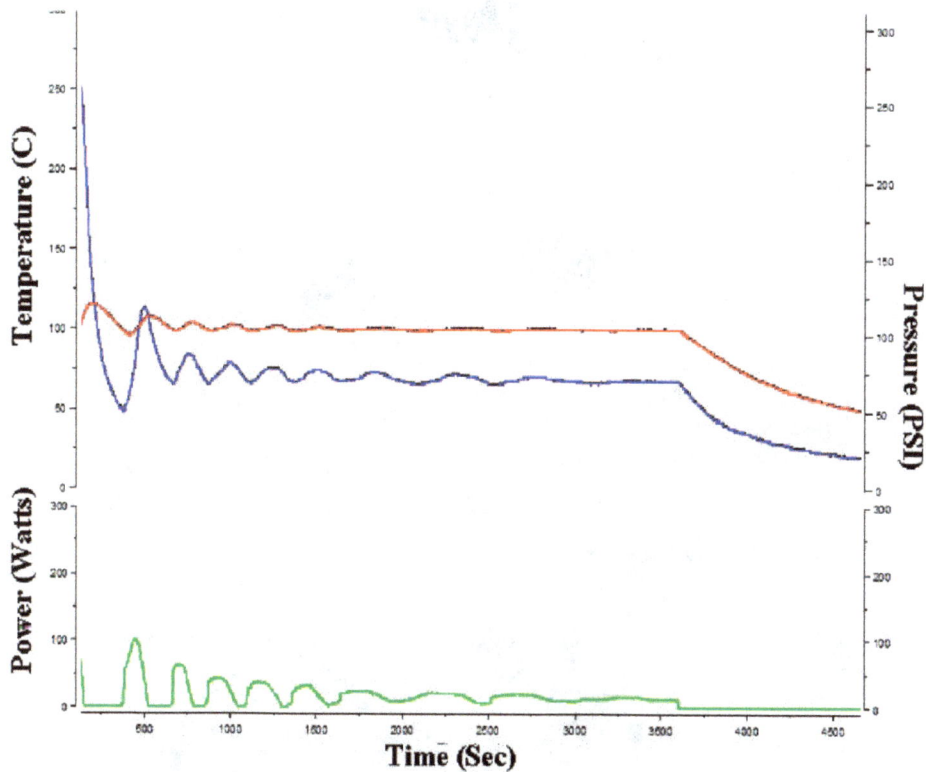

Fig. 6. Reaction profile of 4 mL PEG(300) + 4 mL 0.1 N $AgNO_3$, irradiated at 100 °C for 1 h using MW.

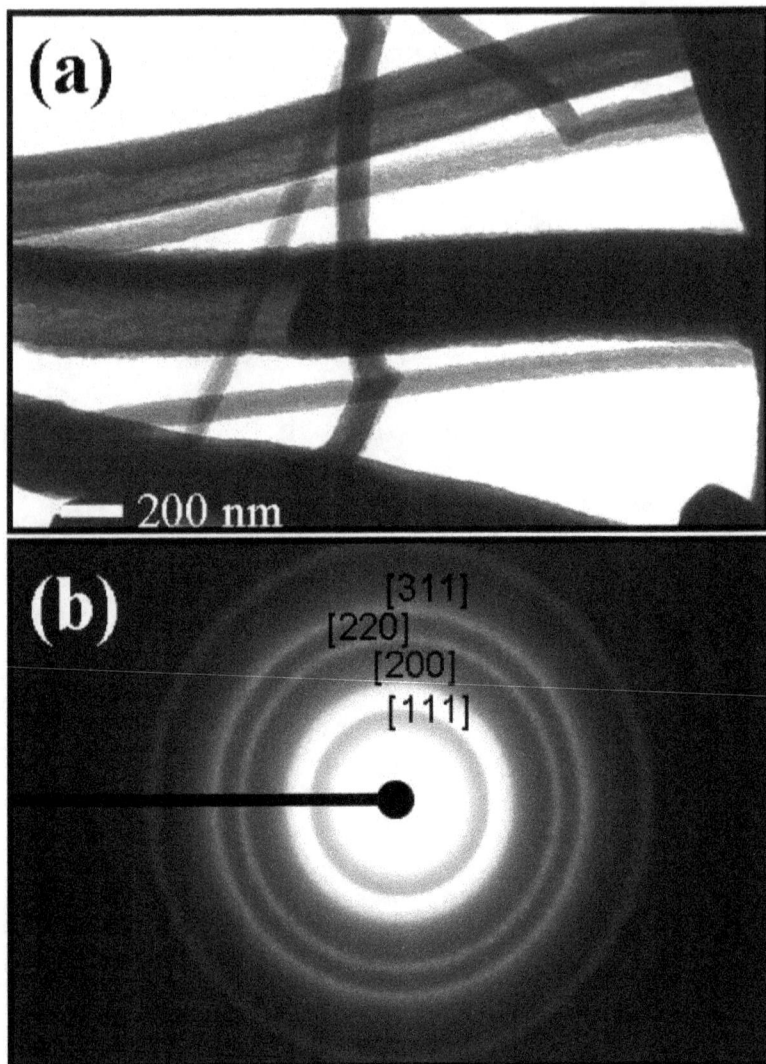

Fig. 7. TEM images of Ag nanorods from (a) 4 mL PEG(300) + 4 mL 0.1 N AgNO₃ under MW conditions, and (b) its SAED pattern obtained, from a bundle of Ag nanorods randomly deposited on the TEM grid.

Shape-controlled synthesis of gold (Au) nanostructures with various shapes such as prisms, cubes, and hexagons was accomplished via the MW-assisted spontaneous reduction of noble metal salts using an aqueous solution of varying concentrations of α-D-glucose, sucrose, and maltose (22). The expeditious reaction was completed under MW irradiation in 30–60s with the formation of different shapes and structures (see Figure 8) and potential application to the generation of nanospheres of Ag, Pd, and Pt. The noble nanocrystals underwent catalytic oxidation with monomers such as pyrrole to generate noble nanocomposites, which have potential functions in catalysis, biosensors, energy storage systems, nano-devices, and other ever-expanding technological applications.

In a typical experiment, an aqueous solution of $HAuCl_4$ (5 mL, 0.01 N) was placed in a 20 mL glass vessel and then mixed with 300 mg of R-D-glucose. The reaction mixture was exposed to high-intensity microwave irradiation (1000 W, Panasonic MW oven equipped with inverter technology) for 30-45 s. Similarly, experiments were conducted using 0.01 N PtCl4, 0.01 N $PdCl_2$, and 0.1 N $AgNO_3$. In the cases of $PdCl_2$ and $AgNO_3$, 300 mg of poly (vinyl pyrrolidinone) (PVP) was added to prevent aggregation and the formation of silver mirror (Tollen's process) on the surface of the glass walls.

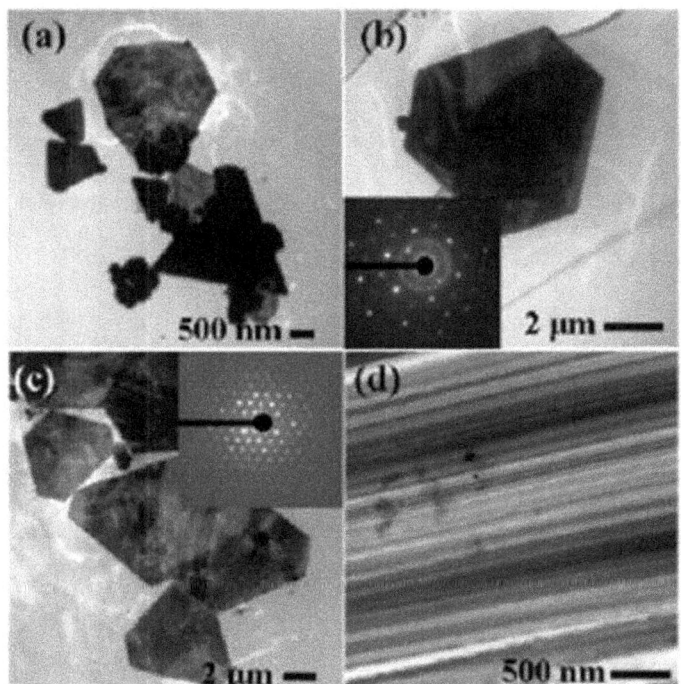

Fig. 8. TEM images of Au nanostructures, synthesized (low concentration of sugar) using MW irradiation with natural polymers such as (a) sucrose, (b) α-D-glucose, or (c, d) maltose. The insets show corresponding electron diffraction patterns.

A green approach was also developed that generated bulk quantities of nanocomposites containing transition metals such as Cu, Ag, In, and Fe at room temperature. A biodegradable polymer, carboxymethyl cellulose (CMC), was reacted with respective metal salts to obtain desired composites (20).

These nanocomposites exhibited broader decomposition temperatures when compared with control CMC and Ag-based CMC nanocomposites, exhibiting a luminescent property at longer wavelengths.

The other noble metals (such as Au, Pt, and Pd) did not react at room temperature with aqueous solutions of CMC, but did react rapidly under MW irradiation (MW) conditions at 100 ^{0}C.

This environmentally-benign approach provides facile entry to the production of multiple-shaped noble nanostructures without using any toxic reducing agents and/or capping/surfactant agents. The method also uses a benign biodegradable polymer, CMC, could find widespread technological and medicinal applications.

Recently, Yu et al. (56) prepared uniform water-soluble silver nanoparticles by reducing silver nitrate with basic amino acids in the presence of soluble starch via MW heating in aqueous medium. Although the fundamental of MW irradiation for this system has yet to be studied completely, the authors believed that MW irradiation plays a major role in the synthesis of the uniform silver nanoparticles. The choice of benign solvent and renewable reacting components and targeted heating approaches amply support the notion that the green chemical synthesis of metal nanoparticles with well-controlled shapes, sizes, and structures is possible.

Microwave irradiation that accomplishes the cross-linking reaction of poly (vinyl alcohol) (PVA) with metallic and bimetallic systems has also been achieved (19).

Nanocomposites of PVA cross-linked metallic systems such as Pt, Cu, and In, and bimetallic systems such as Pt-In, Ag-Pt, Pt-Fe, Cu-Pd, Pt-Pd, and Pd-Fe were prepared expeditiously by reacting the respective metal salts with 3 wt. % PVA under MW irradiation, maintaining the temperature at 100 ^{0}C. This is a radical improvement over the methods used for preparing the cross-linked PVA described in the literature (see Figure 9).

The general preparative procedure is versatile and provides a simple route to manufacturing useful metallic and bimetallic nanocomposites with various shapes, such as nanospheres, nanodendrites, and nanocubes.

Recently, there has been an increasing interest in synthesizing carbon nanotube (CNT)-metal nanoparticle/polymer composites. The larger surface areas and high electric conductivity render them as ideal supporting materials for metal nanoparticle catalysts such as Ag, Au, Pt, and Pd nanoparticles, which have shown great promise in catalysis, surface-enhanced Raman scattering (SERS), and electrochemical and fuel cells. CNTs are also ideal templates for attaching metal nanoparticles and nanoparticle-fused metal nanowires for hydrogen storage and for chemical and biological sensing applications.

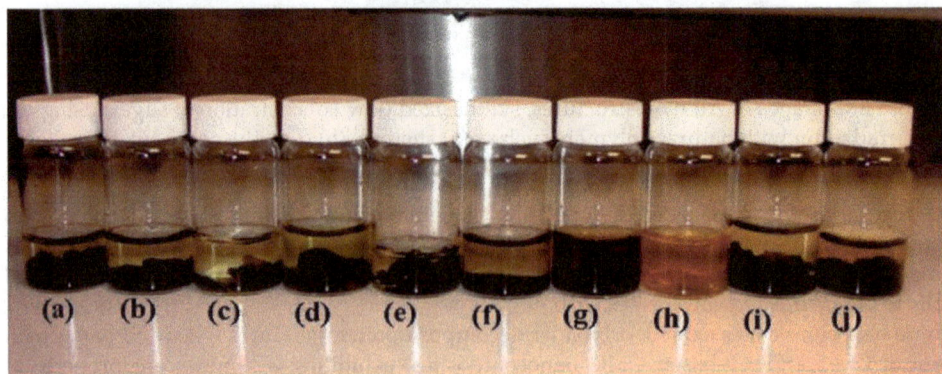

Fig. 9. Photographic image of cross-linked PVA with various metallic and bimetallic systems: (a) Pt, (b) Pt-In, (c) Ag-Pt, (d) Cu, (e) Pt-Fe, (f) Pt with higher concentration ratio, (g) Cu-Pd, (h) In, (i) Pt-Pd, and (j) Pd-Fe.

The cross-linking reaction of PVA with single-walled carbon nanotubes (SWNTs), multi-walled carbon nanotubes (MWNTs), and buckminsterfullerene (C-60) using MW irradiation was achieved with 3 wt. % PVA under MW irradiation, maintaining a temperature of 100 °C, representing a radical improvement over literature methods to prepare such cross-linked PVA composites (Figure 10) (57). This general preparative procedure is versatile and provides a simple route for manufacturing useful SWNT, MWNT, and C-60 cross linked PVA nanocomposites.

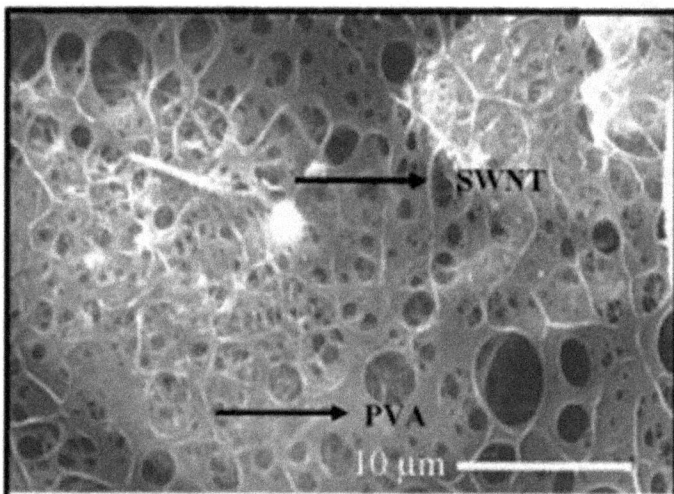

Fig. 10. SEM images of SWNT cross-linked PVA nanocomposites.

Alignment and decoration of noble metals on CNTs wrapped with CMC was also achieved under MW condition. CNTs, such as SWNT, MWNT, and C-60, were well dispersed using the sodium salt of CMC under sonication (58). The addition of respective noble metal salts then generated noble metal-decorated CNT composites at room temperature. However, aligned nanocomposites of CNTs could only be generated by exposing the above nanocomposites to MW irradiation. The general preparative procedure is flexible and provides a straightforward route to manufacturing functional metal coated CNT nanocomposites (Figure 11).

Varma et al. (59) have developed a simple method for the bulk synthesis of monodispersed spinel ferrite nanoparticles with size selectivity using readily available inorganic precursors via a water-organic interface . Hydrothermal as well as MW hydrothermal methods are applicable but the use of MW has the advantage of low temperature, expedient synthesis. The synthesized particles are highly dispersible and are stable in nonpolar organic solvents, which is important in their use as ferrofluids and other magnetic applications. Surface functionalization of the As-synthesized particles with lysine made them water dispersible for possible biological applications (Figure 12).

It has been stated that volumetric and selective heating using MW irradiation may reduce the thermal gradients in the reaction, thereby generating a more homogeneous product with faster consumption of the starting materials. (60)

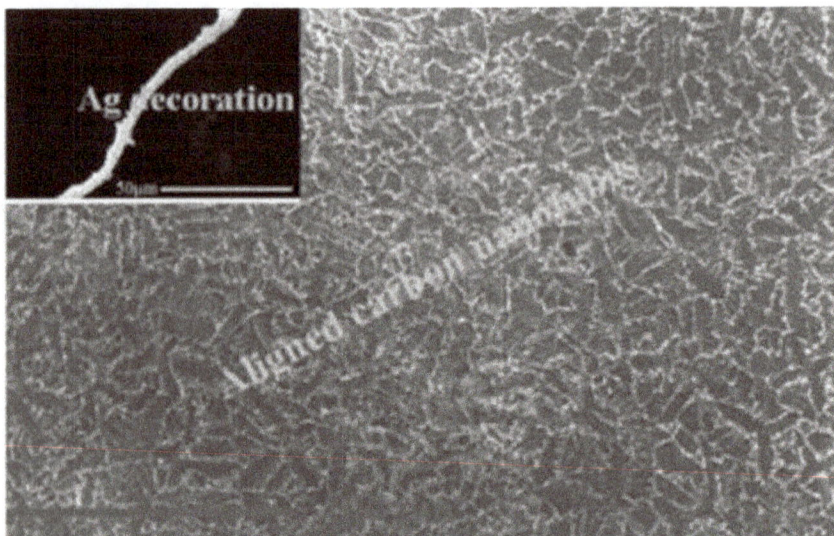

Fig. 11. Aligned CNTs in CMC polymer matrix.

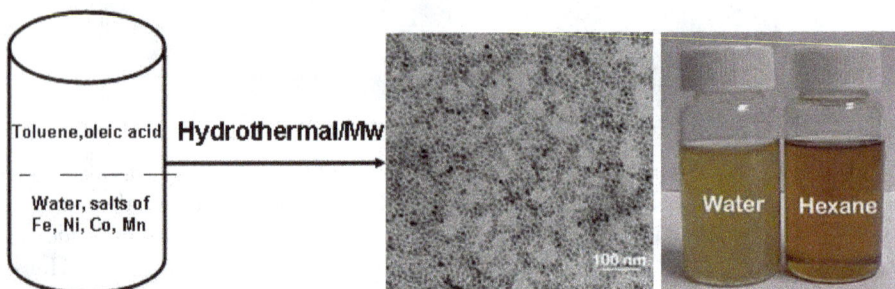

Fig. 12. MW hydrothermal synthesis and fictionalization of nanoferrites.

Varma et al.(61) have synthesized, for the first time under MW irradiation conditions, dendritic ferrites with micro-pine morphology (see Figure 13) without using any reducing or capping reagents. With this adjustment, nano ferrites could then be functionalized (Scheme 2) and coated with Pd metal, which catalyzes various C-C coupling reactions. An assortment of magnetic, nanoparticle-supported metal catalysts have been readily prepared from inexpensive starting materials and shown to catalyze a variety of organic transformations such as oxidation (62), hydration (63), and reduction (hydrogenation) (64). Superior activity and the inherent stability of these catalyst systems coupled with their easy magnetic separation, which eliminates the prerequisite of catalyst filtration after completion of the reaction, are some of the supplementary sustainable attributes of these protocols.

Fig. 13. TEM image of dendritic α-Fe₂O₃.

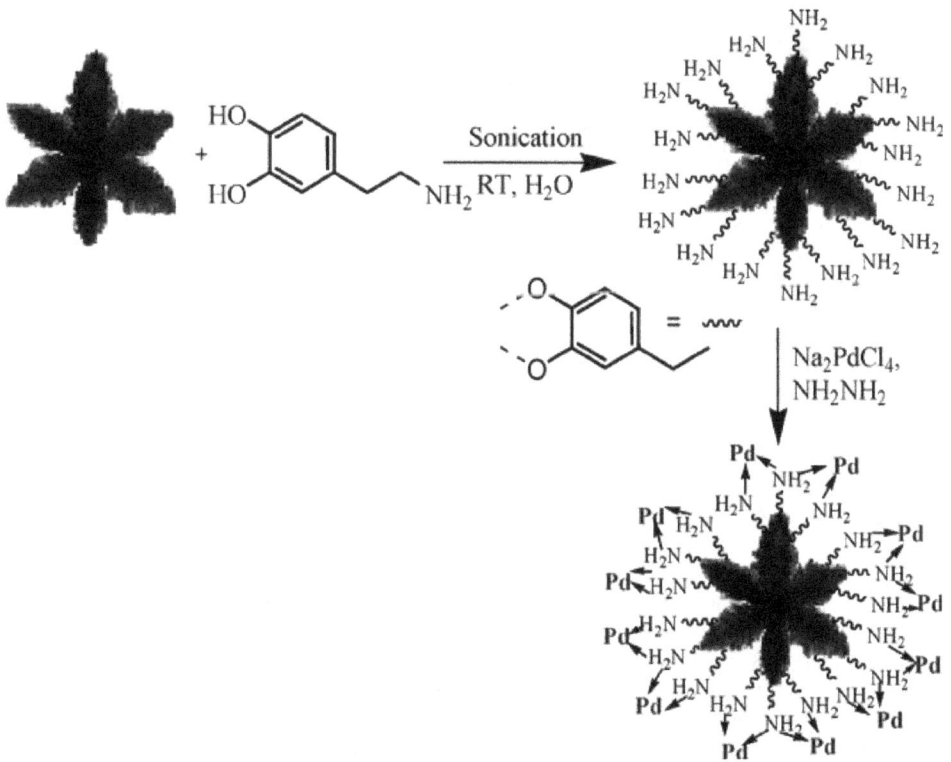

Scheme 2. Schematic diagram of α-Fe₂O₃ functionalization with amino group and Pd.

Microwave strategy can be also be expanded to a solid state reaction. Porous nanocrystalline TiO_2 and carbon coated TiO_2 using sugar dextrose as a template has been achieved through MW and the results were compared with conventional heating furnace (65). Out of three compositions, namely, 1:1, 1:3, and 1:5 (metal: dextrose), 1:3 favors formation of consistent porous structures (see Figure 14). This general and eco-friendly method uses a benign natural polymer, dextrose, to create spongy porous structures and can be extended to other transition metal oxides such as ZrO_2, Al_2O_3, and SiO_2.

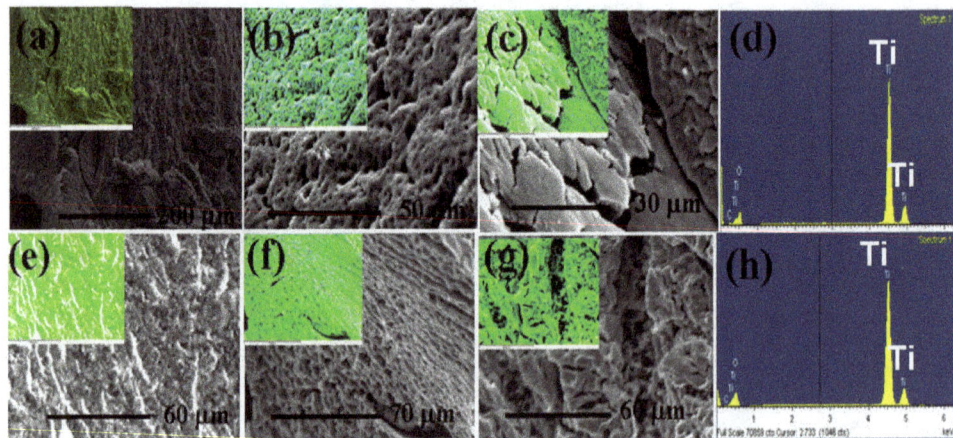

Fig. 14. (a)–(c) SEM images of 1:1, 1:3, and 1:5 (titania: dextrose molar ratio) titania sponges synthesized by microwave combustion and subsequently heated at 850 °C for 1 h by a conventional furnace (the inset shows the x-ray mapping images of the same with the green region showing titania and the red region showing carbon). (d) Representative (1:3 titania: dextrose molar ratio) sample of energy dispersive x-ray analysis (EDX) showing the presence of titania. (e)–(g) SEM images of 1:1, 1:3, and 1:5 (titania: dextrose molar ratio) titania sponges synthesized by heating at 850 °C for 1 h by conventional furnace (the inset shows the x-ray mapping images of the same with the green region showing titania and the red region showing carbon). (h) Representative sample (1:3 titania: dextrose molar ratio) of energy dispersive x-ray analysis (EDX) showing the presence of titania.

4. References

[1] X.Wang andY.Li, Chem. Commun., 2007, 2901.
[2] Y. Sun and Y. Xia, Science, 2002, 298, 2176.
[3] J. Chen, J. M. McLellan, A. Siekkinen, Y. Xiong, Z.-Y. Li and Y. Xia, J. Am. Chem. oc., 2006, 128, 14776.
[4] J. W. Stone, P. N. Sisco, E. C. Goldsmith, S. C. Baxter and C. J.Murphy, Nano Lett., 007, 7, 116.
[5] B. Wiley, Y. Sun and Y. Xia, Acc. Chem. Res., 2007, 40, 1067.
[6] J. Du, B. Han, Z. Liu and Y. Liu, Cryst. Growth Design, 2007, 7, 900.
[7] B. Wiley, T. Herricks, Y. Sun and Y. Xia, Nano Lett., 2004, 4, 2057.
[8] C. J. Murphy, A. M. Gole, S. E. Hunyadi and C. J. Orendorff, Inorg. Chem., 2006, 5, 7544.

[9] B. J. Wiley, Y. Chen, J. M. McLellan, Y. Xiong, Z.-Y. Li, D. Ginger and Y. Xia, Nanoletters, 2007, 4, 1032.

[10] Y.Xiong, H. Cai, B. J. Wiley, J.Wang, M. J. Kimand Y. Xia, J. Am. Chem. Soc., 2007, 129, 3665.

[11] J. Fang, H. You, P. Kong, Y. Yi, X. Song and B. Ding, Cryst. Growth Design, 2007, 7, 864.

[12] A. Narayan, L. Landstrom and M. Boman, Appl. Surf. Sci., 2003, 137, 208.

[13] H. Song, R. M. Rioux, J. D. Hoefelmeyer, R. Komor, K. Niesz, M. Grass, P. Yang and G. A. Somorjai, J. Am. Chem. Soc., 2006, 128, 3027.

[14] C. C. Wang, D. H. Chen and T. C. Huang, Colloids Surf. A, 2001, 189, 145.

[15] M. N. Nadagouda and R. S. Varma, Green Chem., 2006, 8, 516.

[16] P. Raveendran, J. Fu and S. L.Wallen, J. Am. Chem. Soc., 2003, 125,

[17] M. N. Nadagouda and R. S. Varma, Green Chem., 2007, 9, 632.

[18] R. R. Naik, S. J. Stringer, G. Agarwal, S. E. Jones and M. O. Stone, Nat. Mater., 2002, 1, 169.

[19] M. N. Nadagouda and R. S. Varma, Macromol. Rapid Commun., 2007, 28, 465.

[20] M. N. Nadagouda and R. S. Varma, Biomacromolecules, 2007, 8, 2762-2767.

[21] M. N. Nadagouda and R. S. Varma, Cryst. Growth Design, 2007, 7((12)), 2582-2587.

[22] M. N. Nadagouda and R. S. Varma, Cryst. Growth Design, 2007, 7((4)), 686-690.

[23] M. N. Nadagouda and R. S. Varma, Cryst. Growth Design, 2008, 8((1)), 291-295.

[24] J. A. Dahl, L. S.Maddux and J. E. Hutchison, Chem. Rev., 2007, 107, 2228.

[25] P. T. Anastas and J. C. Warner, Green, Chemistry: Theory and Practice, Oxford University Press, Inc., New York, 1998.

[26] V. L. Colvin, M. C. Schlamp, A. P. Alivisatos, Nature 1994, 370, 354.

[27] S. Maeda, S. P. Armes, Chem. Mater. 1995, 7, 171.

[28] P. G. Hill, P. J. S. Foot, R. Davis, Mater. Sci. Forum 1995, 191, 43.

[29] R. E. Schwerzel, K. B. Spahr, J. P. Kurmer, V. E. Wood, J. A. Jenkins, J. Phys. Chem. A. 1998, 102, 5622.

[30] T. Trindade, M. C. Neves, A. M. V. Barros, Scr. Mater. 2000, 43, 567.

[31] J. J. Tunney, C. Detellier, Chem. Mater. 1996, 8, 927.

[32] C. O. Oriakhi, M. M. Lerner, Chem. Mater. 1996, 8, 2016.

[33] L. Ouahab, Chem. Mater. 1997, 9, 1909.

[34] J. H. Choy, S. J. Kwon, S. J. Hwang, Y. H. Kim, W. Lee, J. Mater. Chem. 1999, 9, 129.

[35] C. Sanchez, F. Ribot, B. Lebeau, J. Mater. Chem. 1999, 9, 35.

[36] L. Luo, S. Yu, H. Qian, T. Zhou, J. Am. Chem. Soc. 2005, 127, 2822.

[37] S. Xiong, L. Fei, Z. Wang, H. Y. Zhou, W. Wang, Y. Qian, Eur. J. Inorg. Chem. 2005, 2006, 207.

[38] H. Qian, L. Luo, J. Gong, S. Yu, T. Li, L. Fei, Cryst. Growth Design 2005, 6, 607.

[39] J. Gong, L. Luo, S. Yu, H. Qian, L. Fei, J. Mater. Chem. 2006, 16, 101.

[40] W. Wu, Y. Wang, L. Shi, Q. Zhu, W. Pang, G. Xu, F. Lu, Nanotechnology 2005, 16, 3017.

[41] H. Kong, J. Jang, Chem. Commun. 2006, 3010.

[42] Z. Li, H. Huang, C. Wang, Macromol. Rapid Commun. 2006, 27, 152.

[43] S. Porel, S. Singh, T. P. Radhakrishnan, Chem. Commun. 2005, 2387.

[44] G. A. Gaddy, A. S. Korchev, J. L. McLain, B. L. Slaten, E. S. Steigerwalt, G. Mills, J. Phys. Chem. B 2004, 108, 14850.

[45] Z. H. Mbhele, M. G. Salemane, C. G. C. E. van Sittert, J. M. Nedeljkovic, V. Djokovic, A. S. Luyt, Chem. Mater. 2003, 15, 5019.

[46] M. N.Nadagouda, V. Polshettiwar and R. S. Varma, "J. Mater. Chem., 9, 2026 – 2031, (2009).

[47] N. N. Mallikarjuna, and R. S. Varma Aust. J. Chem., 62, 260–264(2009).

[48] M. N. Nadagouda, R. S. Varma, Green Chem. 2008, 10, 859.

[49] M. N. Nadagouda and R. S. Varma "Bulk and template-free synthesis of narrow diameter reduced polyaniline nanofibers at room temperature" Green Chemistry, 9, 632-637(2007).

[50] J.A. Dahl, B. L. S. Maddux, J. E. Hutchison, *Chem. Rev.* 2007, *107*, 2228.

[51] S. Kundu, H. Liang, Langmuir 2008, 24, 9668.

[52] W. Tu, H. Liu, J. Mater. Chem. 2000, 10, 2207.

[53] R. He, X. Qian, J. Yin, Z. Zhu, *J. Mater. Chem.* 2002, *12*, 3783.

[54] X. Zhang, S. K. Manohar, Chem. Commun. 2006, 2477.

[55] J. M. Campelo, T. D. Conesa, M. J. Gracia, M. J. Jurado, R. Luque, J. M. Marinasa, A. A. Romeroa, Green Chem. 2008, 10, 853.

[56] B. Hu, S. –B. Wang, K. Wang, M. Zhang, S. –H. Yu, J. Phys. Chem. C 2008, 112, 11169.

[57] M. N. Nadagouda, R. S. Varma, Macromol. Rapid Commun. 2007, 28, 842.

[58] M. N. Nadagouda, R. S. Varma, Macromol. Rapid Commun. 2008, 29, 155.

[59] B. Baruwati, M. N. Nadagouda, R. S. Varma, J. Phys. Chem. C. 2008, 112, 18399.

[60] A. Gerbec, D. Magana, A. Washington, G. F. Strouse, J. Am. Chem. Soc. 2005, 127, 15791.

[61] V. Polshettiwar, M. N. Nadagouda, R.S. Varma: Chem. Commun., 2008, DOI: 10.1039/B814715A.

[62] V. Polshettiwar, R. S. Varma, Org. Bio. Chem. *Org. Biomol. Chem.*, 2009, 7, 37–40

[63] V. Polshettiwar, R. S. Varma, Chem. Eur. J. 2009, 15, 1582-1586.

[64] V. Polshettiwar, B. Baruwati, R. S. Varma, Green. Chem. 2009, 11, 127-131.

[65] M. N.Nadagouda and R. S. Varma, Smart Materials and Structures, 5, 1-6 (2006).

6

One-Dimensional Meso-Structures: The Growth and the Interfaces

Lisheng Huang[1,2,3], Yinjie Su[2] and Wanchuan Chen[1]
[1]Department of Physics, National Cheng Kung University, Tainan,
[2]National Laboratory of Solid State Microstructures, Nanjing University, Nanjing,
[3]College of Sciences & College of Materials Science and Engineering,
Nanjing University of Technology, Nanjing,
[1]Taiwan
[2,3]China

1. Introduction

One-dimensional (1D) meso-structures have become the focus of intensive research worldwide due to their unique physics and potential to revolutionize broad areas of device applications. They act as the most basic building blocks of nano-electronic systems, nano-optics and nano-sensors, so the controlled growth of these meso-structures is important for applying them in these fields. Materials properties can be tuned through control of micro-structural characteristics such as the physical size, shape, and the surface. Efforts to explore structures with multiple length scales unite the frontiers of materials chemistry, physics, and engineering. It is in the design and characterization of advanced materials that the importance of new interdisciplinary studies may be realized [1-4]. Recent research focused on well-faceted meso-structures has shown that the shape as well as the hetero-[5, 6] or homo-junctions [7, 8] contribute much to the tuning of properties of structured materials. Many significant properties, including optical, chemical, as well as electronic, have been revealed to be shape- or junction-related. For example, the lasing behaviors of nonlinear optical nano-scale wires or belts derive from the resonance cavity effect functioned by the parallel end-faces of the nano-structures [9-11]. Quantitative characterization of optical waveguiding in straight and bent nanowires is achievable in active devices [12]. Such study has shown that the optimization of surfaces, boundaries, and interfaces in materials with well-faceted structures plays an important role in furthering the application of these materials.

For efficient fabrication and assembly of well-faceted meso-structures, the anisotropy of the crystal can be utilized to control the nucleation and manipulate the surface energy [13, 14]. Macroscopically, a crystal has different kinetic parameters for different crystal planes guided by certain growth conditions. After initial nucleation, a crystallite will commonly develop into a three-dimensional entity with well-defined, low index crystallographic facets. Thus, the growth anisotropy can be advantageously utilized to create crystals with specific desired characteristics through control of the growth conditions. It has been extremely successful in different growth systems, such as solution-based route for growing shaped nanocrystals, vapor-phase growth of quasi-one-dimensional meso-structures with well-defined cross sections and surface polarities as well as some other exotic configurations through vapor-liquid-solid (VLS) or vapor-solid (VS) process [15-20].

In this chapter, we have examined the growth mechanisms and the morphology evolutions of one-dimensional meso-structures systematically based on the experimental and theoretical aspects of crystal growth. The 1D ZnO meso-structures will be selected as an example to show the morphologic evolution at multiple length scales. The quasi-one-dimensional SnO_2 meso-structures are studied to describe the morphological multiformity of crystal growth. The outline of the chapter is as follows. In Sect. 2, the growth of ZnO meso-structures is discussed, which includes the controlled growth (Sect. 2.1), structural characterization and crystal models (Sect. 2.2), the growth process and mechanism (Sect. 2.3), and structure-related optical properties (Sect. 2.4). In Sect. 3, the SnO_2 zigzag meso-structures growth mechanism is discussed, which includes the controlled growth (Sect. 3.1), structural characterization and crystal models (Sect. 3.2), the morphological evolution mechanism (Sect. 3.3). Concluding remarks are given in Sect. 4.

2. ZnO meso-structures

2.1 Growth control

ZnO, a wide direct band-gap semiconductor, is piezoelectric and transparent to visible light [21]. It is attracting much attention for application in UV light-emitters, varistors, transparent high power electronics, surface acoustic wave devices, piezoelectric transducers, gas-sensors, photo-catalysts, and as a window material for display and solar cells [22-31]. The wurtzite structure of the ZnO crystal has pronounced anisotropy. It possesses three fast growth directions of $<2\bar{1}\bar{1}0>$, $<01\bar{1}0>$, and $\langle 0001 \rangle$. Currently much effort has been focused on the fabrication of ZnO nano-/micro-scale structures. A number of methods, based on solid reaction, solution based synthesis, and vapor rout have been developed to grow this material. These methods include the reaction of zinc salt with base, thermal decomposition, pulse laser deposition (PLD), thermal evaporation/vapor phase transport (CVD), metal-organic CVD, molecular beam epitaxy (MBE), electrochemical deposition, chemical bath deposition, aqueous solution decomposition, modified micro-emulsion, and sol-gel methods [32-44]. ZnO nano-/micro-structures of varied geometries, exemplified by wires/rods, belts/ribbons, comb-like structures, tetra-pod whiskers and their various assemblages have all been produced by our group (Figs. 1a-d).

We have also reported a new type of modulated and well-faceted ZnO microfibers, which was synthesized via a convenient CVD process [8]. Considering the decomposition of of $Ni(NO_3)_2$ at high temperature, we used nickel oxide (Ni_2O_3) as a catalyst. This proved to be an efficient way for growing the modulated microfibers. Fig. 2 shows a typical SEM morphology of the as-synthesized product. It is evident that the products are composed of microfibers with periodic junctions at a significant percentage (over 95%) of the yield and over 80% reproducibility from run to run.

The fibers with very thin junctions usually grow parallel to each other, and the roots appear to be compressed and broad. The lengths of the fibers typically range in between 200 and 500 μm. the longest one observed was nearly a millimeter. The spacing between two neighboring junctions normally ranges from 5 to 30 μm. The side surfaces of the fibers are well-faceted. Note that the V-shaped junction derives from the concavo-concave morphology, and the angle between the left and the right facets is exactly of 60 or 120 degrees. The fiber is characteristically decorated by periodically prism-like junction arrays. We refer to this structure as a "junction-prism" structure.

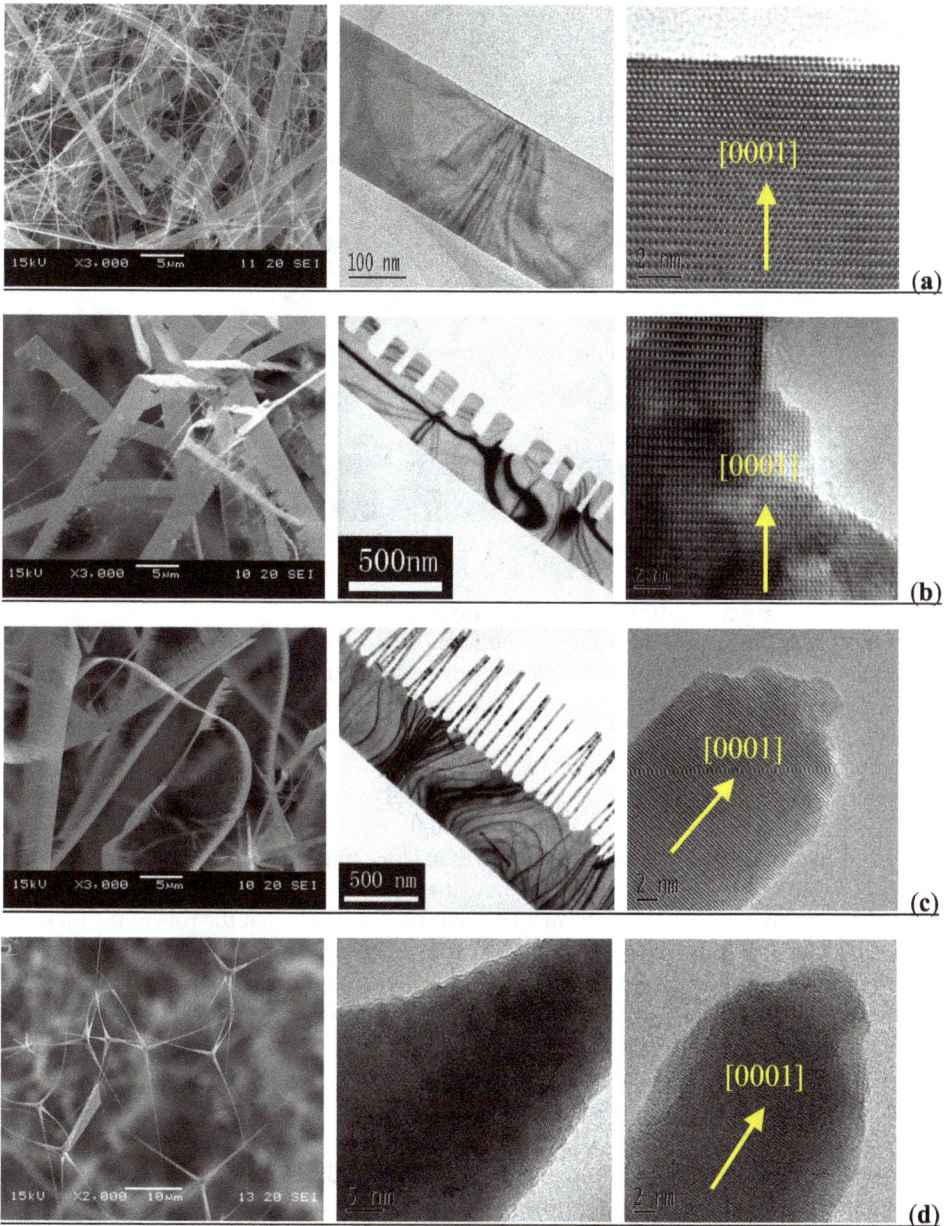

Fig. 1. ZnO meso-structures of belts (a), comb-like structures (b, c) and tetra-pod whiskers (d).

Fig. 2. ZnO junction-prism structure.

2.2 Structural characterization and crystal models

XRD and EDS measurements were performed for element analysis and phase determination. XRD studies show a typical wurtzite structure of ZnO with cell constants of a=0.324 nm and c=0.519 nm (Fig. 3a, JCPDS No. 36-1451). EDS studies (equipped in TEM) at the head, junction and root of a fiber show only peaks belonging to Zn and O without any other impurities (Figs. 3b-d).

In our studies, the presence of a small amount of nickel oxide is critical to synthesizing these modulated and well-faceted ZnO fibers. Although the vapor-liquid-solid (VLS) crystal growth mechanism explains the catalysis growth of some microstructures, no element nickel was observed found in our samples. Of course, the possibility exists that the quantities may be less than can be measured by XRD or EDS analysis. It is likely that the role of the nickel is the same as that of indium oxide and lithium carbonate for nanoring growth [45]. While no catalysts added for their growth, we believe that another important intrinsic factor for growth of the modulated microstructures is the intense anisotropy of the wurtzite-structured ZnO along different axes.

Crystallographic orientations of the fibers were obtained by EBSD. Fig. 4a shows the microfiber with a flat facet upturned, which was automated EBSD mapped for the selected area (Fig. 4b). As EBSD requires a highly tilted surface (near 70° tilt), several microfibers were searched until one was found to give indexable EBSD patterns with illumination corresponding to a flat surface tilted to 70°. The map displayed is corrected for the 70° tilt whereas the SEM image is not tilt-corrected. Pole figures obtained from the EBSD map data show the $[2\bar{1}\bar{1}0]$ direction aligned with the growth direction, the broad surfaces parallel to

{0001} plane and the side surfaces parallel to {01$\bar{1}$0} (Fig. 4c). A schematic unit cell displayed in the orientation was obtained by EBSD (Fig. 4d). The growth direction of the microfiber is [2$\bar{1}$$\bar{1}$0] ($a$ axis) and the side surfaces are ± (0001). The broad top and bottom surfaces are parallel {01$\bar{1}$0} planes.

It was found there are two types of oriented fibers in the production. Fig. 5a shows the crystal models. This is consistent with the crystal structure of the ZnO. The wurtzite structure of the ZnO crystal has pronounced anisotropy, it possesses three fast growth directions of < 2$\bar{1}$$\bar{1}$0 >, < 01$\bar{1}$0 >, and ⟨0001⟩. Generally, [0001] is the fastest based on the kinetic mechanism involved. [0001] growth minimizes the area of exposed {0001} faces (Fig. 5b). Under thermodynamic equilibrium conditions, the surface energy of the polar {0001} planes is larger than that of the nonpolar planes of {01$\bar{1}$0} or {2$\bar{1}$$\bar{1}$0}. Moreover, the surface energies differ less between the {01$\bar{1}$0} and {2$\bar{1}$$\bar{1}$0} planes. Fig. 5c illustrates the basic configurations evolved from ZnO hexagonal unit in Fig. 5b. Changing the growth condition to activate various growth facets, microstructures would be synthesized in shapes with higher complexity than those of the familiar wire, rod, belt, and sphere-like structures. Thus, it is often found that the produced well-faceted ZnO fibers with periodic junction-prisms preferentially grow along [01$\bar{1}$0] as opposed to the [2$\bar{1}$$\bar{1}$0] direction. The structure model shown in Fig. 5a (insert) illustrates a [01$\bar{1}$0] preferred growth axis of the fiber and the geometric relationships between all its outer facets.

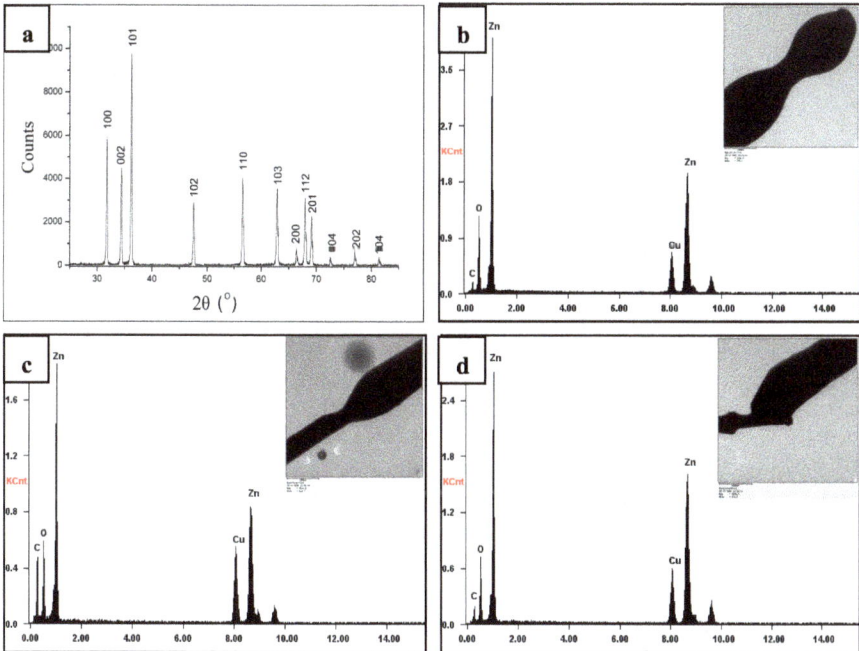

Fig. 3. XRD pattern showing wurtzite structure of ZnO (a), EDS studies at the head, the junction and the root of a fiber (b-d).

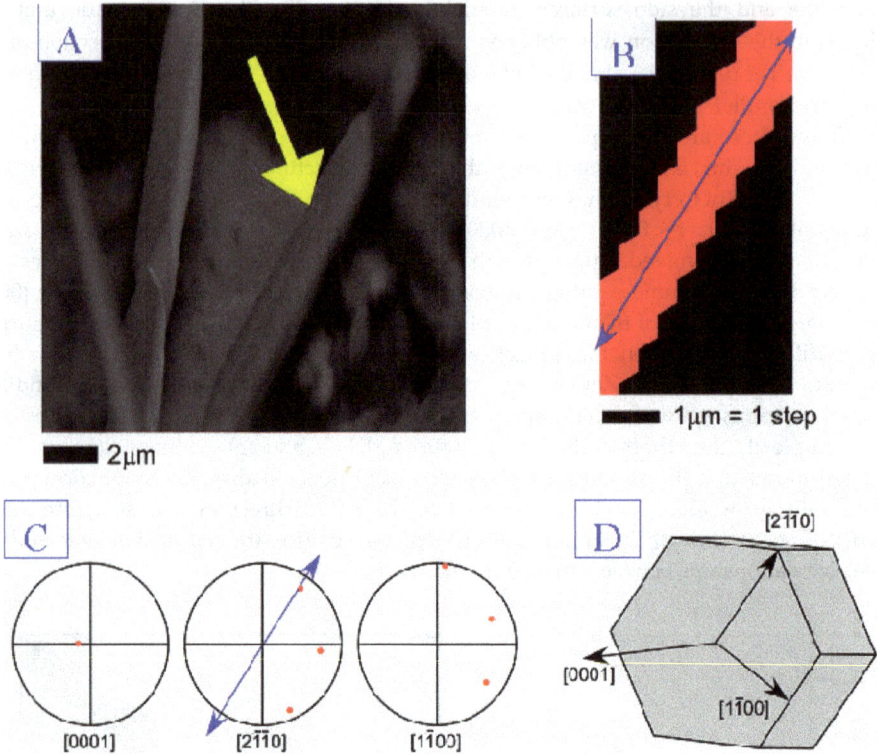

Fig. 4. EBSD measurement for a microfiber.

2.3 Growth process and mechanism

SEM investigations on the microfibers demonstrate growth mechanism of the junction-prisms structures. Fig. 6a shows a newly growing head of a fiber, the growth unit of a nearly hexagonal prism is grown perpendicularly on the nanobelt base. Presumably, the small growth head would develop anisotropically and contact the adjacent unit forming a junction (Fig. 6b). Some similar configurations have been studied by our group [46]. We found that ZnO nano/microcombs are kin to these fibers: every tooth could be considered as a segment unit of the fiber. If the teeth grow short in the [0001] direction and thick in diameter, then they would contact each other and the morphology should be identical to these junction-like fibers (Fig. 6c, d). Additional experiment also showed that large amount necklace-shape structures could be produced. Every microstructure consisted of a row of rhomboids that are equally separated on a straight base of a narrow nanobelt (Fig. 6e, 6f). An anisotropic growth process is shown as follows: a nanobelt base was formed by fast catalysis growth along $[2\bar{1}\bar{1}0]$ or $[01\bar{1}0]$, followed by slow growth along [0001], forming the separate units of nearly hexagonal prisms. Developing on the nanobelt, these small units merged and formed junctions.

Some SEM images of the fibers show segment units quenched at different stages of their growth, and careful examination of the unit's morphology gives insight into the growth process. Fig. 7 (right panel) shows typical units in various stages of growth, along with their schematics (left panel). Although the SEM images are of different fibers, it is presumed that each fiber undergoes a similar sequence of steps during the growth process.

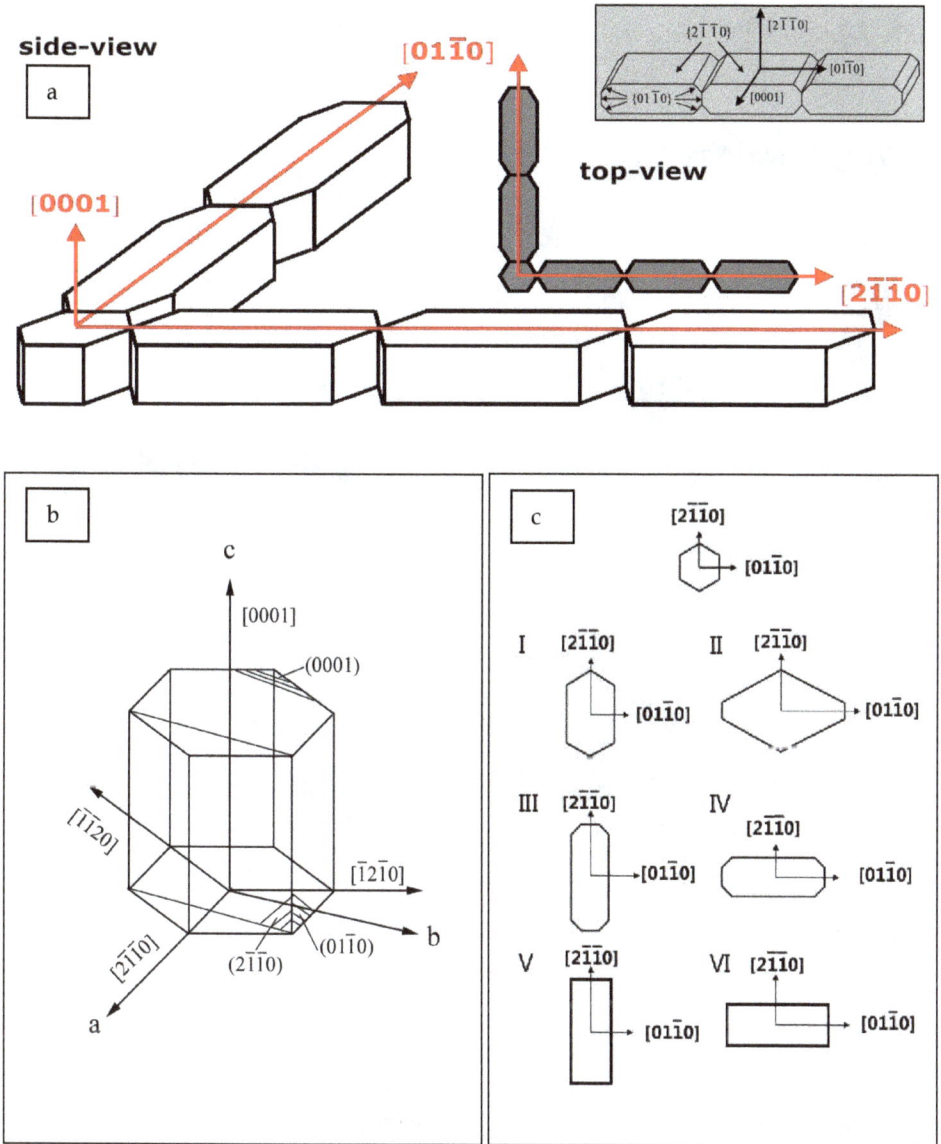

Fig. 5. (a) crystal mode of the two types of oriented ZnO fibers, (b) crystal mode of wurtzite structure of ZnO, (c) the basic configurations evolved from ZnO hexagonal unit.

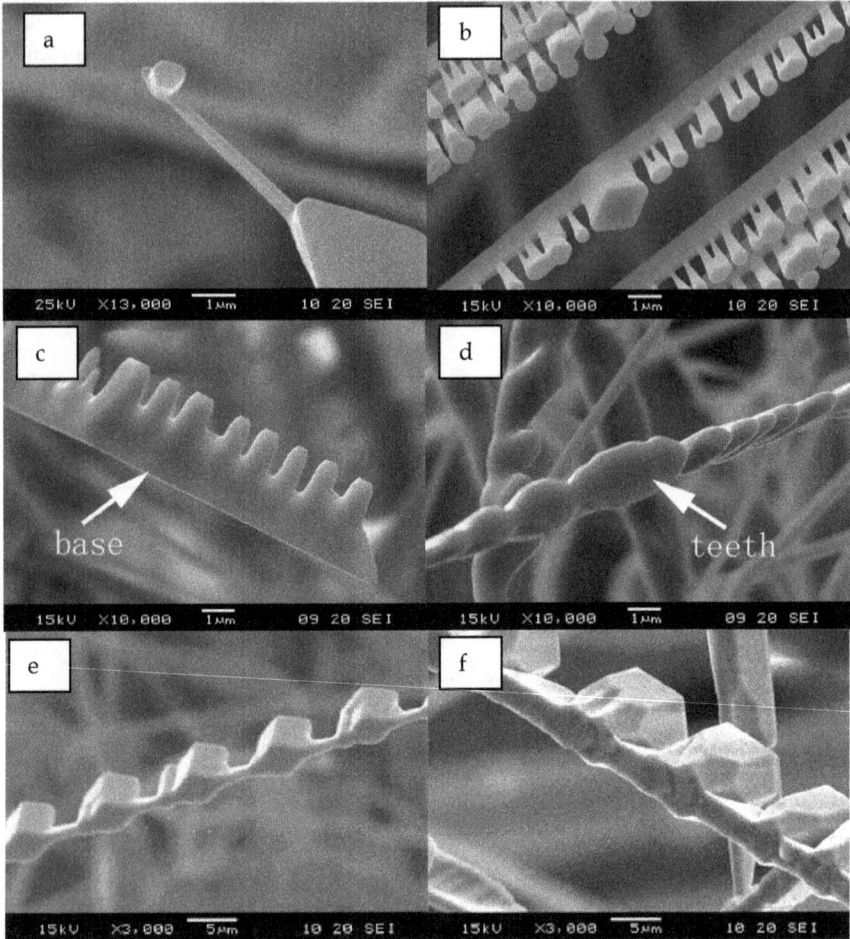

Fig. 6. (a) a newly growing head of a fiber, (b) small growth heads would develop anisotropically and contact the adjacent unit forming a junction, (c, d) ZnO micro-combs are kin to these fibers, (e, f) necklace-shape structures.

For example, consider the growth of $[01\bar{1}0]$ oriented fibers:

1. The first step is to grow a $[01\bar{1}0]$ oriented base with $\{2\bar{1}\bar{1}0\}$ side surfaces and top surfaces of $\{0001\}$. Subsequent grow will be self-modulated by nucleation and growth of the epitaxial pyramids on the c-face, (0001), of the base. The separate units growing along the c-axis on the base have hexagonal shapes (Fig. 7a, left panel are the crystal models). The six sided surfaces are equivalent planes of $\{01\bar{1}0\}$.

2. The units constructing a regular fiber with periodic junctions exhibit a prolonged eight-square shape, where the four profile $\{01\bar{1}0\}$ faces are partly exposed, and two broad $\{2\bar{1}\bar{1}0\}$ faces neighbor them (Fig. 7b). This evolution could be explained by an enhanced growth along $[01\bar{1}0]$ (the base growth direction) and a confined epitaxy perpendicular to $[01\bar{1}0]$.

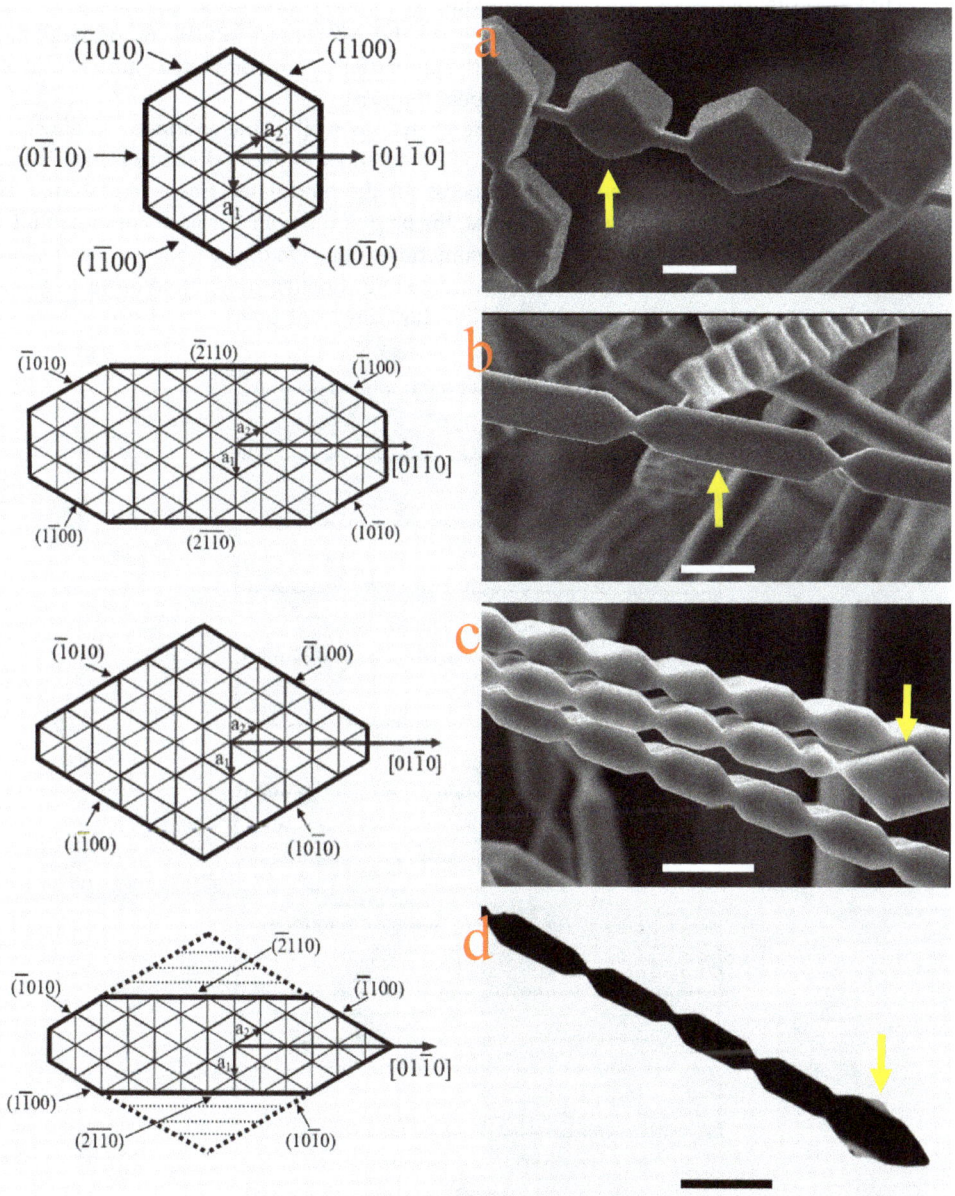

Fig. 7. Typical units in various stages of growth along with their schematics.

3.   A prolonged hexagonal head of a fiber shows evidence for enhanced growth along [01$\bar{1}$0] (Fig. 7c).
4.   TEM observations show that the anti-confined epitaxy process is perpendicular to [01$\bar{1}$0]. Thin epitaxial layers are growing on the two broad {2$\bar{1}$10} facets, and the growth would cease once the four {01$\bar{1}$0} side surfaces are completely grown out of existence and the two {2$\bar{1}$10} facets disappear (Fig. 7d).

Thus, we can deduce that two-step anisotropic growth as well as the confinement effect of the base (substrate) could result from these modulated and well-faceted junction-like fibers. A schematic representation of the growth process for the modulated fiber is illustrated in Fig. 8. Note that in the practical growth process, the base of the fiber has been combined into one united body with the segment units, but sometimes one side of the fiber is thicker than that of the other side. This is evident in the contrast in SEM images (Fig. 2). In order to grow very regular segment units, the growth condition should be controlled.

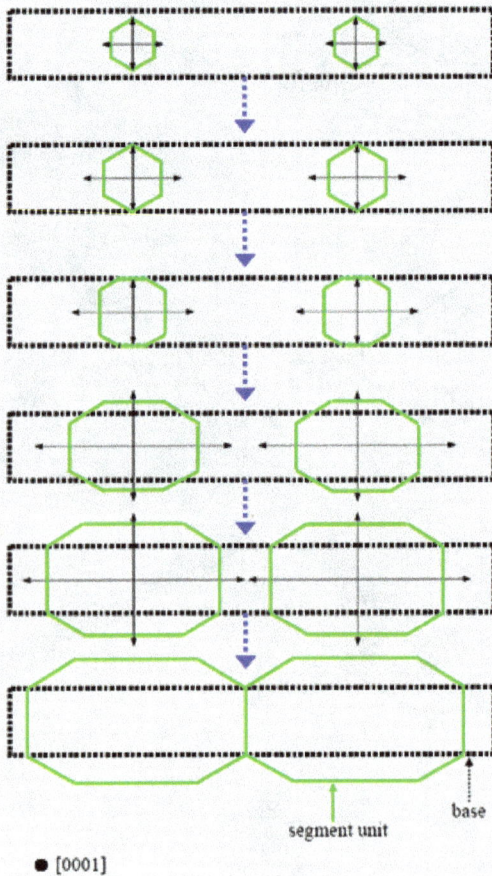

Fig. 8. A schematic representation of the growth process for the modulated fiber.

2.4 Structure-related optical properties

Because it approximately undergoes a thermodynamically equilibrium during the growth process of the fibers, all facets of a fiber are commonly low index crystallographic faces. According to the growth process and the crystal models, all the side surfaces of $\{01\bar{1}0\}$ and $\{2\bar{1}\bar{1}0\}$ should have coordinative crystalline qualities. μ-Raman studies further proved this point of view. The Raman spectra obtained at the $\{01\bar{1}0\}$ facets of a junction and at the $\{2\bar{1}\bar{1}0\}$ facets on the stem are shown in Fig. 9 (the inset shows the sample), respectively. No apparent difference is observed between the spectra from the junction facets and the stem facets.

Since the wurtzite structure of ZnO belongs to the C_{6v}^4 space group, the zone center optical phonons are: $A_1+2E_2+E_1$ [47]. In the spectra, two Raman active E_2 modes were observed at 101 and 437 cm^{-1}, and four Raman active modes--A_1 and E_1 transverse (TO), at 380 and 407 cm^{-1}, and longitudinal (LO), at 574 and 583 cm^{-1}, with second order vibrations observed at 208, 334 and 1050-1200 cm^{-1}. These results can be entirely explained on the basis of the ZnO crystal [48], and signify the good crystalline properties of the junction stem facets of a fiber.

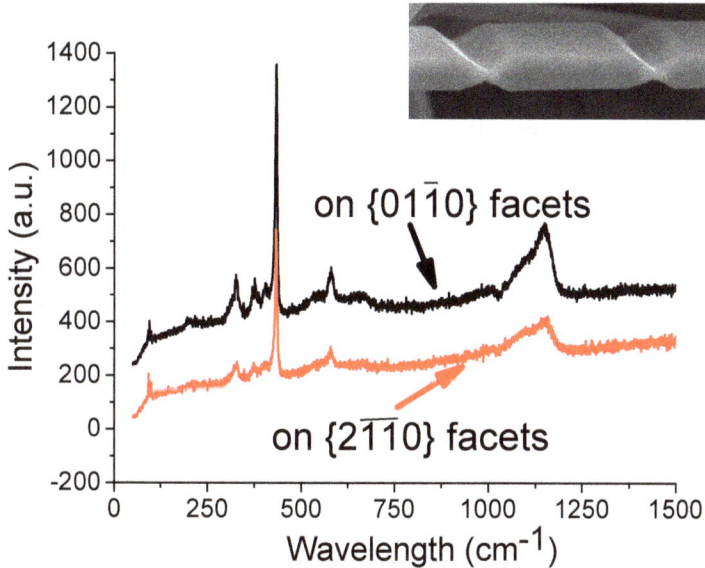

Fig. 9. Raman spectra obtained at the $[01\bar{1}0]$ facets of a junction and at the $\{2\bar{1}\bar{1}0\}$ facets on the stem.

Room temperature micro-PL spectra shown in Fig.10 indicate the enhancement of the green light emission at the junction. The spectrum obtained from the part between the two junctions consists of an intensive UV peak at 383 nm and a weak green band around 510 nm. The spectrum around the junction indicates that the green band is strong. This was further demonstrated using a PL microscope. The PL microscopy images show the fibers emitting strong green light at the junctions (Fig. 11h).

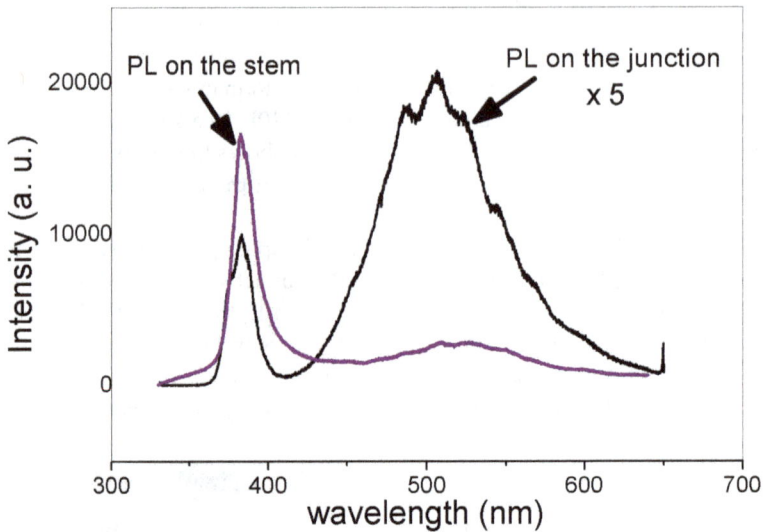

Fig. 10. Room temperature micro-PL spectra on the stem and at the junction.

It is generally accepted that the UV peak at 383 nm resulted from free excitonic emission of ZnO [49], while the green band arises from the recombination of a shallowly trapped electron within a deeply trapped hole [50]. Note that two neighboring units form one thin junction, the V-like slots upon/below the junctions are not suitable as a platform (substrate) for uniform epitaxial growth of the crystals, thus the intrinsic defects such as oxygen vacancies easily develop, resulting in the enhancement of the green light emission [51]. However, the further results of fluorescence microscopy suggest that the inhomogeneous PL emission of green light along the fiber stem, which is characterized by the periodic enhancement at the nearly isometric junctions, should be mainly attributed to the wave-guide property of the well-faceted fibers.

The produced well-faceted ZnO fibers with periodic junctions preferentially grow along $[01\bar{1}0]$ as opposed to the $[2\bar{1}\bar{1}0]$ direction by catalyzing growth. These fibers usually grow broad roots, and the bottom surfaces are $(000\bar{1})$ (Fig. 11a). When the fibers were dispersed onto the quartz substrates by drop-casting, most of the fibers attach to the substrate with broad {0001} facets (Fig. 11c and 11e), the fragments without broad roots (Fig. 11b) attach with {$2\bar{1}\bar{1}0$} facets (Fig. 11d and 11f).

These natural junction-prism arrays as well as the well-faceted surfaces associated with the transparent and homogeneous nature of crystalline ZnO medium offer sharp interfaces between ZnO and air (or other media) for guiding the propagation of light effectively. The optical morphology of the fiber shown in a typical barcode-like black-bright contrast (Fig. 11d) was imaged with a transmission optical microscope. Note in the experimental setup, the parallel light used to illuminate the sample in the microscope came from a lamp underlying the sample, while the camera was located atop the sample. The dark contrast regions correspond to the junction-prisms, while the bright contrast regions correspond to the building blocks of the fiber, which are separated by the junctions. This typical optical phenomenon suggests that refraction and reflection are strongly modulated by the junction-prism arrays within this structural fiber. When parallel light propagates perpendicular to the boundary between the ZnO crystal ($n_{ZnO} \approx 2$) and air medium ($n_{air} \approx 1$), it splits into two parts: light transmitted into ZnO and the light reflected back into air. Considering the reflection and transmission coefficients of ZnO crystal, about 88 % of the incident light was refracted. Moreover, because the ZnO crystal is optically denser than air, no light should enter the air from the V-shape surfaces of the upper junction-prism, where the entire incident light was reflected back due to total reflection.

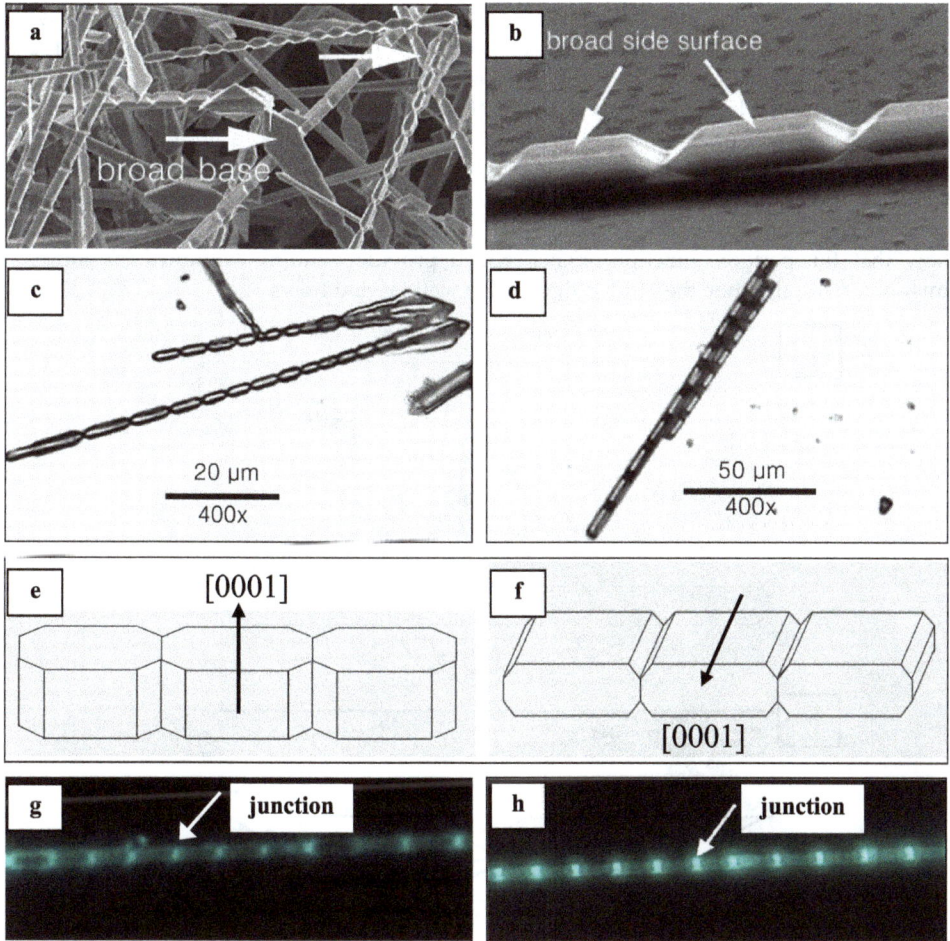

Fig. 11. Most of the fibers attach to the substrate with broad {0001} facets (a, c, e and g), other fragments without broad roots attach with {$2\overline{1}\overline{1}0$} facets (b, d, f and h).

When the fibers were excited by UV light (wavelength: 325-380 nm) with a fluorescence microscope, it is interestingly found that the enhanced green light emits the periodic junctions (Fig. 11h). This result could be explained by the optical waveguide behavior of the well-faceted structure with the junction-prism arrays of ZnO. As to the side surfaces of single building blocks of a fiber, every two parallel broad $\{2\bar{1}\bar{1}0\}$ surfaces and two narrow $\{0001\}$ surfaces could serve as a natural square cavity/waveguides (Fig. 12a). In general, the Vo* centers contributing to the defect-related green emission should be present at the surface region of a given ZnO crystal [50]. An ideal model elucidates the featured enhancement of the green light emitting at the junction-prisms. Analyzing one of these Vo* centers, its emitting light is easily reflected by the two $\{2\bar{1}\bar{1}0\}$ surfaces along the z-axis (i.e., $[01\bar{1}0]$). Note Fig. 12b-1, if the fiber is uniform and cuboid in shape, it should be an ideal bar-like waveguide and the emitting light from the total-reflection widows would be sent out from the ends of the fiber due to total reflection. In this case, it can be considered as an ideal optical fiber. However, the junction-prism arrays of the present fibers destroy the total-reflection condition (Fig. 12b-2). Thus, the junction-prism arrays change the propagation paths of the emitted light and most of the light from the total-reflection windows is guided out of the junction-prism regions directly, resulting in enhanced illumination at the junctions. Moreover, even if the emitted light goes straight though a junction-prism, it would encounter the next junction, and be sent out at last. The thicker the junction of a fiber, the more easily light is arrested by the junction-prism (Fig. 12b-3). All these observations show that the periodic junction-prisms, which provide emitting windows for intrinsic emissions, naturally tune the guided light in the well-faceted fibers.

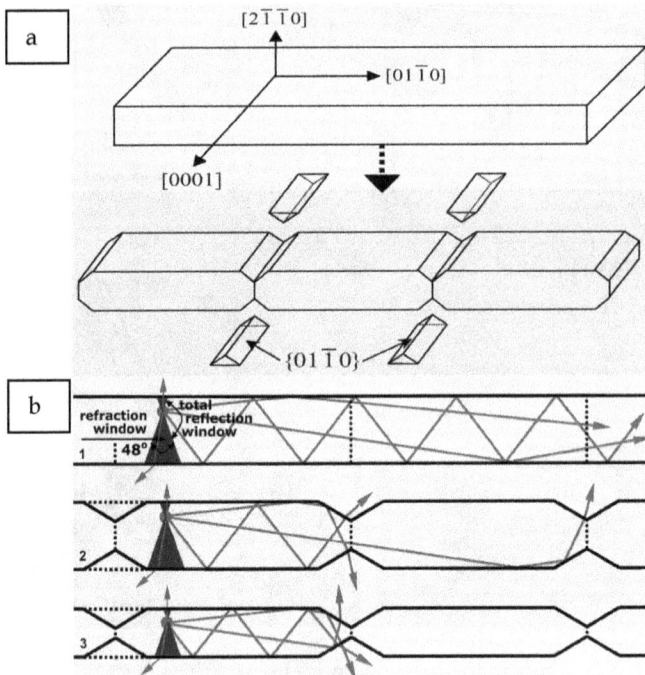

Fig. 12. Schematic illustrations of light reflection at the surface of the junction-prism structure.

3 SnO₂ zigzag shaped meso-structures

3.1 Controlled growth

SnO_2 has been paid attention in a variety of applications in chemical, optical, electronic and mechanical fields, due to its unique high conductivity, chemical stability, gas sensitivity and semiconducting properties [52]. Many syntheses of SnO_2 with different morphologies, such as nano-scale belt, wire, disk and dendrite, have been reported [53-56]. Herein, we report on a kinetics-controlled method to realize selective growth of SnO_2 unconventional zigzag shaped fibers. The morphological evolution process was investigated via SEM and TEM. Previously, the method used to grow SnO_2 single crystals is the high temperature gas phase reaction of evaporating SnO_2 or SnO to lead to SnO in the gas phase, and subsequent re-oxidation [57]. Here we used a lower temperature decomposition of SnO solid powders to produce Sn vapor for deposition, and then to oxidize it to SnO_2. In order to selected deposition of structured products, the growth kinetics was controlled [58].

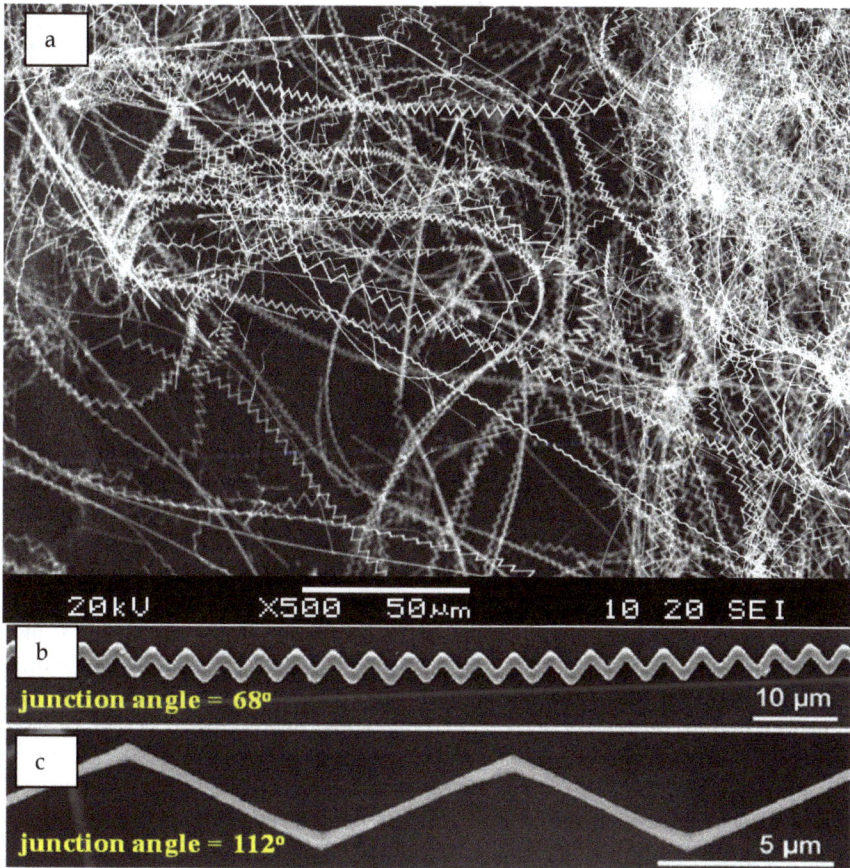

Fig. 13. (a) SEM image of the high yield SnO_2 zigzag fibers, the zigzags with junction angles of about 68º (b) and 112º (c), respectively.

3.2 Structural characterization and crystal models

The SEM image shows the typical growth of zigzag fibers as that shown in Fig. 13a. XRD results show the both structured products are with same crystallography structure: tetragonal rutile SnO_2. The high yield zigzags extend very long and collide with each other. The typical space of one zigzag period ranges from 2 to 10 μm, and the transverse swing is in the range of 5 to 10 μm. The length of the zigzag increases with the growth time, sometimes it can be up to several millimeters. In addition, there are more than three types of the angles of the zigzag junctions. Most of them are about 68° (Fig. 13b), and a few are approximately 112° (Fig. 13c), 90° and 124°, respectively.

The TEM images (Fig. 14a, 14b) give insight of a zigzag angle of 68°. Electron diffractions on the entire junction and on the two blocks reveal that the zigzag is single crystal. The growth directions of the two blocks are parallel to the crystallographic equivalent directions of [101] and [10$\bar{1}$], respectively. High-resolution TEM images (Fig. 14c, 14d) indicate the entire fiber has same lattice arrangements. The structural models are illustrated in Fig. 14b. Structurally, the ±[101] and ±[10$\bar{1}$] in tetragonal SnO_2 are equivalent directions. The angle between the [101] and [10$\bar{1}$] directions and that between the [$\bar{1}$0$\bar{1}$] and [$\bar{1}$01] is 68°, while the angle between the [101] and [$\bar{1}$01] directions and that between the [$\bar{1}$0$\bar{1}$] and [10$\bar{1}$] is 112°. The experimental results of about 68° and 112° correlate well with these values. The formation of a zigzag is mainly accomplished though repeated alternation of growth orientations between the ±[101] and ±[10$\bar{1}$]. Namely, a zigzag could be separated into two types of building blocks, which laterally combine each other periodically. The zigzags with other junction angles should repeatedly shift its growth directions along some other low-index directions, such as from [101] to [001].

3.3 Morphological evolution

Careful examination of the zigzag's morphologies gives insight into the growth habits. As that shown in Fig. 15a, we usually found thin fiber has narrow slab-like morphology with sharp junction corners. The top/down surfaces of the building blocks could be indexed as ±(010) planes, and the side surfaces are ±(10$\bar{1}$) and ±(101), alternately. After further growth, the morphologies of the sample would become well-faceted with some new ±(100) facets present opposite to the junctions (Fig. 15b, 15c). Although the states are of different fibers, it can be presumed that each fiber undergoes a similar sequence of steps during the morphological evolution. The zigzag fiber would be formed by a two-step growth process. The first step is to fast grow to finalize the zigzag frame; the second step is to laterally grow to thicken its diameter. The evolution process illustrated in Fig. 15d reflects the lateral thickening process. In the beginning, the vapor species favor deposition at the V-like slots and it results in some atom steps (Fig. 15e), and then the new arrived species continuously arrange at the steps parallel to the side surface. The epitaxies would cease once the arrange layers meet the ridges of the junctions, due to the higher energy there. At last, these homo-epitaxies equally thicken a fiber in width and some ±(100) facets are constructed at the same time. Note that the transverse swing of the fiber does not change all along, and the original zigzag frame decides the final frame ($A_0 = A_f$). The longer the growth time, the more (100) surface area is present. Ideally, the final morphology could be predicted to be a rectangular crystal bar with long axis parallel to [001] direction and enclosed by lower energy planes of ±(010) and ±(100). This evolution tents to minimize the surface free energy, so the growth should seek thermodynamical equilibrium and be mainly dominated by surface free energy.

Based on this argument, we can explain why fewer zigzags with 112° angels contrast against the zigzags with 68 ° angels in the products. If the zigzag growth changes from [101] to [10$\bar{1}$] periodically, some higher energy facets of ±(001) would be constructed after the lateral thickening process. Thus it is not favorable from the energy point of view.

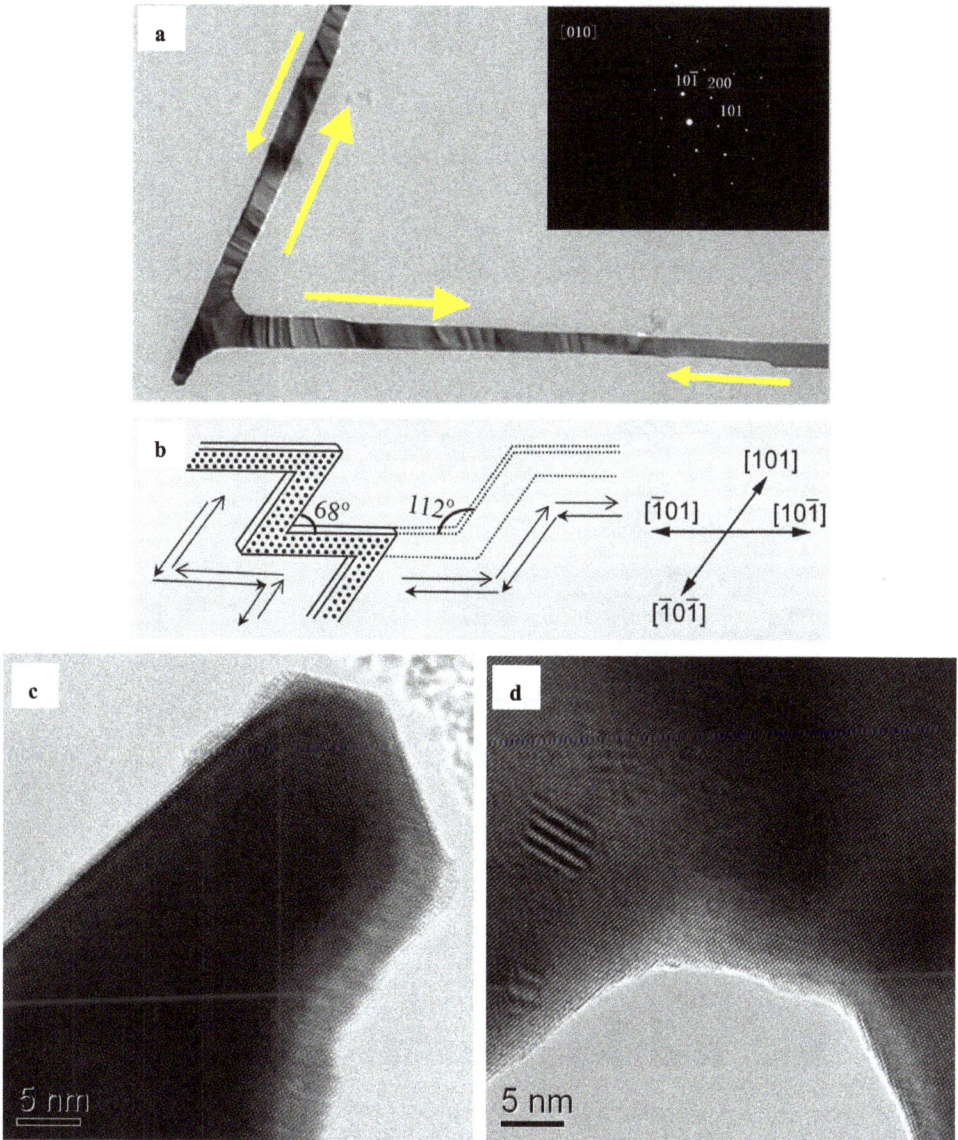

Fig. 14. (a) TEM image showing the junction angles of about 68°, the electron diffractions on the two blocks showing single crystal nature and <101> orientations, (b) crystal structural models for the zigzags, (c, d) high-resolution TEM images taken on the head and V-shape areas.

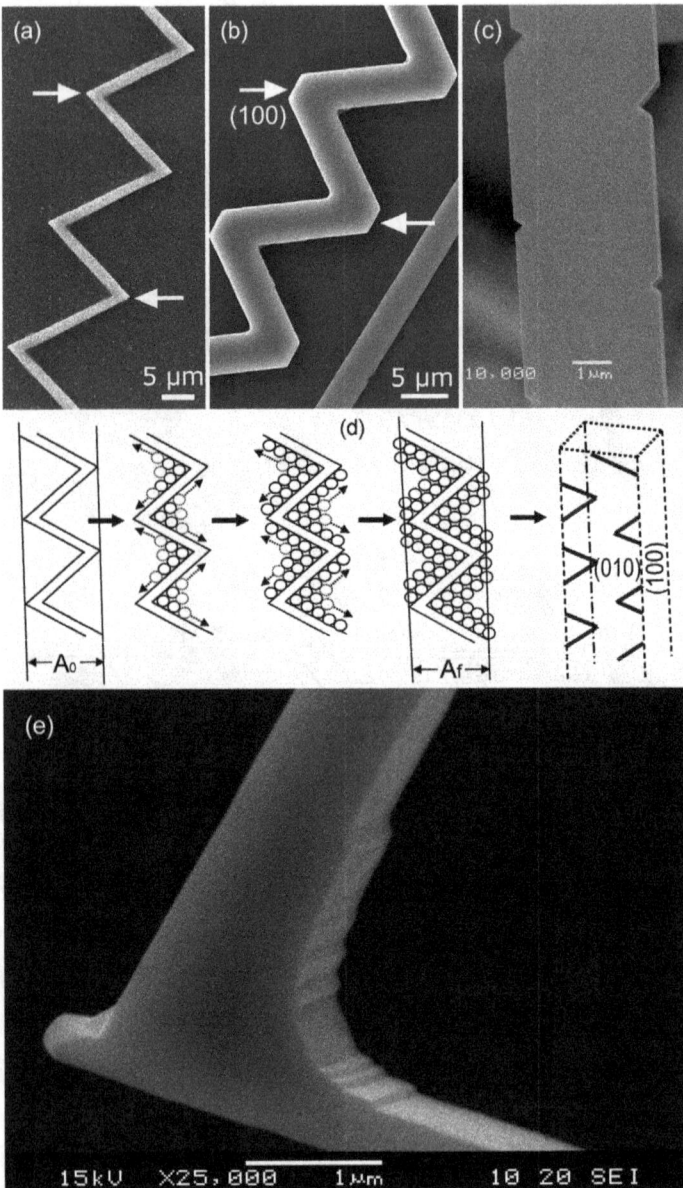

Fig. 15. SEM images of zigzags arranged (a-c) in order to show the evolution process, in (a) the zigzag has narrow slab-like morphology and sharp corners, after epitaxial growth, in (b), (c), the zigzag shows increased width and new ± (100) facets present, schematic illustration for the morphological evolution (d), the vapor species favor deposition at the V-like slots and it results in some atom steps.

4. Concluding remarks

The directional growth of well-faceted ZnO microfibers along different axes could be realized by catalyzing growth. The characterization of the fibers by optical and photoluminescence microscopy showed that the outer facets of the crystalline fibers provide excellent mirror-like surfaces for guiding light propagation along the fiber stem as well as the periodic junction-prisms. The structure-related optical properties of the fibers can be fully explained by a micro-structural model. The model explains several optical properties, such as luminance decreasing at the junction-prisms caused by refraction and total or partial reflection of light, as well as luminance enhancement at the junction-prisms related to waveguiding of the green emission along the ZnO fibers. Further integration of the ZnO junction-prisms into micro-devices could provide micro-scale modulation for light with different wavelengths. Such capability makes such fibers potentially suitable for enhanced light-illumination arrays. Reproducibly high-yield growth of SnO_2 zigzag nanofibers was achieved via controlling the reactant vapor concentration. The formation of the zigzag fiber based on the pre-growing nanobelt is suggested to be in a two-step process: the first is frame growth, which is accomplished through repeated orienting along equivalent directions; the second is lateral epitaxy, which thickens a fiber and results in well-faceted morphology. It is note that the present of intrinsic equivalent directions and the oscillation of external growth kinetics are key roles for producing zigzag structures. The elucidation of the growth mechanism should provide a fully controlled route for reproducibly high-yield growth of zigzag fibers of SnO_2 and give some valuable hints to synthesis other zigzag fibers. This well-faceted zigzag fiber could be studied as optical waveguide in its periodic structure and gas sensor component. These results have shown that the optimization of surfaces, boundaries, and interfaces in 1D meso-scale materials with well-faceted structures plays an important role in furthering the application of these materials.

5. Acknowledgement

We thank Dr. L. Pu, Dr. M. K. Lee, Prof. C. Tien and Prof. L. J. Chang for helpful discussions. L. S. Huang is truly grateful for financial support from NSFC (No. 60606020) and NCKU's "Aim for the Top University Project".

6. References

[1] A. P. Alivisatos, "Semiconductor clusters, nanocrystals, and quantum dots" Science 271, 933-937 (1996).

[2] A. D. Yoffe, "Semiconductor quantum dots and related systems-electronic, optical, luminescence and related properties of low dimensional systems" Advances In Physics 50, 1-208 (2001).

[3] Z. L. Wang, "ZnO nanowire and nanobelt platform for nanotechnology" Mater. Sci. Eng. R. 64, 33-71 (2009).

[4] H. Joyce, Q. Gao, H. Tan, Jagadish, C. "Phase perfection in zinc blende and wurtzite III-V nanowires using basic growth parameters" Nano Lett.10, 908– 915 (2010).

[5] B. M. Wong, F. Lonard, Q. M. Li, T. George "Nanoscale Effects on Heterojunction Electron Gases in GaN/AlGaN Core/Shell Nanowires" Nano Lett. 11, 3074-3079 (2011).

[6] X. F. Qiu, C. Burda, R. L. Fu, L. Pu, H. Y. Chen, J. J. Zhu, "Heterostructured Bi$_2$Se$_3$ nanowires with periodic phase boundaries" J. Am. Chem. Soc. 126, 16276-16277 (2004).

[7] L. Manna, D. J. Milliron, A. Meisel, E. C. Scher, A. P. Alivisatos, "Controlled growth of tetrapod-branched inorganic nanocrystals" Nat. Mater. 2, 382-385 (2003).

[8] L. S. Huang, S. Wright, S. Yang, D. Z. Shen, B. X. Gu, Y. W. Du, "ZnO well-faceted fibers with periodic junctions" J. Phys. Chem. B. 108, 19901-19903 (2004).

[9] M. H. Huang, S. Mao, H. Feick, H. Yan, Y. Wu, H. Kind, E. Weber, R. Russo, P. Yang, "Room-temperature ultraviolet nanowire nanolasers" Science 292, 1897-1899 (2001).

[10] J. Johnson, H. J. Choi, K. P. Knutsen, R. D. Schaller, P. Yang, R. J. Saykally, "Single gallium nitride nanowire lasers" Nat. Mater. 1, 106-110 (2002).

[11] M. Law, D. Sirbuly, J. Johnson, J. Goldberger, R. Saykally, P. Yang, "Ultralong nanoribbon waveguides for sub-wavelength photonics integration" Science 305, 1269-1271 (2004).

[12] C. J. Barrelet, A. B. Greytak, C. M. Lieber, "Nanowire photonic circuit elements," Nano Lett. 4, 1981-1985 (2004).

[13] X. G. Peng, L. Manna, W. D. Yang, J. Wickham, E. C. Scher, A. Kadavanich, A. P. Alivisatos, "Shape control of CdSe nanocrystals," Nature 404, 59-61 (2000).

[14] R. Jin, Y. C. Cao, E. Hao, G. S. Métraux, G. C. Schatz, C. A. Mirkin, "Controlling anisotropic nanoparticle growth through plasmon excitation," Nature 425, 487-490 (2003).

[15] Z. R. Tian, J. Voigt, J. Liu, B. Mckenzie, M. McDermott, M. Rodriguez, H. Konishi, H. F. Xu, "Complex and oriented ZnO nanostructures," Nat. Mater. 2, 821-826 (2003).

[16] T. Schlli, R. Daudin, G. Renaud, A. Vaysset, O. Geaymond, A. Pasturel, "Substrate enhanced supercooling in AuSi eutectic droplets" Nature 464, 1174– 1177 (2010).

[17] M. Tabuchi, A. M. Y. Ohtake, Y. Takeda, "X-ray CTR scattering measurement to investigate the formation process of InP/GaInAs interface" J. Phys.: Conf. Ser. 83, 012031 (2007).

[18] M. Verheijen, R. Algra, M. Borgstrm, G. Immink, E. Sourty, W. van Enckevort, E. Vlieg, E. Bakkers, "Three dimensional morphology of GaP-GaAs nanowires revealed by transmission electron microscopy tomography" Nano Lett. 7, 3051–3055 (2007).

[19] D. E. Perea, N. Li, R. M. Dickerson, A. Misra, S. T. Picraux, "Controlling Heterojunction Abruptness in VLS-Grown Semiconductor Nanowires via in situ Catalyst Alloying" Nano Lett. 11, 3117–3122 (2011).

[20] O. Shpyrko, R. Streitel, V. Balagurusamy, A. Grigoriev, M. Deutsch, B. Ocko, M. Meron, B. Lin, P. Pershan, "Surface crystallization in a liquid AuSi alloy" Science 313, 77-80 (2006).

[21] D. C. Look, "Recent advances in ZnO materials and devices" Mat. Sci. Eng. B. 80, 383-387 (2001).

[22] D. M. Bagnall, Y. F. Chen, Z. Zhu, S. Koyama, M. Y. Shen, T. Goto, "Optically pumped lasing of ZnO at room temperature" Appl. Phys. Lett. 70, 2230-2232 (1997).

[23] P. M. Martin, M. S. Good, J. W. Johnston, L. J. Bond, S. L. Crawford, "Piezoelectric films for 100-MHz ultrasonic transducers" Thin Solid Films 379, 253–258 (2000).

[24] J. Q. Xu, Q. Y. Pan, Y. A. Shun , Z. Z. Tian, "Grain size control and gas properties of ZnO gas sensors" Sensors Actuators B. 66, 277–279 (2000).

[25] M. H. Huang, S. Mao, H. Feick, H. Q. Yan, Y. Y. Wu, H. Kind, E. Weber, R. Russo, P. D. Yang, "Room-temperature ultraviolet nanowire nanolasers" Science 292, 1897–1899 (2001).

[26] J. H. Song, Y. Zhang, , C. Xu, W. Z. Wu , Z. L. Wang, "Polar Charges Induced Electric Hysteresis of ZnO Nano/Microwire for Fast Data Storage" Nano Lett. 11, 2829–2834 (2011).

[27] H. Kind, H. Yan, B. Messer, M. Law, P. Yang, "Nanowire ultraviolet photodetectors and optical switches" Adv. Mater. 14, 158–160 (2002).

[28] Y. Wu, H. Yan, P. Yang, "Semiconductor nanowire array: potential substrates for photocatalysis and photovoltaics" Top. Catal. 19, 197–202 (2002).

[29] G. Gordillo, "New materials used as optical window in thin film solar cells" Surf. Rev. Lett. 9, 1675–1680 (2002).

[30] W. I. Park, G. C. Yi, J. W. Kim, S. M. Park, "Schottky nanocontacts on ZnO nanorod arrays" Appl. Phys. Lett. 82, 4358–4360 (2003).

[31] A. F. Yu, H. Y. Li, H. Y. Tang, T. J. Liu, P. Jiang, Z. L. Wang, "Vertically integrated nanogenerator based on ZnO nanowire arrays" Physica Status Solidi. 5, 162-164 (2011).

[32] N. Audebrand, J. P. Auffredic, D. Louer, "X-ray diffraction study of the early stages of the growth of nanoscale zinc oxide crystallites obtained from thermal decomposition of four precursors" Chem. Mater. 10, 2450-2461 (1998).

[33] A. M. Morales, C. M. Lieber, "A laser ablation method for the synthesis of crystalline semiconductor nanowires" Science 279, 208–211 (1998).

[34] Z. W. Pan, Z. R. Dai, Z. L. Wang, "Nanobelts of semiconducting oxides" Science 291, 1947-1949 (2001).

[35] M. H. Huang, Y. Y. Wu, H. Feick, N. Tran, E. Weber, P. D. Yang, "Catalytic growth of zinc oxide nanowires by vapor transport" Adv. Mater. 13, 113–116 (2001).

[36] Y. G. Wei, W. Z. Wu, R. D. Guo, J. Yuan, D. Suman, Z. L. Wang, "Wafer-Scale High-Throughput Ordered Growth of Vertically Aligned ZnO Nanowire Arrays" Nano Lett. 11, 2829–2834 (2011).

[37] G. D. Yuan, W. J. Zhang, S. T. Lee "p-Type ZnO Nanowire Arrays" Nano Lett. 8, 2591–2597 (2008).

[38] Y. Li, D. W. Meng, L. D. Zhang, F. Phillip, "Ordered semiconductor ZnO nanowire arrays and their photoluminescence properties" Appl. Phys. Lett. 76, 2011-2013 (2000).

[39] D. S. Boyle, K. Govender, P. O'Brien, "Novel low temperature solution deposition of perpendicularly orientated rods of ZnO: substrate effects and evidence of the importance of counter-ions in the control of crystallite growth" Chem. Commun. 1, 80-81 (2002).

[40] L. Vayssieres, K. Keis, S. Lindquist, A. Hagfeldt, "Purpose-built anisotropic metal oxide material: 3D highly oriented microrod array of ZnO" J. Phys. Chem. B 105, 3350-3352 (2001).

[41] L. Vayssieres, "Growth of arrayed nanorods and nanowires of ZnO from aqueous solutions" Adv. Mater. 15, 464–466 (2003).

[42] L. Guo, S. H. Yang, C. L. Yang, J. N. Wang, W. K. Ge, "Synthesis and Characterization of Poly (vinylpyrrolidone)-Modified Zinc Oxide Nanoparticles" Chem. Mater. 12, 2268-2274, (2000).

[43] Z. R. Tian, J. A. Voigt, J. Liu, B. Mckenzie, M. J. Mcdermott, "Biomimetic arrays of oriented helical ZnO nanorods and columns" J. Am. Chem. Soc. 124, 12954–12955 (2002).

[44] L. E. Greene, M. Law, D. H. Tan, P. D. Yang "General Route to Vertical ZnO Nanowire Arrays Using Textured ZnO Seeds" Nano Lett. 5, 1231–1236 (2005).

[45] X. Y. Kong, Y. Ding, R. S. Yang, Z. L. Wang, "Single-crystal nanorings formed by epitaxial self-coiling of polar nanobelts" Science 303, 1348-1351 (2004).

[46] Y. J. Su, Y. J. Zhou, L. S. Huang, Y. F. Liu, S. Z. Shi, Y. N. Lv, "The morphology evolution mechanism of ZnO quasi-one-dimensional nanostructures" J. Inor. Mater. 24, 6 (2009).

[47] C. A. Arguello, D. L. Rousseau, S. P. S. Porto, "First-Order Raman Effect in Wurtzite-Type Crystals" Phys. Rev. 181, 1351-1363 (1969).

[48] T. C. Damen, S. P. S. Porto, B. Tell, "Raman Effect in Zinc Oxide" Phys. Rev. 142, 570-574 (1966).

[49] L. E. Greene, M. Law, J. Goldberger, F. Kim, J. C. Johnson, Y. F. Zhang, R. J. Saykally, P. D. Yang, "Low-temperature wafer-scale production of ZnO nanowire arrays" Angew. Chem. Int. Ed. 42, 3031-3034 (2003).

[50] A.Van Dijken, E. A. Meulenkamp, D.Vanmaekelbergh, A. Meijerink, "Identification of the transition responsible for the visible emission in ZnO using quantum size effects" J. Lumin. 90, 123-128 (2000).

[51] L. S. Huang, L. Pu, Y. Shi, R. Zhang, B. X. Gu, Y. W. Du, Y. D. Zheng, "Light propagation tuned by periodic junction-prisms within well-faceted ZnO fibers" Opt. Express, 13, 5263-5269, (2005).

[52] Z. M. Jarzebski, J. P. Marton, "Physical properties of SnO2 materials" J. Electrochem. Soc. 123, 199c-205c (1976).

[53] Z. R. Dai, Z. W. Pan, Z. L. Wang, "Ultra-long single crystalline nanoribbons of tin oxide" Solid State Commun. 118, 351-354 (2001).

[54] C. Y. Wen, J. Tersoff, K. Hillerich, M. C. Reuter, J. H. Park, S. Kodambaka, E. A. Stach1,and F. M. Ross "Periodically Changing Morphology of the Growth Interface in Si, Ge, and GaP Nanowires" Phys. Rev. Lett. 107, 025503 (2011).

[55] J. Q. Hu, Y. Bando, D. Golberg, "Self-catalyst growth and optical properties of novel SnO2 fishbone-like nanoribbons" Chem. Phys. Lett. 372, 758-762 (2003).

[56] A. Beltran, J. Andres, E. Longo, E. R. Leite, "Thermodynamic argument about SnO2 nanoribbon growth" Appl. Phys. Lett. 83, 635-637 (2003).

[57] B. Thiel, R. Helbig, " Growth of SnO2 single crystals by a vapour phase reaction method" J. Crystal Growth 32, 259-264 (1976).

[58] L. S. Huang, L. Pu, Y. Shi, R. Zhang, B. X. Gu, Y. W. Du, S. Wright, "Controlled growth of well-faceted zigzag tin oxide mesostructures" Appl. Phys. Lett., 87, 163124 (2005).

7

Crystal Habit Modification Using Habit Modifiers

Satyawati S. Joshi
University of Pune,
India

1. Introduction

The synthesis of inorganic materials with a specific size and morphology has recently received much attention in the material science research area. Morphology control or morphogenesis is more important for the chemical industry than size control. Many routes have been reported to control the crystal growth and eventually modify the morphology of the crystals. For crystal-habit modification, crystals are grown in the presence of naturally occurring soluble additives, which usually adsorb or bind to the crystal faces and influence the crystal growth or morphology. A number of recent investigations show that such type of crystal-habit modifiers can be used to obtain inorganic crystals with organized assemblies. (Xu, et al. 2007, Yu & Colfen 2004, & Colfen, 2001).

The crystal-habit modifiers may be of a very diverse nature, such as multivalent cations, complexes, surface active agents, soluble polymers, biologically active macromolecules, fine particles of sparingly soluble salts, and so on. (Sarig et al.,1980) These crystal modifiers often adsorb selectively on to different crystal faces and retard their growth rates, thereby influencing the final morphology of the crystals. (Yu & Colfen, 2004) The strategy that uses organic additives and/or templates with complex functionalization patterns to control the nucleation, growth, and alignment of inorganic crystals has been universally applied for the biomimetic synthesis of inorganic materials with complex forms. (Qi et al., 2000) The biomimetic process uses an organized supramolecular matrix and produces inorganic crystals with characteristic morphologies. (Xu et al., 2007& Loste &Meldrum, 2001) Understanding the mechanism involved in such a matrix-mediated synthesis has a great potential in the production of engineering materials. Thus, catalyst particles of controlled size and morphology, magnetic materials with appropriate anisotropy, highly porous materials, composites, and well-organized crystallite assemblies can be produced by this synthesis method. (Sinha etal.,2000) Using water-soluble polymers as crystal modifiers for controlled crystallization is widely expanding and becoming a benign route for controlling and designing the architectures of inorganic materials. (Yu & Colfen, 2004) Investigators have used different double hydrophilic block copolymers, such as poly(ethylene glycol)-block-poly(methacrylic acid), to control the morphology of a number of inorganic salts, namely, $CaCO_3$, (Sedlak & Colfen, 2001, Rudolff et al., 2002, Meng et al., 2007, Guo et al., 2006, He et al., 2006, Wang et al., 2005, Meldrum et al., 2007, Gorna et al.,2007, & Colfen &Qi, 2001) $BaCO_3$, $CdCO_3$, $MnCO_3$, $PbCO_3$, (Yu et al.,2003) $BaCrO_4$, (Liu et al.2005& Yu et al.,

2002) $BaSO_4$, (Qi et al., 2000, Robinson et al.,2002, Wang et al., 2005,& Yu et al., 2005) tolazamide, Pb- WO_4, (Kuldipkumar et al., 2005) and so forth. In the early stages, gel matrices have been used for the control of nucleation and morphology in aqueous solution-based crystal growth. (Yu et al., 2007 Oaki &Imai, 2003) Investigators have used poly(vinyl alcohol) (PVA)-, agar-, gelatin-, and pectin-based gel matrices to control the morphology of inorganic crystals such as PbI_2, AgI, $Ag_2Cr_2O_7$, $PbSO_4$, $PbCl_2$, and so forth. (Henisch, 1988) The advantage of a gel medium is believed to be the reduction of the nucleation rate and suppression of convection. (Yu & Colfen, 2004) The functional groups, such as amine, amide, carboxylic acid and so forth, are known to significantly influence the mineralization process. Among the reported common gel matrices used as crystal-habit modifiers, PVA is a water soluble synthetic polymer with excellent film-forming and emulsifying properties. PVA is a crystalline polymer with a monoclinic structure and is known for its biological activities. (Merrill & Bassett, 1975) Also, PVA is reported to have been used for the morphology control of $K_2Cr_2O_7$, AgBr, and $CaCO_3$, (Sinha, 2001) and even for the selective nucleation of $CaCO_3$ polymorphs. (Lakshminarayanan, 2003) In this chapter, the results on morphological changes using polymers as habit modifiers are discussed on the basis of nucleation theory and growth process.

2. Crystal habit

Although crystals can be classified according to seven crystal systems, the relative sizes of the faces of a particular crystal can vary considerably. This variation is called a modification of habit.

2.1 Crystal habit modifications
2.1.1 Crystal morphology and structure
The morphology of a crystal depends on the growth rates of the different crystallographic faces. Some faces grow very fast and have little or no effect on the growth form; while slow growing faces have more influence. The growth of a given face is governed by the crystal structure and defects on one hand and by the environmental conditions on the other. (Mullin 2002)

A number of proposed mechanisms and theories have been put forth to predict the equilibrium form of a crystal. According to the Bravais rule, the important faces governing the crystal morphology are those with the highest reticular densities and greatest interplanar distances, d_{hkl}. Or in simpler terms, the slowest growing and most influential faces are the closest packed and have the lowest Miller indices. The surface theories suggest that the equilibrium form should be such that the crystal has a minimum total surface free energy per unit volume.

The crystals may grow rapidly, or be stunted, in one direction; thus an elongated growth of a prismatic habit gives a needle shaped crystal (acicular habit) and a stunted growth gives a flat plate-like crystal (tabular, platy or flaky habit). The relative growths of the faces of a crystal can be altered and often controlled by a number of factors. Rapid crystallization, produced by the sudden cooling or seeding of a supersaturated solution, may result in the formation of needle crystals. The growth of a crystal may be stunted in certain directions due to presence of impurities in the crystallizing solution. A change of solvent often changes the crystal habit.

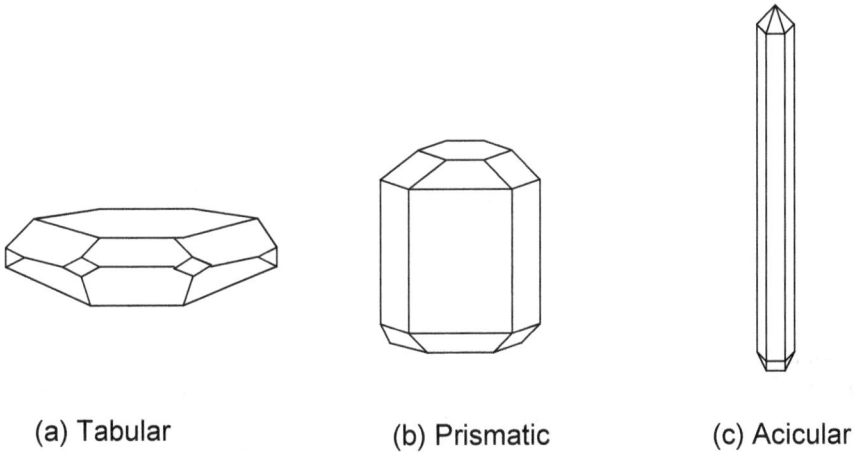

(a) Tabular   (b) Prismatic   (c) Acicular

Fig. 1. Crystal habit illustrated on a hexagonal crystal

2.1.2 Crystal surface structure

The structure of a growing crystal face at its interface with the growth medium has been characterized by a quantity as surface roughness or surface entropy factor or the alpha factor defined by

$$\alpha = \xi \, \Delta H / kT \qquad (1)$$

Where ξ is an anisotropy factor related to the bonding energies in the crystal surface layers, ΔH is the enthalpy of fusion and k is the Boltzmann constant. Values of α less than 2 are indicative of rough crystal surface which will allow continuous growth to proceed. The growth will be diffusion controlled and the face growth rates, v, will be linear with respect to the supersaturation, σ, i.e.

$$v \, \alpha \, \sigma \qquad (2)$$

For $\alpha > 5$, a smooth surface is indicated

2.2 Effect of crystal size

In order for crystallization to occur, there must exist in a solution a number of minute solid bodies, nuclei or seeds that act as a centre of crystallization, the classical theory of nucleation stemming from the work of Gibbs (1948) Volmer (1939) and others is based on the condensation of vapor to liquid and this treatment may be extended to crystallization from melts and solutions. Crystallization process can be explained on the basis of nucleation and growth process.

2.2.1 Nucleation

Schematically the nucleation steps are as shown below:

Nucleation

Primary

Secondary
(induced by crystals)

Homogeneous
spontaneous

Heterogeneous induced by
(foreign particles)

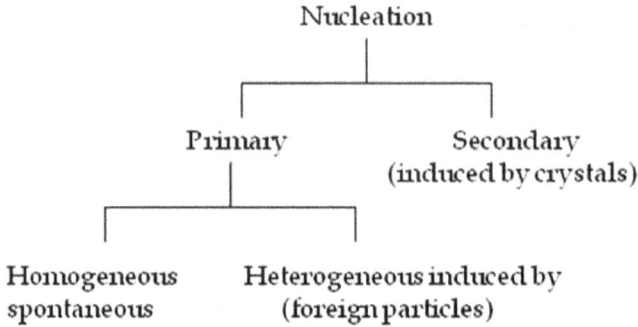

Let us consider the free energy changes associated with the process of homogenous nucleation. The overall excess free energy, ΔG, between a small solid particles of solute (assume here a sphere of radius r for simplicity) and the solute in solution is equal to the sum of surface excess free energy ΔGs i.e. excess free energy between the surface of the particles and the bulk of the particles, and the volume excess free energy, ΔGv, i.e. the excess free energy between very large particles ($r = \infty$) and the solute in the solution , ΔGs is a positive quantity, the magnitude of which is proportional to r^2. In a supersaturated solution Gv is a negative quantity proportional to r^3. Thus

$$\Delta G = \Delta Gs + \Delta Gv \tag{3}$$

$$= 4\pi r^2 \gamma + \frac{4}{3} \pi r^3 \Delta Gv \tag{4}$$

ΔGv also can be understood as free energy change of formation per unit volume
γ is the interfacial tension or surface energy.
ΔGs and ΔGv are opposite in sign and depend differently on r. the free energy of formation, ΔG, passes through a maximum (ΔG_{crit}) corresponds to critical nucleus, r_c. For a spherical cluster, it is obtained by setting $d\Delta G / dr = 0$

$$\frac{d\Delta G}{dr} = 8\pi r\gamma + 4\pi r^2 \Delta G_v = 0 \tag{5}$$

$$r_c = \frac{-2\gamma}{\Delta G_v} \tag{6}$$

From equations 1 & 3 we get

$$\Delta G_{crit} = \frac{16\pi\gamma^3}{3(\Delta G_v)^2} = \frac{4\pi\gamma r_c^2}{3} \tag{7}$$

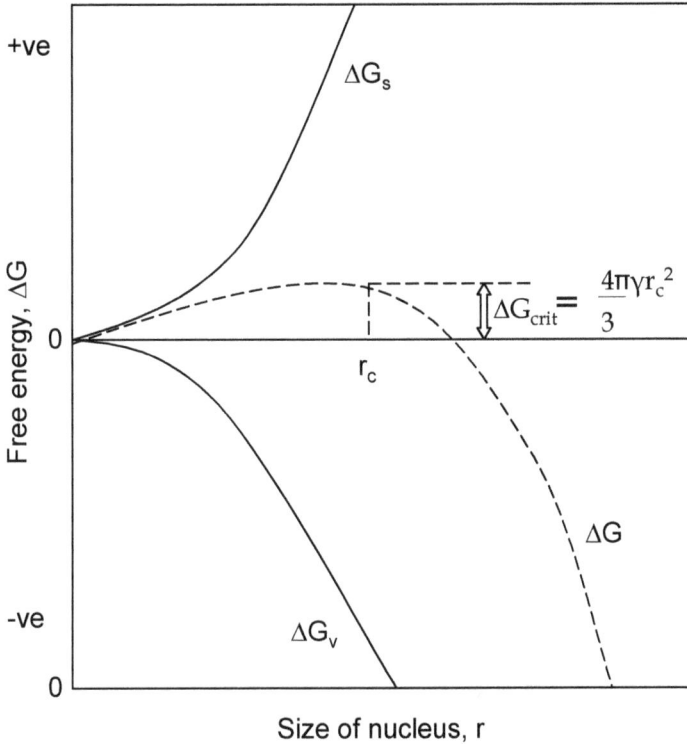

Fig. 2. Free energy diagram for nucleation explaining the existence of a 'critical nucleus'.

The behavior of a newly crystalline lattice structure in a supersaturated solution depends on its size, the crystal may grow or redissolve and it undergoes decrease in free energy of the particle. Particles smaller than r_c will dissolve if present in a liquid in order to achieve reduction in free energy. Similar particles larger than r_c will continue to grow.

There will be fluctuations in the energy about the constant mean value i.e. there will be a statistical distribution of energy, or molecular velocity, in the molecules constituting the system, and in those supersaturated regions where the energy level rises temporarily to a high value, nucleation will be favored.

The rate of nucleation J e.g. the number of nuclei formed per unit time per unit volume can be expressed in the form of Arrhenius reaction velocity equation

$$J = A \exp(-\Delta G / kT) \tag{8}$$

k = Boltzmann constant, the gas constant per molecule

The basic Gibbs-Thomson relationship for a non-electrolyte may be written as

$$\ln S = \frac{2\gamma v}{kTr} \tag{9}$$

Where S is defined by equation 10

$$S = \frac{C}{C^*} \tag{10}$$

Where C is the solution concentration and C* is equilibrium saturation at given temperature and v is the molecular volume; this gives

$$-\Delta G_v = \frac{2\gamma}{r} = \frac{kT \ln S}{v} \tag{11}$$

Hence, from equation 7

$$\Delta G_{crit} = \frac{16\pi \gamma^3 v^2}{3 (kT \ln S)^2} \tag{12}$$

And from equation 8

$$J = A \exp \left(\frac{16\pi \gamma^3 v^2}{3k^3 T^3 (\ln S)^2} \right) \tag{13}$$

This equation indicates that three main variables govern the rate of nucleation: temperature, T; degree of supersaturation, S; and interfacial tension, γ. Equation 13 may be rearranged to give

$$\ln S = \left(\frac{16\pi \gamma^3 v^2}{3k^3 T^3 \ln(A/J)} \right)^{1/2} \tag{14}$$

And if, arbitrarily, the critical supersaturation, S_{crit}, is chosen to correspond to a nucleation rate, J, of say one nucleus per second per unit volume, then equation 14 becomes

$$\ln S_{crit} = \left(\frac{16\pi \gamma^3 v^2}{k^3 T^3 \ln A} \right)^{1/2} \tag{15}$$

From equation 6 and 11, the radius of a spherical critical nucleus at a given supersaturation can be expressed as

$$r_c = \frac{2\gamma v}{kT \ln S} \qquad (16)$$

For the case of non-spherical nuclei, the geometrical factor $16\pi/3$ in equations 7 and 12-14 must be replaced by an appropriate (e.g. 32 for cube).

Critical reviews of nucleation mechanism have been made by Nancollos and Purdie (1964), Nielsen (1964), Walton (1967), Strickland-Constable (1968), Zettlemoyer (1969), Nyvlt et al (1985) and Sohnel, Garside (1992) and Kashchiev (2000).

2.2.2 Mechanism of growth

Nucleation occurs over some time with constant precursor concentration. Eventually surface growth of clusters begins to occur which depletes the initial supply. When the initial concentration falls below the critical level for nucleation (critical supersaturation level), nucleation ends. A general analysis of the growth process is then important to understand nanocrystal synthesis. In general, the surface to volume ratio in smaller particles is quite high. As a result of the large surface area present, it is observed that surface excess energy becomes more important in very small particles, constituting a non-negligible percentage of the total energy. Hence, for a solution that is initially not in thermodynamic equilibrium, a mechanism that allows the formation of larger particles at the cost of smaller particles reduces the surface energy and hence plays a key role in the growth of nanocrystals. A colloidal particle grows by a sequence of monomer diffusion towards the surface followed by reaction of the monomers at the surface of the nanocrystal. Coarsening effects, controlled either by mass transport or diffusion, are often termed the Ostwald ripening process. This diffusion limited Ostwald ripening process is the most predominant growth mechanism and was first quantified by Lifshitz and Slyozov [Lifshitz and Slyozov, 1961], followed by a related work by Wagner [Wagner and Elektrochem, (1961)] known as the LSW theory.

The diffusion process is dominated by the surface energy of the nanoparticle. The interfacial energy is the energy associated with an interface due to differences between the chemical potential of atoms in an interfacial region and atoms in neighboring bulk phases. For a solid species present at a solid/liquid interface, the chemical potential of a particle increases with decreasing particle size, the equilibrium solute concentration for a small particle is much higher than for a large particle, as described by the Gibbs–Thompson equation. The resulting concentration gradients lead to transport of the solute from the small particles to the larger particles. The equilibrium concentration of the nanocrystal in the liquid phase is dependent on the local curvature of the solid phase. Differences in the local equilibrium concentrations, due to variations in curvature, set up concentration gradients and provide the driving force for the growth of larger particles at the expense of smaller particles [Sugimoto, (1987)].

2.3 Habit modification by polymers
2.3.1 Habit modification by polymers of inorganic materials

All crystal growth rates are particle size dependent and size range. For microscopic, submicroscopic particles, the size effect becomes significant.

The morphology-controlling effect of PVA is long known and utilized as a capping agent during the synthesis of nanoparticles. Using radiation chemical reduction, we have successfully synthesized morphology controlled copper and silver metal nanoparticles by using PVA as a capping agent. (Joshi et al.,1998, & Temgire & Joshi,2004) In the presence of crystal habit- modifying polymers, the crystal growth or nucleation is diverted from the non-uniform to a uniform shape. In most of the earlier studies using PVA as a crystal-habit modifier, a gel matrix made out of PVA has been used for the control of nucleation and morphology in aqueous-solution-based crystal growth. (Merrill & Bassett, 1975, Sinha et.al., 2001, Lakshminarayanan et al., 2003, Joshi et al.,1998 & Temigre & Joshi,2004) Ammonium perchlorate (AP) is one of the most extensively used solid propellant oxidizers in the propellant industry. The percentage of oxidizer in the propellant formulation varies from 70 to 80% by weight, depending on the energetic requirements and compatibility with the other ingredients. Because of the high percentage in the propellant formulation, the performance of the propellants (specific impulse and burning rate) varies with the oxidizer properties, and in turn, the performance of the oxidizer varies with the particles' size and morphology. (Sutton & Oscar, 2001) Hence, in the present investigation, PVA has been used as a supermolecular matrix to control the morphology of AP.

AP is the most commonly used rocket propellant oxidizer and one of the extensively studied ammonium compounds. The morphology of the oxidizer has an important role in the formulation and performance of solid propellants, and the AP crystallized from its saturated solution gives needle-shaped crystals. The nucleation of the crystals was observed immediately after drying began. The crystals grown in the PVA showed entirely different morphologies, such as rectangular prism and rectangular wedge, in comparison to the morphologies of the AP crystals grown in the absence of PVA. The SEM images obtained are shown in Figures 3–8. Three different sets of SEM images were chosen for different concentrations, such as a low salt concentration, equal salt-polymer concentration, and high salt concentration (crystals grown from mixtures A, C, and E) (Vargeese et al.2008)

The images of the crystals grown immediately after mixing the solution and after 24 h of reaction time are shown in panels a and b of each figure, respectively. Figures 3–5 show the images of the crystals grown from PVA 14000. Figure 3 shows that the crystals have an irregular morphology and do not have any growth orientation toward a particular plane. The images also indicate that the crystals have an irregular shape, although they tend to grow in an organized manner. This could be due to the polymer-substrate interaction that prevents the crystals from growing in an organized manner. At low salt concentrations, there is too much hydrogen bonding between the hydroxyl groups of the polymer and the hydrogen of the ammonium ion. Adsorption characteristics of polymers are different from those of other systems because of the polymers' flexibility. In addition to the usual adsorption factors, such as adsorbate–adsorbent and adsorbate-solvent interactions, a major aspect is the conformation of molecules at the interface and its role in dispersion. PVA is a flexible linear molecule with no charge and which can potentially adsorb on the surface. Bridging is considered to be a consequence of the adsorption of individual intermolecular polymer molecules on the surface. This happens through hydrogen bonding. Because of the high polymer concentration, not all the segments of the polymer are in direct contact with the surface. Also, the diffusion of ions is slow at high polymer concentrations. The possibility that the solution does not contain enough AP to grow in an organized manner cannot be ruled out. The viscous nature of the solutions containing a large quantity of PVA polymer leads to rectangular wedge- shaped crystals because diffusion is predominant and convection is suppressed for the transformation of solutes.

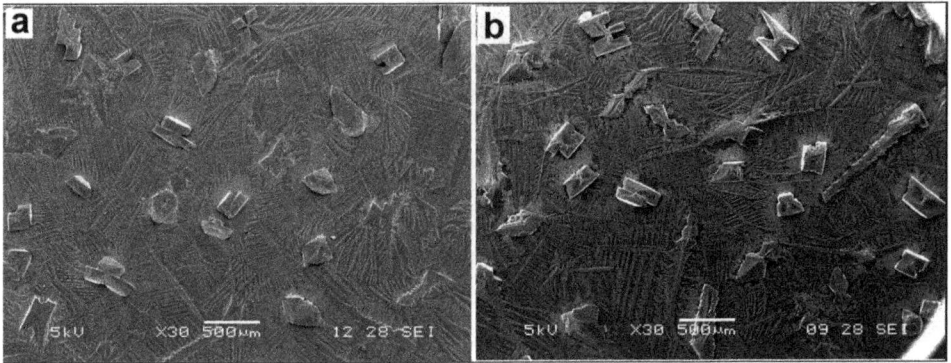

Fig. 3. SEM images of crystals grown from mixture A (PVA 14000) after (a) 0 h and (b) 24 h.

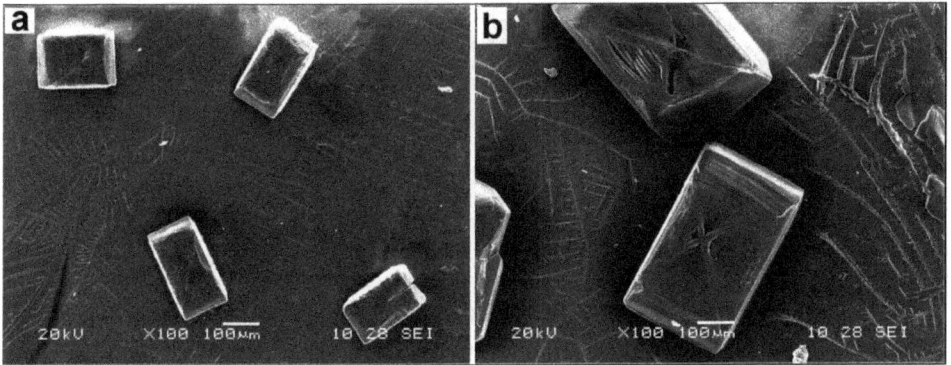

Fig. 4. SEM images of crystals grown from mixture C (PVA 14000) after (a) 0 h and (b) 24 h.

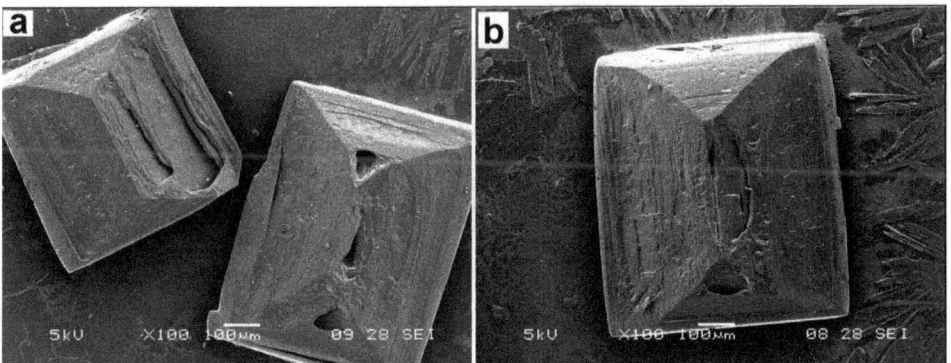

Fig. 5. SEM images of crystals grown from mixture E (PVA 14000) after (a) 0 h and (b) 24 h.

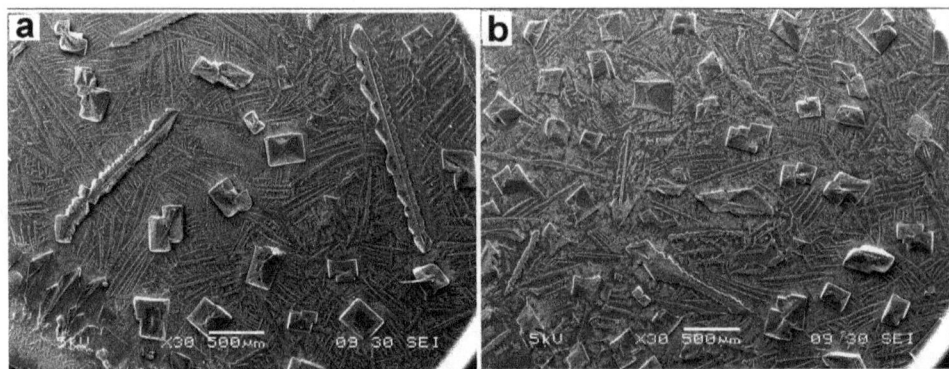

Fig. 6. SEM images of crystals grown from mixture A (PVA 125000) after (a) 0 h and (b) 24 h.

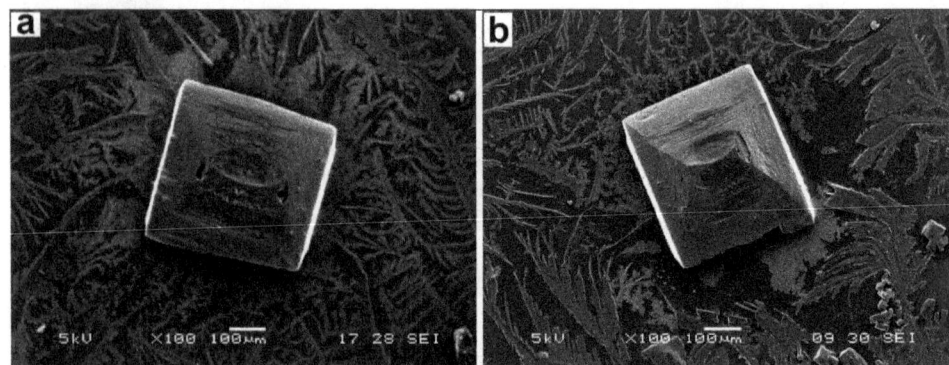

Fig. 7. SEM images of crystals grown from mixture C (PVA 125000) after (a) 0 h and (b) 24 h.

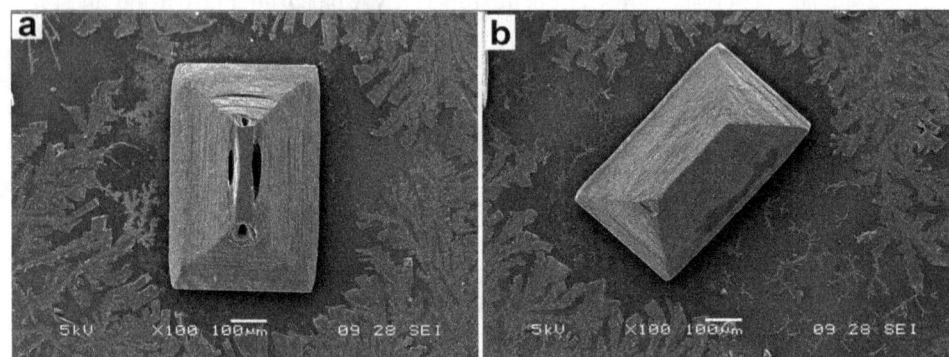

Fig. 8. SEM images of crystals grown from mixture E (PVA 125000) after (a) 0 h and (b) 24 h.

As seen from the SEM images, the salt-to-polymer solution ratio change is reflected in the morphology of the crystals. Although the crystals grown from mixture A have an irregular morphology, the particles grown from the other compositions contain individual crystals with a specific morphology. The crystals grown from mixture C have a rectangular prism shape, whereas the crystals grown from the solution containing a high salt concentration (mixture E) have a rectangular wedge shape. The crystals grown from mixtures B and D also show a more or less similar morphology, comparable to the crystals grown from mixtures C and E, respectively. Hence, the crystals grown from mixtures B and D are not discussed here. The crystals obtained from mixture E have grown in another plane, leading to a rectangular wedge-like morphology in response to the change in polymer-to-salt ratio. This outgrowth to another plane of crystals obtained from mixture E is one of the observed modifications from the crystals grown from mixture C. The morphological evolution of the crystals from rectangular prism to wedge-like shape clearly shows the dependency of the particles' morphology on the polymer concentration and the minimum salt concentration requirement.

The advantage of using PVA as a habit modifier is a reduced agglomeration, leading to samples containing only individual particles. Studies show that, for a chemical system in which PVA and AP molecules are involved, the PVA induces the crystallization of individual AP crystals irrespectively of the polymer-substrate concentration variations. Here, the colloidal action of the PVA or the surfactant activity of the polymer could prevent the particles' agglomeration. The PVA could possibly get adsorbed on the surface, thereby preventing the agglomeration and leading to the formation of individual particles of AP. That is, immediately after the nucleation has started, the polymer might isolate the salt solution into packets and force them to grow as individual particles.

It has also been speculated that the densification of the gel medium increases the random noise for crystal growth because the polymer gel matrices disturb the progress of the growing surface. (Oaki & Imai, 2003) The PVA gel forms an organized matrix under the optimum concentration of components, which ensures the homogeneous distribution of the cations in the polymeric network structures. The chain alignment and interchain separation of PVA, which depend on temperature and concentration, lead to the formation of a polymeric matrix with complex structures. These structures chelate the cations through a process of weak chemical bonding (such as van der Waals, hydrogen bonding) and steric entrapment. (Sinha, 2001) The polymer matrix not only provides an organized surface of mineralization, but also induces a vector growth on the polymeric surface, the direction of which differs from the characteristically preferred direction of the unit cell. (Lee et al., 1999, Addadi &Weiner, 1992, Walsh & Mann, 1995, Mann et al,1988) This results in the formation of extend and nonequilibrium morphologies, as well as metastable phases with lattice parameters on the order of the spacing available in the polymeric matrix. However, the nucleation of a particular space group on a charged polymeric surface not only depends on the lattice geometry, but also includes spatial charge distribution, hydration, defect sites, and surface relaxation. (Mann, 1988) These factors affect the collision frequency and in turn the activation energy for nucleation; hence, the transition state theory might be considered to explain the nucleation of biominerals. (Sinha et.al., 2001)

The shape of inorganic crystals is normally related to the intrinsic unit cell structure, and the crystal shape is usually the outside embodiment of the unit cell replication and amplification. (Yu and Colfen, 2004, Colfen & Mann, 2003, & Jongen et al., 2000)The diverse crystal morphologies that a mineral, identical to that for calcium carbonate, can have are due to the different surface energies and external growth environments of the crystal faces. (Wulff &Kristallogr 1901) The morphological evolutions (from irregular to organized crystal assemblies) of the AP crystals are seen in the SEM images. The polymer-substrate interaction is clearly seen in the polymer pattern observed near the crystals. The polymer seems to have grown in the form of dendrites surrounding the crystals with primary and secondary branches. Usually, the rate of nucleation is governed by the temperature, the degree of supersaturation, and the interfacial tension. Crystals often grow from the center of the face and spread outward toward the edges in layers, and these layers may have a thickness of several 1000 Å. During this growth, dissolved impurities may affect the thickness and shape of the layers, which in turn change the morphology of the crystals. Usually, the effect of these impurities is highly specific and depends upon a number of parameters. The growth rate of a crystal face is usually related to its surface energy, if the same growth mechanism acts on each face. The fast growing faces have high surface energies, and they will vanish in the final morphology, and vice versa. This treatment assumes that the equilibrium morphology of a crystal is defined by the minimum energy resulting from the sum of the products of the surface energy and the surface area of all exposed faces (Wulff rule). (Yu & Colfen 2004) The driving force for this spontaneous oriented attachment is that the elimination of the pairs with a high surface energy will lead to the substantial reduction of the surface free energy from a thermodynamic viewpoint. (Banfield et al., 2000 & Alivisatos, 2000) The surface roughness on the molecular level is governed by energetic factors arising from fluid-solid interactions at the interface between the crystal and its growth environment. A change in the solvent often changes the crystal habit, and this may sometimes be explained in terms of interface structure changes. The structure of the growing crystal surface at its interface with the growth medium has an important effect on the particular mode of crystal growth adopted. A functional group with a high affinity ensures the anchoring of the molecule on a particular phase, and the polymeric chain protects the surface from coalescing with the next one through electrostatic repulsion or steric hindrance. (Joshi et al.1998) This result suggests a significant interaction between the polymeric hydroxyl groups and the crystallizing AP, resulting in the considerable influence on both the primary crystallization and the superstructure. (Qi et. al.2000)

2.3.2 Habit modification by polymers of nanomaterials

2.3.2.1 Zinc oxide nanoparticles

In the studies of nanomaterials, it has been observed that the size and shape of a nanomaterial depends on nature of stabilizer i.e. surfactant, ligand, polymer to salt ratio, reaction temperature and time. The synthetic method applied also plays a role. The systematic adjustment of the reaction parameters can be used to control the size and shape, a quality of nanocrystals and inorganic crystals as well. Nanoparticles are small and thermodynamically unstable. After the primary nucleation, the particles grow via molecular addition. Particles can grow by aggregation with other particles called secondary growth. Their growth rates may be arrested during the reaction either by adding surface protecting agents. Nanocrystal dispersions are stable if interaction between the capping groups and solvent is favorable providing an energetic barrier to counter act van der Waals' forces.

In our recent studies, we have synthesized flower-like ZnO nanostructures comprising of nanobelts of 20 nm width by template and surfactant free low-temperature (4 ^{0}C) aqueous solution route. The ZnO nanostructures exhibit flower-like morphology, having crystalline hexagonal wurtzite structure with (001) orientation. The flowers with size between 600 and 700 nm consist of ZnO units having crystallite size of 40 nm. Chemical and structural characterization reveals a significant role of precursor: ligand molar ratio, pH, and temperature in the formation of single-step flower-likeZnO at low temperature. Plausible growth mechanism for the formation of flower like structure has been discussed in detail. Photoluminescence studies confirm formation of ZnO with the defects in crystal structure. The flower-like ZnO nanostructures exhibit enhanced photochemical degradation of methyleneblue (MB) with the increased concentration of ligand, indicating attribution of structural features in the photocatalytic properties. (Vaishampayan et.al.2011)

ZnO exhibits a varied range of novel structures. The relative surface activities of various growth facets under given conditions determine the surface morphology of the grown structure. Macroscopically, a crystal has different kinetic parameters for different crystal planes, which are emphasized under controlled growth conditions. Thus, after an initial period of nucleation and incubation, a crystallite will commonly develop into a three-dimensional object with well-defined, low-index crystallographic faces.

Wurtzite ZnO being a polar crystal, Zn forms a positive polar plane and O forms a negative polar plane. Zn^{2+} and O^{2-} ions are tetrahedrally coordinated and stack alternatively along the c-axis thus, ZnO grows along the c-axis. When EA is added in the aqueous solution, it gets hydrolyzed and forms EAH+ molecule. Thus, by Coulomb interaction EAH+ molecule gets adsorbed on the negative polar plane retarding the growth of ZnO along the negative polar plane. Therefore, when appropriate amount of EA is used, it covers the side surfaces of ZnO crystal, enhancing growth along the (0 0 1) direction. When EA concentration is lower, i.e. not enough to cover the whole surface, the Oswald ripening takes place and thereby role of EAH+ in the growth of ZnO crystal results in the formation of flower-like structure where individual petal is formed by the overlay of nanobelts.

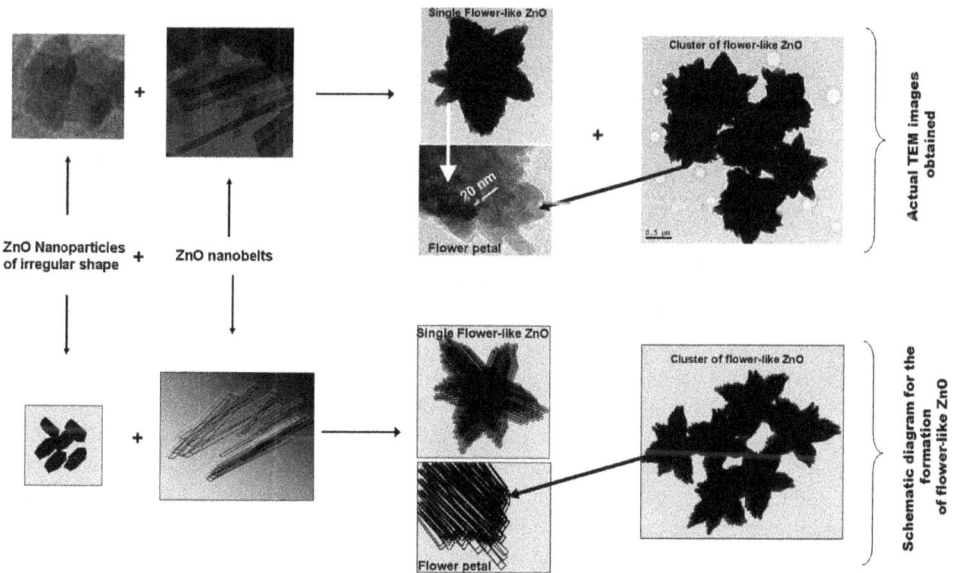

Fig. 9. Diagrammatic representation of formation of flowerlike structure.

Also, some particles of crystallite size 30 nm are seen on the nanobelts. The nanobelts and nanoparticles are formed by conventional nucleation followed by crystal growth process. The thermal energy released during the hydrolysis of EA facilitates nanoparticles to arrange themselves in between the nanobelts so as to form a compact flower-like structure. Also, as the concentration of EA increases, the nanobelts appear to have tapering feature. Thus, the ZA:EA molar ratio plays an important role in framing the morphology of the final product. The EA chelates the cations through a process of weak interactions such as van der Waals forces, hydrogen bonding, and steric entrapment. The ligand not only provides organized surface for structure formation but also induces a vector growth on the surface, the direction of which differs from characteristically preferred direction of the unit cell. This results in non-equilibrium morphologies as well as metastable phases.

We have also synthesized ZnO by aqueous thermolysis method. (Patil & Joshi, 2007) PVA of two different molecular weights was used as a capping agent and as a fuel. A TEM study of ZnO nanoparticles was undertaken to highlight the shape, size and size distribution as well as the crystallinity of the particles. Fig. 10 shows the TEM image of zinc oxide nanocrystals after TGA. The micrograph of sample A (PVA 14,000) showed uniform distribution with nearly spherical morphology (Fig. 10a). All the particles are separated from each other. While sample B (PVA 125,000) synthesized with higher molecular weight exhibits cuboid like morphology and few particles appear to be close to spherical shape (Fig. 10b). In this micrograph the crystals are structurally perfect and attached like beads due to cross-linking of the polymer. This may be due to migration of defects to the surface of crystal during the calcination and growth of particles (Gu et.al, 2004) Fig. 10c shows very small particles of ZnO mostly of spherical shape for sample C (PVP 40,000). PVP does not form gel at room temperature. Therefore, it directly gets solidified while heating the precursor. As the solution is not viscous, particles formed are not cross-linked in the polymer matrix. Hence due to lack of steric interactions as compared to PVA, particles synthesized by PVP are dense but quite separated from each other. The particle sizes of XRD and TEM are comparable. The surface morphology of zinc oxide nanoparticles changes greatly with an increase in oxidation temperature. This can be clearly observed in scanning electron micrographs (SEM). Fig. 10 also shows SEM micrographs for samples A and B after TGA (500 ℃) in air. At annealing temperature of 500 ℃ the ZnO nanoparticles consist of fine grains. Since the grains are agglomerated together, grain boundaries cannot be distinguished clearly. In sample A (Fig.10d) numerous micropores were observed compared to sample B (Fig.10e). This porosity may be generated by the evolution of gases and removal of organic matter, which was loaded in the polymer network and also due to heat generated during combustion of polymers. A significant change in surface morphology is observed in the ZnO annealed at 1000 ℃ (Fig.10 f&g), well facet grains are observed acquiring dumbbell morphology; their size becomes larger, with a wide range of distribution. At such high temperature, migration of grain boundaries occurs causing the coalescence of small grains and the formation of large grains. In order to understand the process occurring during thermolysis, we have to consider the 'cross-linking' of the polymeric network, which depends on the average molecular weight, degree of polymerization and solubility of the polymer in water. Our observation shows that PVA of molecular weight 14,000 takes almost 24 h to dissolve in water while higher molecular weight PVA, 125,000 dissolves in 2 h. Both have a very good gelling property. Poly-vinyl alcohol is linear flexible molecule with no charge. Therefore, it adsorbs non-specifically on the surface of oxides. The interaction with the surface takes place through hydrogen bonds between polar functional groups of the polymer chain and hydroxylated and protonated groups on the surface. Though the interaction energy between surface and each chain segment is smaller than kT, chains adsorb very well because of large number of contact points. The affinity of the macromolecule for the surface usually increases with its molecular weight. The conformation of the adsorbed polymer remains similar to that of free macromolecule and exhibit tails and loops between contact points. The adsorbed layer provides an excellent steric protection against aggregation (Gennes, 1987, Dickson & Ericsson, 1991). This can be schematically depicted as below.

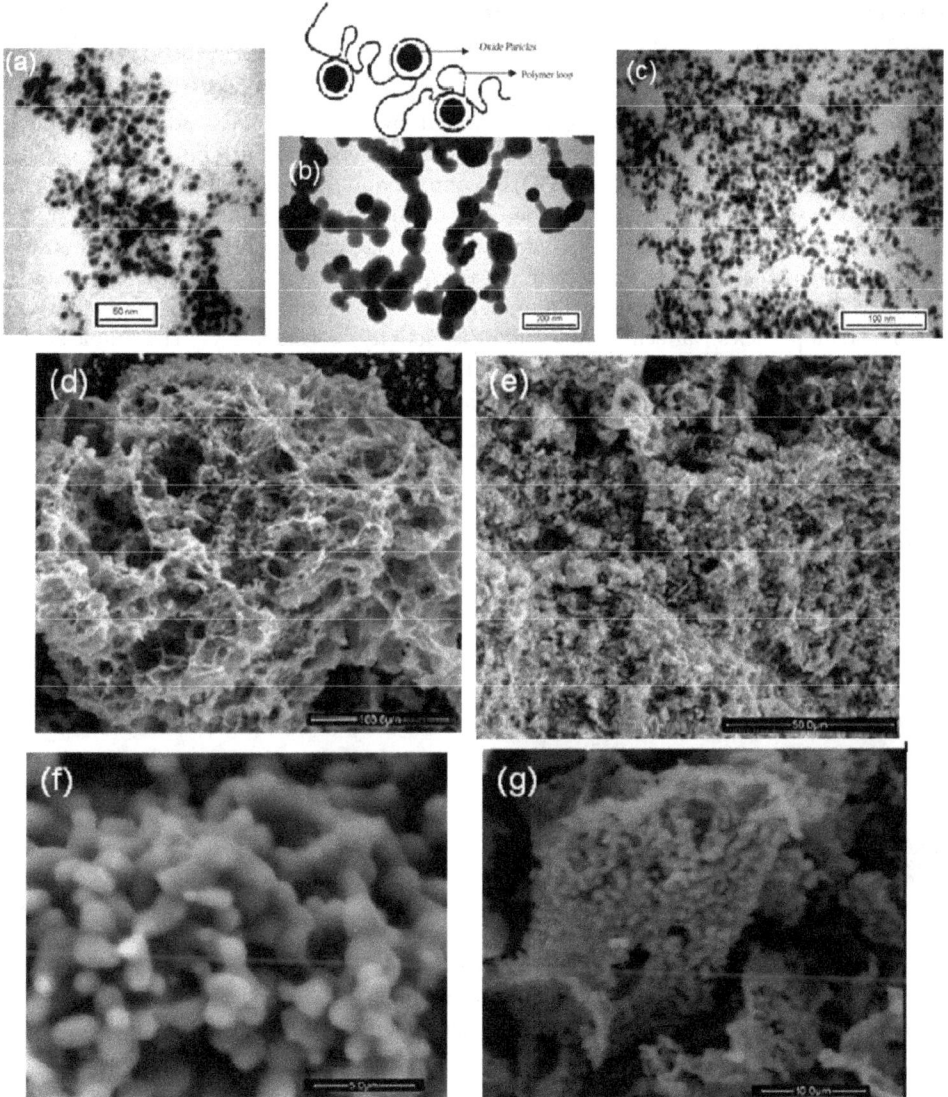

Fig. 10. TEM of ZnO for: (a) sample A, (b) sample B and (c) sample C and SEM of ZnO after TGA in air: (d) sample A at 800× magnification; (e) sample B at 2000× magnification and SEM of ZnO after annealing at 1000 ∘C: (f) sample B at 12000× magnification; (g) sample C at 6000× magnification.

Here polymer is adsorbed and acting as a bridge between particles. The linear chains of PVA can be cross-linked in aqueous medium, i.e. water (Kirk-Othmer, 1983). The cross-linking between the chains may provide small cages wherein the 'sol' of the reactant mixture gets trapped. During thermolysis, the 'sol' trapped in the cages may get converted to ultrafine particles of zinc oxide. Thus, the cages formed by the cross-linking may offer resistance to the agglomeration of the particles and the particle growth. The degree of polymerization can also affect the formation and morphology (Temgire & Joshi, 2004).

2.3.2.2 Palladium nanoparticles

Stable palladium nanocluster catalysts prepared by chemical and -radiolytic reduction methods were found to give very high turn-over frequency numbers in hydrogenation of styrene oxide and 2-butyne-1,4-diol (B3D) as compared to the conventional catalysts. (Telkar et al.2004) A systematic study was carried out on the effects of different transition metals, their reduction methods, types of polymer used as a capping agent, and the concentration and composition of solvent used during catalyst preparation on the size and shape of nanoparticles. The reduction method of metal precursor directly influenced the morphology of the nanoparticles, affecting the catalyst activity considerably. The cubic-shaped nanoparticles (5-7 nm) were obtained in chemical reduction, while radiolytic reduction method gave spherical nanoparticles (1-5 nm).

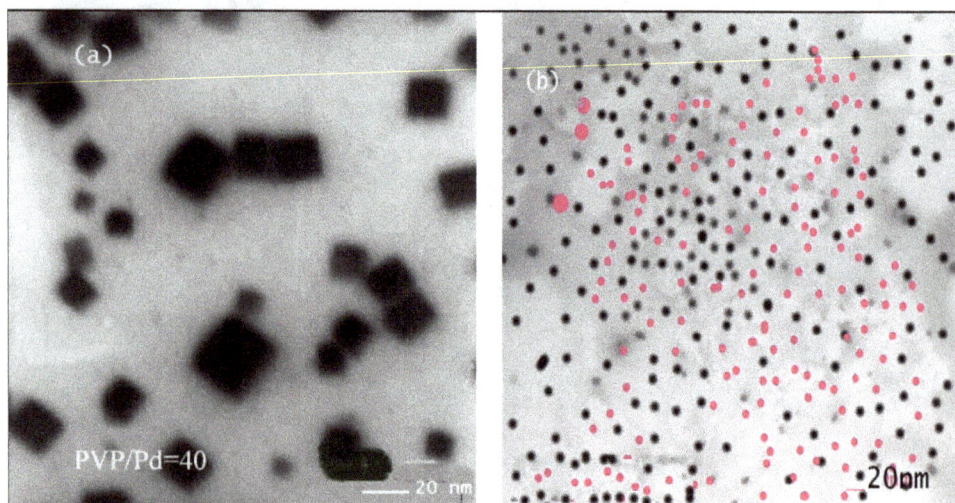

Fig. 11. TEM photograph for Pd nanoparticles prepared by (a) chemical method (AV = 80 kV; magnification = 40,000×) and (b) radiolytic method (PVP/Pd: 40) (AV = 80 kV; magnification = 80,000×). (Applied Catalysis A: General 273 (2004) 11–19)

The activity results, along with the particle sizes of various nanocluster catalysts, are presented in Table 1. The catalyst activity of RRPd (Radiolytic Reduction) catalyst was higher than that of CRPd (Chemical Reduction) catalyst when PVP/Pd ratio was 1, which was in accordance with the fact that the particle size of RRPd was less (5.2 nm) than that of CRPd sample (7.1 nm). As the PVP/Pd ratio was increased from 10 to 40, the activity trend was reversed with respect to particle sizes, thus CRPd samples showed higher catalyst activities than that of RRPd samples. The catalyst activity of CRPd was maximum for PVP/PD ratio of 40, in spite of the fact that particle size of RRPd sample was one-fifth of that of CRPd sample. Such a trend was consistent for both 2-butyne-1, 4-diol and styrene oxide hydrogenation. However, the extent of activity enhancement was dramatically higher for styrene oxide hydrogenation. Besides the reduction in the size, the polymer concentration seems to have a significant effect on the adsorption of reactants. It is reported that, for polymer concentration higher than 50 mg/l, fully developed steric layers are formed around the particle; these act as an effective diffusion barrier that blocks further growth of the metal particles. Pd particles are considered to adsorb onto the polymer. At higher PVP/Pd ratios, beads on string-type of complexes may thus be formed, adsorbing multi-particle complexes (Boonekamp&Kelly, 1994), which is also one of the reasons for higher activity of nanoparticles at higher PVP concentration. The selectivity to B_2D obtained was more than 98% for the Pd catalysts having PVP/Pd ratio in the range of 1–30. A further increase in PVP/Pd ratio to 40 caused a marginal decrease in 2-butene-1,4-diol selectivity from 98 to 91% due to the formation of butane-1,4-diol. In the case of catalysts prepared by the radiolytic method, the particle size reduced to 1 nm with increase in concentration of PVP. Surprisingly, the activity was found to reduce with decrease in particle size. This observation was consistent for both (2-butyne-1, 4-diol (B3D)) as well as styrene oxide hydrogenation (Table 1). This trend can be attributed to differences of shape of the Pd nanoparticles formed by chemical and radiolytic reduction methods. TEM photographs of Pd nanoparticles (Fig.11) showed a distinct morphological change depending on the method of preparation of nanoparticles. Pd particles prepared by ethanol reduction showed particles with a square outline, from which the three-dimensional shapes determined, were found to be cubic. Similar morphology was observed for Pt colloids prepared by H_2 gas reduction. (Ahmadi et al.1996) However, they also obtained a mixture of tetrahedral, polyhedra and irregular-prismatic particles. Milligan and Morris also observed cubical gold nanoparticles for hydroxylamine hydrochloride as the reducing agent (Milligan & Morriss, 1964). In contrast to this, the radiolytic reduction of $PdCl_2$ gave colloidal Pd particles of spherical and oval shapes, 1–5 nm diameter. The final structure and size of the clusters depend on the mechanism of growth process. In the case of radiolytic reduction, the solvated electrons and H• are strong reducing agents and with high rate of reduction, the free metal ions are generally reduced at each encounter (Belloni et al., 1998). In the chemical reduction method, an adsorption of excess of metal ions on the reduced metal clusters, get reduced at a slower rate. This difference in reduction mechanism of radiolytic and chemical reduction may give rise to two distinct shapes of the nanoparticles. It is known that the active sites are more concentrated on the edges of the catalyst these sites may be formed in chemically reduced Pd nanoparticles, leading to higher catalyst activity for these samples. This clearly indicates that not only the particle size but also the shape of the nanoparticles influences the activity of the catalyst (Chen et al., 2000). Nanoparticles of other metals reduced by ethanol also showed cubic shapes while the radiolytic reduction gave spherical particles. As mentioned earlier, the concentration of a stabilizer influenced the nanoparticle size dramatically in case of radiolytically reduced Pd colloids. However, the shape remained spherical, thus confirming that the stabilizer concentration did not contribute to the shape of the nanoparticles. In order to further understand why the catalyst activity decreased in spite of considerable size reduction of radiolytically reduced Pd catalysts; stabilizing polymer alone was irradiated in a separate experiment. There was an increase in the viscosity of irradiated PVP, which indicates increased cross-linking of the polymer (Wang et al.1997). With increase in concentration of polymer, (PVP/Pd = 40), the polymer cross-linking may hinder the access of substrate to the Pd metal particles thereby decreasing the activity of radiolytic nanosize particles.

Method of preparation	PVP/Pd	Particle size (nm)	B_3D hydrogenation			Styrene oxide hydrogenation	
			TOF $(x10^{-5}h^{-1})^a$	Selectivity (%)		TOF $(x10^{-4}h^{-1})a$	Selectivity PEA (%)
				B_1D	B_2D		
Chemical method (CRPd)	1	7.1	3.0	1.0	99	1.5	99.9
	10	6.1	3.2	1.8	98.2	1.4	99.8
	20	6.0	4.3	1.6	98.4	3.4	99.6
	30	5.5	4.6	1.6	98.4	7.0	99.8
	40	5	5.7	8.8	91.2	10.2	99.9
Radiolytic method (RRPd)	1	5.2	3.6	1.0	99.0	1.9	99.5
	10	5.0	3.2	1.7	98.3	1.8	99.5
	20	4.0	3.0	1.6	98.4	1.5	99.5
	30	3.0	2.8	1.5	98.5	1.3	99.4
	40	1	2.4	1.4	98.6	1.0	99.8

Table 1. Effect of polymer to Pd ratio prepared by chemical and radiolytic methods for hydrogenation reactions.

2.3.2.3 Copper chromite nanoparticles

Amorphous and monodispersed copper chromite nanoparticles were prepared by aqueous thermolysis method using PVA and different ratios of urea-PVA as fuel in air (Hrishikeshi and Joshi, unpublished results, 2011). Morphology and size of nanoparticles were measured by SEM and TEM analysis. Copper chromite ($CuCr_2O_4$) is a tetragonally distorted normal spinel; this distortion is due to Jahn Teller effect of Cu^{+2} (d^9) ions in tetrahedral sites. It is a p-type semiconductor which is widely used as a catalyst for the oxidation of CO (Hertl et al., 1973), hydrocarbons (Mc Cabe & Mitchell, 1983) alcohols (Solymosi & Krix 1962) and as a burn rate catalyst in composite solid propellants, (Prince, 1957, Solymosi & Krix 1962, Patil et al., 2008) Well resolved square bipyramidal morphology was seen in all copper chromite samples using PVA alone. The habit modification of copper chromite was observed due to presence of urea. The urea molecule is planar in the crystal structure, but the geometry around the nitrogens is pyramidal in the gas-phase minimum-energy structure. In solid urea, the oxygen center is engaged in two N-H-O hydrogen bonds. The resulting dense and energetically favorable hydrogen-bond network probably changes the morphology after combustion process.

Figure 12 shows scanning electron micrograph of Copper chromite (a) using only PVA and (b) using Urea and PVA after annealing at 800°C. As obtained as well as annealed samples show uniform and compact distribution of copper chromite $CuCr_2O_4$ nanoparticles. There is almost no porosity in the as obtained as well as in annealed samples. Polymer is adsorbed and acting as bridge between particles. The linear chains of PVA can be cross linked in aqueous medium (Kirk-Othmer, 1983). The cross linking between the chains may provide small cages wherein the "sol" of the reactant mixture gets trapped. During combustion, the "sol" trapped in the cages may get converted to ultrafine particles of copper chromite. Thus cages formed by the cross linking may offer resistance to the agglomeration of the particles and particle growth. Perfect square bipyramidal morphology is seen in PVA capped and orthorhombic in annealed samples. Sharpness of edges decreases gradually with increase in urea content in the fuel mixture.

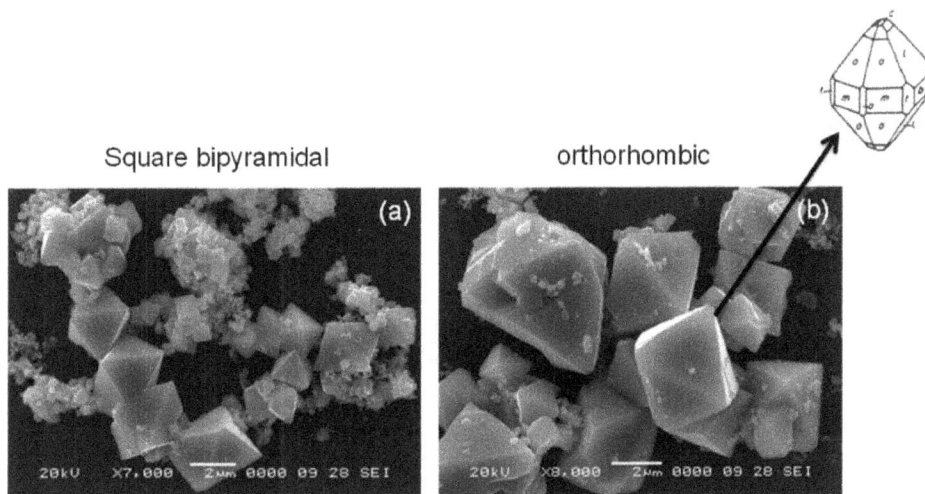

Fig. 12. SEM micrographs of Copper chromite (a) using only PVA and (b) using Urea and PVA after annealing at 800°C.

I have tried to discuss the morphological changes and habit modification of some of the materials studied in our group, on the basis of the theories put forth and the literature.

3. Conclusion

In this Chapter, the habit modification and morphological changes of some inorganic materials in microsize and nanosize are discussed. In most of these studies polymers play multiple roles as a fuel in combustion synthesis, encapsulating agent and as a habit modifier in other synthesis method applied. We have observed that the size, shape, morphology of the synthesized material depends on various factors like nature of polymer, its degree of polymerization, molecular weight, reaction time, synthetic method applied and also on heat of reaction. In the methods applied at high temperature, rapid nucleation time gives rise to short burst of nuclei which might react with intermediate species and the reactions are more kinetically controlled. When the synthesis was carried out at low temperatures, nucleation process is slow and thermodynamically driven process. With aging, growth process becomes more favorable. Final morphology of the material depends on equilibrium conditions related to minimum surface energy, rate of nucleation and growth.

4. Acknowledgement

My sincere thanks are due to Ms. Tajana Jevtic, Publishing Process Manager, In Tech, for inviting me to write a chapter on the work related to crystal growth. It is rewarding to be a contributory of the book, "Crystal Growth". I take this opportunity to thank all my research students, who have worked hard and contributed to the field of Nanoscience and related area. I need to mention the effort of Ms. Shubhangi Borse, in editing the manuscript, as per the requirements of the prescribed format. I acknowledge my family members, for their co-operation, wholehearted support and constant encouragement during the preparation of this chapter.

5. References

Addadi, L., Weiner, S. (1992) Preparation of portland cement components by poly(vinyl alcohol) solution polymerization *Angew.Chem., Int. Ed.*, *31*, 153–169.

Ahmadi T.S., Wang Z.L., Green T.C., Henglein A., El-Sayed M.A. (1996) Shape-Controlled Synthesis of Colloidal Platinum Nanoparticles, *Science* 272, 1924.

Alivisatos, A. P. (2000) Naturally Aligned Nanocrystals *Science,* 289, 736–737.

Banfield, F., Welch, S. A., Zhang, H., Ebert, T. T. Penn, R. L. (2000) Aggregation-Based Crystal Growth and Microstructure Development in Natural Iron Oxyhydroxide Biomineralization Products Banfield, *Science* 2000, *289*, 751–754.

Boonekamp E.P, Kelly J.J, Fokkink L.G.J (1994) Adsorption of nanometer-sized palladium particles on Si(100) surfaces, *Langmuir*, 10, 4089–4094.

Colfen, H. (2001) Double-Hydrophilic Block Copolymers: Synthesis and Application as Novel Surfactants and Crystal Growth Modifiers, *Macromol. Rapid Commun., 22*, 219–252.

Colfen, H., Qi, L. A. (2001) Systematic Examination of the Morphogenesis of Calcium Carbonate in the Presence of a Double-Hydrophilic Block Copolymer, *Chem. Eur. J.*, 7, 106–116.

Colfen, H., Mann, S. (2003) Higher-Order Organization by Mesoscale Self-Assembly and Transformation of Hybrid Nanostructures, *Angew. Chem., Int. Ed.*, 42, 2350–2365.

De Gennes P.G. (1987) Polymers at an interface; a simplified view, *Adv. Colloid Interface Sci.*, 27 189

Dickson E., Ericsson, L. (1991) *Adv. Colloid Interface Sci.*, 143 9.

Gibbs,J.W (1948) *Collected works*, Vol.1, *Thermodynamics,* Yale University Press,New Haven.

Gorna, K., Munoz-Espi, R., Grohn, F., Wegner, G. (2007) Bioinspired Mineralization of Inorganics from Aqueous Media Controlled by Synthetic Polymers, *Macromol. Biosci.*, *7*, 163–173.

Gu, F., Wang,S.F., Lu, M.K.,. Xu, D, Yuan. D.R. (2004) Structure evalution & highly enhanced luminescence of DY^{+3} –doped ZnO nanocrystals by Li doping via combustion, *Langmuir* 20, 3528.

Guo, X., Yu, S., Cai, G. (2006) Crystallization in a Mixture of Solvents by Using a Crystal Modifier: Morphology Control in the Synthesis of Highly Monodisperse $CaCO_3$ Microspheres, *Angew. Chem., 118*, 4081–4085.

He, L., Zhang, Y., Ren, L., Chen, Y., Wei, H., Wang, D. (2006) Double-Hydrophilic Polymer Brushes: Synthesis and Application for Crystallization Modification of Calcium Carbonate *Macromol., Chem. Phys. 207*, 684–693.

Henisch, H. K. (1988) In *Crystals in Gels and Liesegang Rings*; Cambridge University Press: Cambridge,; Chapter 2, 29–47.

Jongen, N., Bowen, P., Lemaitre, J., Valmalette, J. C.; Hofmann, H. J. (2000) Precipitation of Self-Organized Copper Oxalate Polycrystalline Particles in the Presence of Hydroxypropylmethylcellulose (HPMC): Control of Morphology, *Colloid Interface Sci., 226*, 189–198.

Joshi, S. S., Patil, S. F., Iyer, V., Mahamuni, S. (1998) Radiation induced synthesis and characterization of copper nanoparticles, *Nanostruct. Mater., 10*(7) 1135–1144.

Kashchiev, D. (2000) Nucleation, Butterworth-Heinemann Oxford.

Kirk-Othmer Encyclopedia of Chemical Technology, vol. 23, John Wiley and Sons, New York, 1983, p. 856.

Kuldipkumar, A., Tan, Y. T. F., Goldstein, M., Nagasaki, Y., Zhang, G. G. Z., Kwon, G. S. (2005) Amphiphilic Block Copolymer as a Crystal Habit ModifierKuldipkumar, *Cryst. Growth Des., 5*, 1781–1785.

Lakshminarayanan, R., Valiyaveettil, S., Loy, G. L., (2003) Selective Nucleation of Calcium Carbonate Polymorphs: Role of Surface Functionalization and Poly(Vinyl Alcohol) Additive, *Cryst. Growth Des., 3*, 953–958.

Lee, S. J., Benson, E. A., Kriven, W. M. (1999) Preparation of portland cement components by poly(vinyl alcohol) solution polymerization, *J. Am. Ceram. Soc., 82*, 2049–2055.

Lifshitz, I.M., Slyozov, V. V. (1961) *J. Phys Chem. Solids, 19*, 35

Liu, S., Yu, J., Cheng, B., Zhang, Q. (2005) Controlled Synthesis of Novel Flower-shaped $BaCrO_4$ Crystals, *Chem. Lett., 34*, 564–565.

Loste, E., Meldrum, F. C. (2001) Control of calcium carbonate morphology bytransformation of an amorphous precursor in a constrained volume, *Chem. Commun.*, 901–902.

Makoto, K. (2001) *Mem. Nat. Def. Acad., Math., Phys., Chem. Eng., 1*, 1–8.

Mann, S. (1988) Molecular recognition in biomineralization, *Nature, 332*, 119–124.

Mann, S., Heywood, R., Rajam, S., Birchall, J. D. (1988) Controlled crystallization of $CaCO_3$ under stearic acid monolayers, *Nature, 334*, 692–695.

Meldrum, F. C., Ludwigs, S. (2007) Template-Directed Control of Crystal Morphologies *Macromol. Biosci., 7*, 152–162.

Meng, Q., Chen, D., Yue, L., Fang, J., Zhao, H., Wang, L. (2007) Hyperbranched Polyesters with Carboxylic or Sulfonic Acid Functional Groups for Crystallization Modification of Calcium Carbonate, *Macromol. Chem. Phy., 208*, 474–484.

Merrill, L. Bassett, W. A. (1975) The crystal structure of $CaCO_3$ (II), a high-pressure metastable phase of calcium carbonate, *Acta Crystallogr. B., 31*, 343–349.

Mullin, J. W. (2002) In *Crystallization*; Butterworth-Heinemann: Oxford,; Chapter 6, 216–314.

Mullin. J.W., (1961) *Crystallization* (4th Edition Reprinted) Butterworth-Heinemann, ISBN O 7506 4833 3, Oxford.

Nanocallas, G.H. and Purdie, N. (1964) The kinetics of crystal growth. Quarterly Reviews of the chemical Society, 18,1-20.

Nielsen, A.E (1964) Kinetics of precipitation, Pregamon, Oxford.

Nyvlt, J. (1995) The ostwale Rule of Stages. Crystal Research and Technology, 30, 445–451.

Oaki, Y., Imai, H., (2003) Experimental Demonstration for the Morphological Evolution of Crystals Grown in Gel Media, *Cryst. Growth Des,*. 2003, 3, 711–716.

Patil P. R., Joshi S. S., (2007) Polymerized organic–inorganic synthesis of nanocrystalline zinc oxide, *Materials Chemistry and Physics*,105, 354–361.

Patil P. R., Krishnamurty V. N., Joshi S. S. (2008) Effect of Nano-Copper Oxide and Copper Chromite on the Thermal Decomposition of Ammonium Perchlorate, *Propellants, Explosives, Pyrotechnics*, 4, 33, 266 – 270.

Prince, E., (1957)*Acta.Crystallogr.* 10, 554

Qi, L., Colfen, H., Antonietti, M. (2000) Control of Barite Morphology by Double-Hydrophilic Block Copolymers, *Chem. Mater.*, 12, 2392–2403.

Qi, L., Colfen, H., Antonietti, M. (2000) Crystal Design of Barium Sulfate using Double-Hydrophilic Block Copolymers, *Angew. Chem., Int. Ed.*, 39, 604–607.

Robinson, K. L., Weaver, J. V. M., Armes, S. P., Marti, E. D., Meldrum, F. C. (2002) Synthesis of controlled-structure sulfate-based copolymers *via* atom transfer radical polymerisation and their use as crystal habit modifiers for BaSO₄, *J. Mater. Chem.*, 12, 890–896.

Rudolff, J., Antonietti, M., Colfen, H., Pretula, J., Kaluzynski, K., Penczek, S. (2002) Double-Hydrophilic Block Copolymers with Monophosphate Ester Moieties as Crystal Growth Modifiers of CaCO₃, *Macromol. Chem. Phys.*, 203, 627–635.

Said A.A., (1991) The role of Copper-chromium oxide catalyst in the thermal decomposition of ammonium perchlorate, *J. Therm. Anal.*, 37,959.

Sarig, S., Mullin, J. W. (1980) Size Reduction of Crystals in Slurries by the Use of Crystal Habit Modifiers, *Ind. Eng. Chem. Process Des. DeV.*, 19, 490–494.

Sedlak, M., Colfen, H. (2001) Synthesis of Double-Hydrophilic Block Copolymers with Hydrophobic Moieties for the Controlled Crystallization of Minerals, *Macromol. Chem. Phys.*, 202, 587–597.

Sinha, A., Agrawal, A., Das, S. K., Ravi Kumar, B., Rao, V., Ramachandrarao, P. (2001) On the growth of monoclinic calcium carbonate in poly(vinyl alcohol), *J. Mater. Sci. Lett.*, 20, 1569–1572.

Sinha, A., Kumar Das, S., Rao, V., Ramachandrarao, P. (2000) Synthesis of organized inorganic crystal assemblies, *Curr. Sci.*, 79, 646–648.

Sohnel, O. and Garside, J. (1992) Precipitation: Basic Principle and Industrial Applications, Butterworth-Heinemann Oxford.

Solymosi F., Krix E., (1962) Catalysis of solid phase reactions effect of doping of cupric oxide catalyst on the thermal decomposition and explosion of ammonium perchlorate, *J.Catal.*, 1, 468.

Strickland-Constable, R.F (1968) Kinetics and Mechanism of crystallization, Academic Press London.

Sugimoto, T. (1987) *Adv. Colloid Interface Sci.*, 28, 165.

Sutton, G. P., Oscar, B. (2001) In *Rocket Propulsion Elements*; John Wiley & Sons: New York,; Chapter 12. 474–519.

Telkar M.M., Rode C.V., Chaudhari R.V., Joshi S.S., Nalawade A.M., (2004) Shape-controlled preparation and catalytic activity of metal nanoparticles for hydrogenation of 2-butyne-1,4-diol and styrene oxide, *Applied Catalysis A: General*, 273, 11–19.

Temigre, M. K., Joshi, S. S. (2004) Optical and structural studies of silver nanoparticles, *Radiat. Phys. Chem.*, 71, 1039–1044.

University Press: Cambridge, (1988) Chapter 2, 29–47.

Vaishampayan M., Joshi. S. S., Mulla I. S., (2011) Low temperature pH dependent synthesis of flower-like ZnO nanostructures with enhanced photocatalytic activity *Material Research Bulletin*, 46(5), 771–778.

Vargeese. A. A., Joshi. S. S., Krishnamurthy V. N. (2008) Role of Poly(vinyl alcohol) in the Crystal Growth of Ammonium Perchlorate, *Crystal Growth & Design, 8, 3*, 1060-66.

Volmer,M. (1939) *Kinetic der Phasenbildung*, Steinkopff, Leipzig.

Wagner, C. (1961) *Elektrochem.Z*, 65, 581

Walsh, D., Mann, S. (1995) Molecular recognition in biomineralization, *Nature, 377*, 320–323.

Walton.A.G. (1967) The formation and properties of precipitates, Interscience, New York.

Wang, F., Xu, G., Zhang, Z., Xin, X. (2005) Morphology control of barium sulfate by PEO-PPO-PEO as crystal growth modifier, *Colloids Surf., A, 259*,151–154.

Wang, T., Rother, G., Colfen, H. (2005) A New Method to Purify Highly Phosphonated Block Copolymers and Their Effect on Calcium Carbonate Mineralization, *Macromol. Chem. Phys.*, 206, 1619–1629.

Wulff, G. Z. (1901) On the question of the rate of growth and dissolution of crystal surfaces, *Kristallogr. Mineral.*, 34, 449–530.

Xu, A., Ma, Y., Colfen, H. (2007) Biomimetic mineralization, *J. Mater. Chem.*, 17, 415-449

Yu, J. G., Zhao, X. F., Liu, S. W., Li, M.; Mann, S., Ng, D. H. L. (2007) Poly(methacrylic acid)-mediated morphosynthesis of $PbWO_4$ micro-crystals, *Appl. Phys. A: Mater. Sci. Process.*, 87, 113–120.

Yu, J., Liu, S., Cheng, B. (2005) Effects of PSMA additive on morphology of barite particles, *J. Cryst. Growth.*, 275, 572–579.

Yu, S., Colfen, H. (2004) Bio-inspired crystal morphogenesis by hydrophilic polymers, *J. Mater. Chem.*, 14, 2124–2147.

Yu, S., Colfen, H., Antonietti, M. (2002) Control of the Morphogenesis of Barium Chromate by Using Double-Hydrophilic Block Copolymers (DHBCs) as Crystal Growth Modifiers, Chem. *Eur. J.*, 8, 2937–2945.

Yu, S., Colfen, H., Antonietti, M. (2003) Polymer-Controlled Morphosynthesis and
    Mineralization of Metal Carbonate Superstructures, *J. Phys. Chem. B, 107,* 7396–
    7405.
Zettlemoyer, A.C (ed.) (1969) Nucleation, Dickker, New York.

Part 2

Growth of Organic Crystals

8

Protein Crystal Growth

Igor Nederlof[1], Eric van Genderen[1], Flip Hoedemaeker[2],
Jan Pieter Abrahams[1] and Dilyana Georgieva[1]
[1]Leiden University,
[2]Kabta Consultancy Ltd.,
The Netherlands

1. Introduction

The biological activity of most proteins is determined by their 3D structure. For instance, a substantial number of molecular diseases are caused by protein structural alterations, which are genetically encoded. Drugs operate by binding to proteins, inducing alteration of their functional structure and thereby affecting their biological activity. Hence the design and improvement of drugs is greatly facilitated by knowledge of the 3D structures of their macromolecular targets. In the light of these considerations, it is clear that elucidation of the 3D structure of proteins is of prime importance for understanding the underlying mechanisms of molecular diseases. It was initially believed that any protein that could be made soluble and could be purified would be relatively easy to crystallize. However, the results have indicated that solubility and purity of proteins, although being important factors, do not secure a yield of useful crystals. The crystallization behavior of proteins turns out to be very complex.

In an effort to identify the naturally occurring protein folds, large structural genomics consortia were set up. The somewhat disappointing outcome of these efforts is that only about 3% of all proteins that were targeted by these consortia yielded a crystal structure (http://targetdb.pdb.org/statistics/TargetStatistics.html), despite massive investments in high-throughput, automated protein production, purification and crystallization. It is clear that in order to improve the current situation, better strategies for protein crystallization are required, combined with techniques that allow the use of smaller nano-crystalline material.

2. Crystallization of bio-macromolecules

Biocrystallization involves the three classical steps of nucleation, growth, and cessation of growth, even though the protein crystals contain on average 50% of disordered solvent (Figure 1) (Matthews et al., 1968). However, crystal growth of biological molecules differs substantially from small molecule crystalogenesis. The reason is the much larger number of parameters involved in biocrystallization, as well as the specific physico-chemical properties of the biological compounds. The main difference from small molecule crystal growth is the conformational flexibility and chemical versatility of macromolecules and their greater sensitivity to external factors. An overview of different parameters affecting the crystallization of biomacromolecules is presented in table 1 (Bergfors T, 2009).

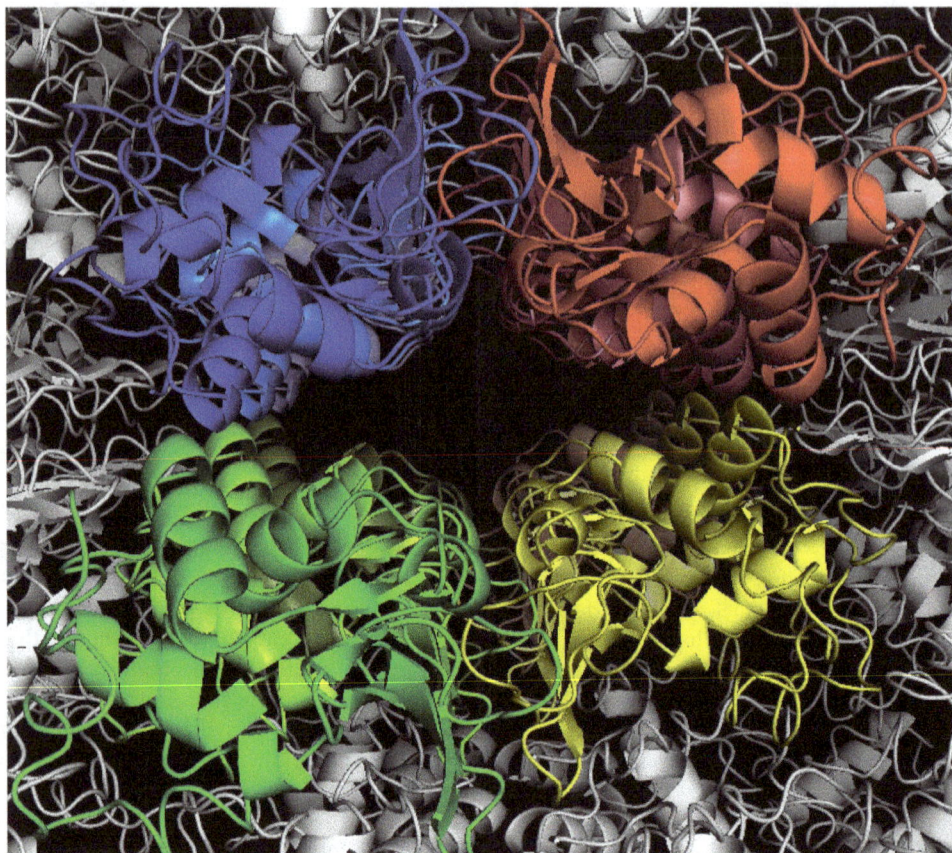

Fig. 1. Crystal packing in lysozyme crystals (pdb:1Lyz) shows large cavities. These cavities are filled with disordered solvent (not shown).

Intrinsic physico-chemical properties	Biochemical and biophysical parameters
• Supersaturation • Temperature, pH • Ionic strength and purity of chemicals • Pressure, electric and magnetic fields • Vibration and sound	• Sensitivity of conformation to physical parameters • Binding of ligands • Specific additives • Aging of samples
Biological parameters	**Purity of macromolecules**
• Rarity of biological macromolecules • Bacterial contaminants • Biological sources of organisms and cells	• Macromolecular contaminants • Sequence (micro) heterogeneity • Conformational (micro) heterogeneity • Batch effects

Table 1. Overview of parameters affecting bio-macromolecular crystallization.

Another important prerequisite for successful crystallization is the quality of the macromolecular samples. Bio-macromolecules are extracted from living cells or synthesized *in vitro* and they are frequently difficult to prepare at a high degree of purity and homogeneity. Besides traces of impurities, the different treatments proteins are subjected to may decrease their stability and activity through different kinds of alterations. As a general rule, purity and homogeneity are regarded as conditions of prime importance. Accordingly, purification, stabilization, storage and handling of macromolecules are other essential steps prior to crystallization.

3. Purity of bio-macromolecular samples

The concept of purity has a special meaning when biological crystallogenesis is concerned. Molecular samples need to be not only chemically pure, but they must also be conformationally uniform (Giege et al., 1986). This concept is based on the fact that the best crystals are grown from solutions containing well-defined entities with identical physico-chemical properties. For X-ray crystallographic studies, the aim is to grow 'single crystals' diffracting to high resolution with a low mosaicity and prolonged stability in the X-ray beam. It is therefore understandable that contaminants may compete for sites on the growing crystals and generate lattice errors leading to internal disorder, dislocations, poor diffraction or early cessation of growth (Vekilov et al., 1996). Because of the high molecular weight of molecules in a single crystal (up to millions of daltons), and hence low molarity of their solutions even relatively small amounts of contaminant may induce formation of non-specific aggregates, alter macromolecular solubility, or interfere with nucleation and crystal growth (Skouri et al., 1996; McPherson et al., 1996). Successful crystallization of rare proteins and nucleic acids support the importance of purity and homogeneity (Wierenga et al., 1987; Thegesen et al., 1996; Aoyama et al., 1996; Douna et al.,: 1993). Usually most of the contaminants are eliminated during the different purification steps, however traces of polysaccharides, lipids or proteases may still be present and hinder crystallization. Small molecules, like peptides, oligonucleotides, amino acids, as well as uncontrolled ions should also be considered as contaminants. Buffering molecules remaining from a purification step can be responsible for irreproducible crystallization. For instance, phosphate ions are relatively difficult to remove and may crystallize in the presence of divalent cations (Ca^{2+}, Mg^{2+}). Counterions play a critical role in the packing of biomolecules. Often macromolecules do not crystallize or yield different habits in the presence of various buffers adjusted at the same pH.

Bio-molecular samples containing traces of contaminants can further be subjected to purification through recrystallization, column chromatography, ultra-centrifugation, fractionated precipitation, affinity purification and other techniques. Microheterogeneity in pure macromolecules can only be revealed by very sensitive methods. The most common causes for heterogeneity are uncontrolled fragmentation and post-synthetic modification.

Proteolysis normally takes place in many physiological processes and represents a major difficulty that needs to be overcome during protein extraction from the living cells that produce the desired protein (Achstetter et al., 1985; Barrett et al., 1986; Dalling et al., 1986; Bond et al., 1987; Arfin et al., 1988; Wandersman et al., 1989). The reason is that proteases are localized in various cellular compartments or excreted in the extracellular medium. Upon cell disruption, cellular compartments are mixed with extracellular proteases and control over proteolysis is lost. Decrease of protein size and stability, modification of their charge or hydrophobicity, partial or total loss of activity are usually signs of proteolysis. Traces of protease may not be detectable even when overloading electrophoresis gels, but they can cause damage during concentration or storage of samples.

Co- or post-translational enzymatic modifications generate microheterogeneity in proteins when different groups, for instance oligosaccharide chains, occupy specific modification sites on the protein, or when correct modifications are unevenly distributed. Only certain modifications are reversible, for instance phosphorylation, but others like glycosylation or methylation are not. Microheterogeneity can also appear during storage, for instance by deamidation of asparagines or glutamine residues is a well-documented phenomenon.

Pure, chemically uniform macromolecules can be fully functional in a biochemical activity assay even though they are microheterogeneous. Conformational heterogeneity may have several origins: binding of ligands, intrinsic flexibility of molecular backbones, oxidation of cysteine residues or partial denaturation. In the first case, macromolecules should be prepared in both forms, the one deprived of and the other saturated with ligands. In the second case, controlled fragmentation may be helpful. In the last one, oxidation of a single cysteine residue leads to complex mixtures of molecule species for which the chances of growing good crystals are low (Van der Laan et al., 1989).

Although macromolecules may crystallize readily in an impure state (Holley et al., 1961), this is an exception and it is always preferable to achieve a high level of purity before starting crystallization experiments. In order to gain more information about the quality of the protein samples, different techniques can be used. For instance, spectrophotometry and fluorometry give information about the quality of samples if macromolecules or their contaminants have special absorbance or emission properties. SDS-PAGE indicates the size of protein contaminants, but not that of non-protein contaminants. Isoelectric focusing gives an estimate of the pI of protein components in a mixture and electrophoretic titration shows the mobility of individual proteins as a function of pH. The latter method can also suggest the type of chromatography suitable for further purification. Capillary electrophoresis is well adapted for purity analysis (Karger et al., 1996). Amino acid composition and sequencing of N- and C- termini verify in part the integrity of primary structure.

Electrospray ionization and matrix-assisted laser desorption/ionisation mass spectrometry are also powerful tools in the analysis of recombinant protein chemistry. Nuclear magnetic resonance can detect small size contaminants and gives structural information on biomolecules (Wuthrich , 1995).

It is widely believed that the success of crystal trials is largely dependent on various, not very well identified, properties of the protein. For example, a positive correlation has been established between the degree of protein monodispersity in solution and the ability of the protein to crystallize. On the other hand, it's thought to be a negative correlation between the degree of disorder in the protein and its ability to crystallize (Mikol et al., 1989).

A number of biophysical techniques and methods are employed to evaluate the quality and stability of protein solutions. Dynamic light scattering is a useful tool for non-invasive *in situ* monitoring of crystallization trials because it detects the formation of aggregates or nuclei before they become visible under a light microscope (Berne et al., 1976). Fluorescence and light scattering are helpful to rapidly identify stabilizing conditions compromising simple agents (salts, co-factors etc.). Emission fluorescence is used to measure changes if the protein unfolds or undergoes other conformational changes (Konev et al. 1967).

4. Solubility, supersaturation and phase transition

Biological macromolecules follow the same thermodynamic rules as inorganic or organic small molecules concerning supersaturation, nucleation and crystal growth. However, protein macromolecules are organized in tertiary and quaternary structures. The intra-molecular interactions responsible for their tertiary structure, the intermolecular interactions involved in the crystal contacts, and the interactions necessary to solubilise them in a solvent are similar.

To crystallize a biological macromolecule, its solution must have reached supersaturation, which is the driving force for crystal growth. The under- and supersaturated states are defined by the solubility of the macromolecules. When the concentrations of the crystallization agent and the macromolecules correspond to the solubility condition, the saturated macromolecule solution is in equilibrium with the crystallized macromolecules. Below the solubility curve (fig. 2) the solution is under saturated and the system is thermodynamically stable. In this case, phase transition (crystallization) will not occur. Above the solubility curve, the concentration of the biological macromolecules is higher than the concentration at equilibrium. A supersaturated macromolecular solution contains an excess of macromolecules that will appear as a solid phase until the macromolecular concentration reaches the solubility value in the solution. The higher the supersaturation, the faster this solid phase appears. However, at very high supersaturation precipitation, not crystallization occurs, but insoluble macromolecules rapidly separate from the solution in an amorphous state.

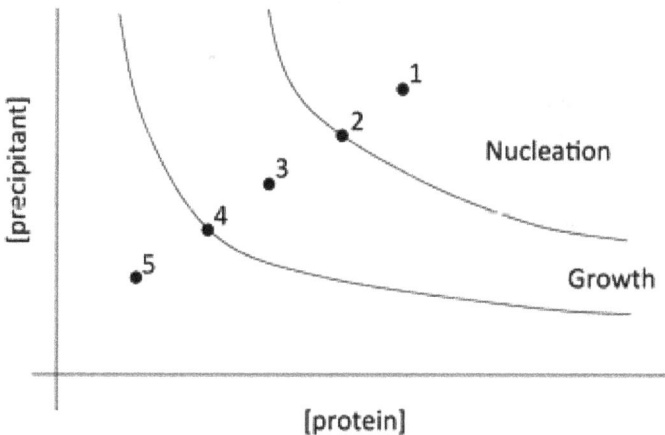

Fig. 2. Solubility curve of a protein, where the phase state of the protein is plotted against the concentration of both protein and precipitant. At the point (1), the protein may precipitate so fast that an amorphous precipitate or at best shower of microcrystals is formed. At (2) the conditions are just right for the protein to form a stable crystal nucleus, which will start to grow – passing (3) – into a stable protein at equilibrium with the mother liquor (4). At (5), the concentration of protein and precipitant are too low for crystal nucleation or growth, and the solution will remain clear. Note that the true solubility curve of any protein is highly multidimensional, with every parameter affecting protein solubility (cf. Table 1) representing a different independent axis.

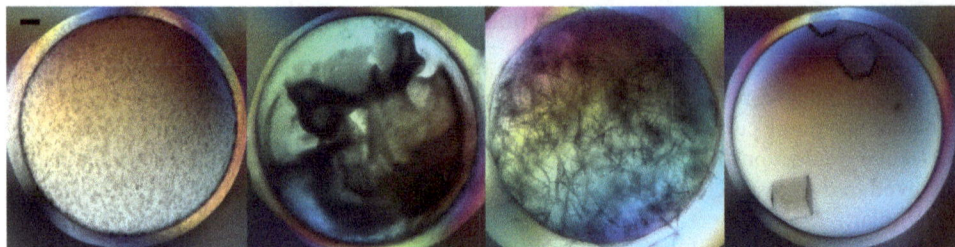

Fig. 3. Glucose Isomerase crystallization condition yielding phase separation (far left) amorphous precipitation (near left) micro-crystals (near right) and macro-crystals (far right) bar on the top left represents 200 micrometer.

5. Crystallization strategies

5.1 Crystallization screens

Finding crystallization conditions for a new protein target is largely based on a trial and error method. The first step is to set up screening trials, exposing the protein to a variety of agents in order to find useful "leads", which can be crystals, crystalline precipitates and phase separation that point to conditions that are conductive to crystallization.

The most popular screens to perform the initial screening step are called sparse-matrix screens. These screens rely on a compilation of conditions that had previously led to successful crystallization. Systematic screens sample the crystallization parameter space in a balanced, rational way using information on the protein properties. Systematic screens are usually used as second remedy or in order to optimize the crystallization conditions.

5.2 Choosing the crystallization method

There are different methods to crystallize biological macromolecules. However, all of them aim at bringing the solution of macromolecules to a supersaturation state (McPherson, 1985; Giege, 1987). It's important to keep in mind that not only the various chemical and physical parameters influence protein nucleation and crystallization, but also the method of crystallization. Therefore, it's wise to try different methods when searching for optimal crystallization conditions. As solubility is dependent on temperature (it could increase or decrease depending on the protein), it's highly recommended to perform crystallization trials at constant temperature unless temperature variation is part of the experiment. Solubility of most chemicals is given in Merck Index. The chemical nature of the buffer is an important parameter for protein crystal growth. It must be kept in mind that the pH of buffers is often temperature dependent, this is particularly significant for Tris buffers. Buffers, which must be used within one unit from their pK value, are well described in textbooks (Perin et al., 1974).

Protein samples often contain large amount of salts of unknown composition when first obtained. Thus it's wise to dialyse a new batch of a macromolecule against a large volume of well-characterised buffer of given pH, to remove unwanted salts and to adjust the pH. Starting from known conditions helps to increase the reproducibility.

Whatever the crystallization method used, it requires high concentration of biological macromolecules as compared to normal biochemistry conditions. Before starting a crystallization experiment, a concentration step is generally needed. It's also important to keep pH and ionic strength at desired value, since pH may vary when the concentration of macromolecules increases. Also, low ionic strength could lead to early precipitation. Many commercial devices are available based on Different concentration principles such as concentration under pressure, using centrifugation, or lyophilisation. The choice of method for concentration depends on the quantity and the stability of the macromolecules.

Before a crystallization experiment, solid particles such as dust, denatured proteins, and solids coming from purification columns or lyophilization should be removed. This could be achieved by centrifugation or filtration, depending on the available quantity.

The most common method to measure macromolecular concentration is to sample an aliquot, dilute it with buffer, and measure absorbance at 280 nm for proteins within the linear range of a spectrophotometer. Proper subtraction with the reference cell should be made especially when working with additives absorbing in the 260-300 nm wavelength range. When working with enzymes, an alternative method to measure the concentration of protein is to perform activity test, otherwise colorimetric methods can be performed.

5.2.1 Vapour diffusion

The most widely used method of crystallization is vapour diffusion. The protein solution is a hanging, sitting or sandwich drop that equilibrates against a reservoir containing crystallizing agents at either higher or lower concentration than in the drop. Equilibration proceeds by diffusion of the volatile species (water or organic solvent) until vapour pressure in the droplet equals the one of the reservoir. If equilibration occurs by water exchange from the drop to the reservoir, it leads to a droplet volume decrease. Consequently, the concentration of all constituents in the drop will increase. For species with a vapour pressure higher than water, the exchange occurs from the reservoir to the drop. In such a 'reverse' system, the drop volume will increase and the concentration of the drop constituents will decrease. The same principle applies for hanging drops, sitting drops and sandwich drops. Most people use a ratio of 1:1 between the concentration of the crystallizing agent in the reservoir and in the droplet. This is achieved by mixing a droplet of protein at twice the desired final concentration. When no crystal or precipitate is observed, either supersaturation is not reached or one has reached the metastable region. In the latter case changing the temperature by a few degrees is generally sufficient to initiate nucleation. Although unique in this respect, vapour diffusion permit easy variation of physical parameters during crystallization, and many successes were obtained by modifying supersaturation by temperature or pH changes. With ammonium sulphate as the crystallizing agent, it has been shown that the pH in the droplets is imposed by that of the reservoir. Consequently, varying the pH of the reservoir permits gentle adjustments of that in the droplets. From another point of view sitting drops are well suited for attempting epitaxial growth of macromolecule crystals on appropriate mineral matrices. In other words vapour diffusion provides a way to sample the crystallization space with the conditions continuously varying, as the equilibration proceeds. The kinetics of water evaporation determines the kinetics of supersaturating and accordingly affects nucleation rates. Evaporation rates from hanging drops have been determined experimentally in the presence of ammonium sulphate, PEG, MPD and NaCl as crystallizing agents. The main parameters that determine the rate of water equilibration are temperature, initial drop volume, water pressure of the reservoir, and the chemical nature of the crystallization agent. Theoretical modelling has shown in addition the pivotal role of the drop to reservoir distance. It was shown that the effect of this parameter is negligible in classical set-ups and becomes only noticeable when drop to reservoir distance is more than 2 cm. From practical point of view, the time for water evaporation to reach 90% completion can vary from about 25 hours to more than 25 days. The fastest equilibration occurs in the presence of ammonium sulphate and the slowest in the presence of PEG. Equilibration rates are significantly slowed down by increasingly appropriately the distance between the drop and the reservoir. An alternative solution to decrease equilibration rates is to apply a layer of oil over the reservoir.

Fig. 4. Schematic drawing of sitting drop (left), hanging drop (middle), and batch crystallization (right). Well solution is blue, protein mixed with well solution is brown and oil is green.

5.2.2 Batch crystallization methods

Another routinely used method for crystallization is the batch method. The biological macromolecule to be crystallized is mixed with the crystallizing agent at a concentration such that supersaturation is instantaneously reached. Crystallization trials are dispensed and incubated under low-density paraffin oil. The crystallization drops remain under oil, where they are protected from evaporation, contamination and shock. Since supersaturation is reached at the start of the experiment, nucleation tends to be higher, if compared to the vapour diffusion method. However, in some cases fairly large crystals can be obtained when working close to the metastable region. Although the microbatch method has not been compared in a statistically significant scale against hanging drop-vapour diffusion method, a comparison on a small scale has been performed (Baldok et al., *1999*). The study demonstrated that the methods are not entirely identical, but are equally effective. The results suggest that vapour diffusion method and the microbatch technique will probably produce similar numbers of crystals, but may not produce crystals for the same conditions. Microbatch and vapour diffusion methods are both suitable for high throughput crystallization experiments where all the steps of dispensing, mixing and sealing are automated and performed by a robot. Other crystallization methods worth mentioning, although with more limited success and use are crystallization in gel, dialysis, microfluidics, free interface diffusion. Microfluidic chips are also being used for high throughput crystallization screening.

5.2.3 Crystallization in gels

Special attention has been paid to crystallization in gels (Robert at al., 1987). The protein crystallization process consists of two main steps – the transport of growth units towards the surface of the crystals and second, the incorporation of the growth units into a crystal surface position of high bond strength. The whole growth process is dominated by the slower of these two steps and is either transport controlled or surface controlled. The ratio between transport to surface kinetics, which can be tuned by either enhancing or reducing transport processes in solution, was shown to control the amplitude of growth rate fluctuations. These are the reasons why gels if properly designed are expected to enhance the quality of crystals. It's worth mentioning that crystals growing in gel do not sediment as they do in free solution. They develop at the nucleation site, sustained by the gel network. For small molecule crystals grown in silica gel, the gel often forms cusp-like cavities around the crystal and a thin liquid film that reduces contamination risk, separates the crystal from the gel. Such cavities have not been seen in macromolecular crystals. Recent studies have shown that silica gel can be incorporated in the crystal network almost without disturbing the crystal lattice. Such crystals that still diffract to a high resolution, are mechanically reinforced and are more resistant to dehydration, because the silica gel framework embedded in the crystal slows down water loss due to its hygroscopic properties. Although seeding can be used, it appears that most of the gel-grown crystals are obtained by spontaneous nucleation inside a macroscopically homogeneous gel. When the gel adheres to the walls of the container, no nucleation occurs on the cell walls, neither on dust. So, heterogeneous nucleation is strongly reduced, if not suppressed. Another type of nucleation, namely secondary nucleation, is due to attrition of a previous crystal by the solution flux. When nucleation occurs inside the gel, one observes that all the crystals appear at the same time and have about the same size. They are homogenously distributed in the whole volume.

5.2.4 Dialysis methods

Crystallization by dialysis methods allow for an easy variation of the different parameters that influence the crystallization of biological macromolecules. Different types of dialysis cells are used, but follow the same principle. The macromolecule is separated from a large volume of solvent by a semi-permeable membrane that gives small molecules free passage, but prevents macromolecules from circulating. The kinetics of equilibration will depend on the membrane cut-off, the ratio of the concentration, the temperature and the geometry of the cell.

The method of crystallization by interface diffusion was developed (Salemme, 1972) and used to crystallize several proteins. In the liquid/liquid diffusion method, equilibration occurs by diffusion of the crystallization agent into the biological macromolecule volume. To avoid rapid mixing, the less dense solution is poured gently on the most dense (salt in general) solution. Sometimes, the crystallizing agent is frozen and the protein layered above to avoid rapid mixing.

5.3 The role of heterogeneous substrates in the process of protein nucleation and crystallization

In general, additives play an important role in protein crystallization. Heterogeneous substrates are usually regarded as additives when they are purposefully added to the solution in order to obtain a desired effect (inhibition of nucleation, habit change of crystals). However, impurities of foreign substances may also exist in the solution originating from other sources (the solvent, crystallization agent, etc.). Heterogeneous crystallization which is induced by a properly chosen additive may allow better control of nucleation and growth. The first report of a nucleant inducing nucleation of macromolecules was the epitaxial growth of protein crystals on minerals (McPherson et al., 1988). Other candidate nucleants followed like zeolites, silicates, charged surfaces, porous materials etc. and have been tested for multiple proteins (Sugahara et al., 2008, Takehara et al., 2008). Previous results showed that horsehair and dried seaweed showed increased hits when added to sparse-matrix crystallization trials. The increase in crystallization was 35% when horsehair was added to 10 test proteins (Thakur et al., 2008). The underlying mechanism is explained with epitaxial nucleation in the case of minerals, electrostatic interactions if the nucleants contain charged surfaces, nucleation through specific favourable protein-protein interactions or physical entrapment in the caves of porous materials.

Seeding techniques can be advantageous in both screening of crystallization conditions to obtain crystals as well in the later optimisation steps. The streak seeding technique may provide a fast and effective way to facilitate the optimization of growth conditions without the uncertainty that is intrinsic in the process of spontaneous nucleation (Bergfors, 2003). A probe for analytical seeding is easily made with an animal whisker mounted with wax to the end of a pipette tip. The end of the fibre is then used to touch an existing crystal and dislodge seeds from it. Gentle friction against the crystal is normally sufficient. The probe is then used to introduce seeds into pre-equilibrated drop by rapidly running the fibre in a straight line across the middle of the drop containing protein and precipitant. Sitting drop set-ups are preferable since hanging drops tend to evaporate more quickly.

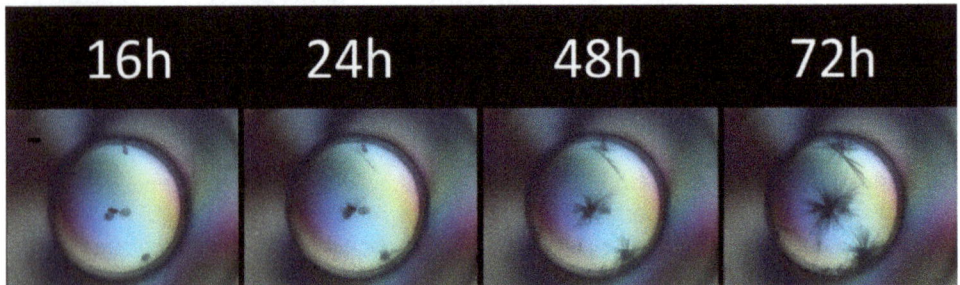

Fig. 5. Lysozyme needle crystals growing on sliced human hair as a nucleant, the black bar in the left picture represents 200 micrometer.

6. Combining heterogeneous crystallization and high throughput methods

A method for the introduction of heterogeneous nucleants in high throughput crystallization experiments has recently been developed (Nederlof et al., 2011). The method includes preparing of crystallization plates that are locally coated with fragments of human hair, allowing automated, high throughput crystallization trials in a fashion entirely compatible with standard vapour diffusion crystallization techniques. The effect of the nucleants was assessed on the crystallization of 11 different proteins in more than 4000 trials. Additional crystallization conditions were found for 10 out of 11 proteins when using the standard JCSG+ screen. In total, 34 additional conditions could be identified. The increase in crystallization conditions ranged between 33.3% (two additional conditions were identified for myoglobin on top of four homogeneous crystallizations) to 1.2% (we identified a single additional condition for insulin, which crystallized in 85 out of 96 conditions); the median increase in crystallization hits was 14%. The method is straightforward, inexpensive and uses materials available in every crystallization lab.

7. Lab automation

In recent years, setting up protein crystallisation trials and analysis of the results has become largely automated. More and more of the crystallisation methods mentioned in section 5 have been made amenable to automation, with the sitting drop method still the most popular experiment type in this respect. Lab automation includes the use of dispensing robots, imaging robots, in situ crystal analysis as well as automated diffraction analysis (Stevens et al., 2000, Berry et al., 2006).

7.1 Automation in dispensing

Dispensing robots that are used routinely are either specialized for dispensing well recipes (e.g. Formulator, MatrixMaker) or drop-setting (NT8, Phoenix, Mosquito), but there are also more generic robots that can do both (Hamilton Star, Tecan Evo). In general, the experimenter will start crystallisation trials with a set of pre-defined conditions, contained in one or more screens. Over 150 of these screens can be bought from commercial vendors in a wide range of formats. A number of these are designed on the basis of statistical analysis of results obtained at structural genomics initiatives. When initial hits are found with screens like these, secondary optimisation experiments need to be performed to produce diffracting crystals. In this stage, interaction with a Lab Information Management System (LIMS), where experiment design can be coupled to experiment preparation and analysis, greatly enhances the potential throughput in a lab and thereby the success rate. There are a number of these software packages that can be used to create grid experiments around an initial hit condition, as well as randomized sparse-matrix screens based on initial successes.

7.2 Automated experiment imaging

Automated experiment analysis is an essential part of the lab setup. Due to the increase in throughput obtained by using dispensing robots it is impossible to routinely scan the results manually under a microscope. The dynamic nature of these experiments can cause the crystallographer to miss events, even crystals. Imaging robots vary from semi-automated microscopes with a moving plate stage and camera to fully automated incubators that are capable of following all lab experiments from start to finish without human intervention. Ideally, images are displayed to the user in the context of the experiment design, so that the results are easily interpreted. If this functionality is integrated with the experiment design and dispensing the optimisation circle is complete. Such LIMS systems (Bard et al., 2004) can be further expanded to follow up on harvested crystals, to assess their diffraction quality and finally the structural data derived. (see 7.4)

7.3 *In situ* crystal analysis

When crystals are found an assessment needs to be made whether the crystals are indeed protein crystals or just salt crystals. And the quality of the crystal needs to be established as well as their usefulness for collecting diffraction data. It has always been difficult to distinguish protein crystals from salt crystals without actually collecting diffraction data. Historically, destructive methods have been used like the "crunch" method and protein dyes, the idea being that crystals similar to the ones destroyed will have the same properties. These methods have not always been conclusive and often the true nature of the crystal was only revealed on the X-ray beam. In recent years, three new techniques have been developed in this field; in situ diffraction analysis, UV detection and second harmonic microscopy.

7.3.1 *In situ* diffraction analysis

A number of years ago, Oxford Diffraction has come up with a device for X-ray diffraction analysis of crystals in the plate where they were grown (Skarzynski 2009, le Maire, et al., 2011). The idea is fairly simple, you center a crystal in the X-ray beam using a visual alignment tool and you subsequently take a single or a small number of X-ray diffraction images to assess whether a crystal is indeed protein, and to get some idea about the diffraction quality (mosaicity, resolution). An advantage is that the method is non-invasive (bearing in mind potential radiation damage) and fast. The method is not suitable for complete diffraction analysis, as the sample can only be rotated by 6°. It is also possible to automatically screen a complete crystallisation plate for potential diffraction. When suitably diffracting crystals are found they will still needed to be harvested and frozen for complete diffraction analysis.

7.3.2 UV detection of protein crystals

An increasing number of imaging devices (see 7.2) make use of a secondary light path in the UV range to detect protein crystals. These imagers make use of the fluorescence in UV by proteins, mostly caused by tryptophan (Judge et al., 2005). Since the protein concentration in the crystal will be much larger than in solution, any protein crystal will light up under UV, provided that the protein contains tryptophan. This is a relatively fast and non-invasive method, UV illumination can cause some ionisation in the drop, but this effect is much less than with X-ray illumination. In order to maximize its use, the experiment media (plates, seals) have to be chosen with care, some plastics are not sufficiently translucent in UV, or fluoresce themselves, adding noise to the image. One also has to bear in mind that some non-protein crystals (ATP, other co-factors added), might also fluoresce in UV. Having visible light and UV cameras integrated in a single imaging device greatly enhances its usefulness to distinguish protein from salt crystals.

7.3.3 Second harmonic microscopy

Fairly recently, a new development in the field of in situ crystal analysis has been reported. The technology makes use of a phenomenon called second harmonic generation (SHG), more often referred to as "frequency doubling" (Wampler et al., 2008). When an intense laser pulse travels through a highly polarizing, non-centrosymmetric material, light emerges with exactly half of the wavelength of the incident beam. The explanation is that two photons of the incident beam merge, creating a single photon with twice the energy. If the incident beam is in the near-infrared, the emerging beam, will be in the visible range. As mentioned, the technology requires intense laser light, delivered in femtosecond pulses. Most chiral crystal classes, with the exception of octahedral and icosahedral crystals, allow for SHG, thus encompassing over 99% of all protein crystals grown so far. A first commercial device using this technology, called SONICC, is available since 2011. When combined with a LIMS and a visible light imaging station, SHG can be used to automatically score and pre-sort the results for the experimenter.

7.4 Automated diffraction analysis

In parallel with automation taking hold of many crystallisation labs, the last part of the protein structure analysis pipeline, diffraction analysis in the X-ray beam, is increasingly automated as well. Not so long ago, a crystallographer would either measure his/her crystals at a home X-ray source or would travel to a synchrotron facility to do so. The process involved manually harvesting of the crystals, preparing them for the X-ray beam (mounting in a capillary or in a cryoloop, freezing), mounting them manually in the X-ray beam, gathering a few trial images to determine optimal settings for exposure, distance etc. and finally recording a set of diffraction images to solve the structure. The most time consuming steps have now been automated (Cork et al., 2006, Song et al 2007). Most notably, crystal mounting robots will now automatically take samples out of a liquid nitrogen dewar and place them in the X-ray beam, eliminating the need for the user to enter the X-ray hub of the synchrotron after every crystal. At the home lab, the mounted crystals are packed in specific dewar compatible with the robot arms at the beamline, and they are mailed to the synchrotron. In many synchrotrons, the user now has a choice of having a local operator collecting the data, or to drive the computers at the facility remotely from their own lab, there is no need to travel to a remote synchrotron anymore. In the near future, the automation can be improved by automatic crystal centering routines (Vernede et al., 2006). With e.g. the use of SHG (see 7.3) crystals can automatically be located inside the cryo loops, and this information can be used to automatically center the crystal in the beam.

8. Conclusions

The chapter covers some of the main aspects of protein nucleation and crystallization. Different diagnostic tools, crystallization techniques and strategies are explained. New tendencies in the field such as combining heterogeneous nucleants and high throughput methods are also presented.

9. References

Abbot, A. (2000) Nature 408, 130-132

Aoyama, K. (1996). J. Cryst. Growth, 168, 198

Achstetter, T. and Wolf, D. H. (1985). Yeast, 1, 139

Arfin, S. M. and Brandshaw, R. A. (1988). Biochemistry, 27, 7979

Baldock, P., Mills, V., Stewart, P.S. (1996). J. Cryst.Growth 168, 170-174

Bard, J., Ercolani, K., Svenson, K., Olland, A., Somers, W. (2004) Methods 34, 329-347

Barrett, A. J. and Salvesen, G. (1986) Research monographs in cell and tissue physiology, Vol 12, pp. 1-661. Elsevier. Amsterdam

Berry, I. M., Dym, O., Esnouf, R. M., Harlos, K., Meged, R., Perrakis, A., Sussman, J. L., Walter, T. S., Wilson, J. & Messerschmidt, A. (2006). Acta Cryst. D62, 1137-1149.

Bergfors T. 2003, J Struct Biol; vol. 142(1): pp 66-76.

Bergfors, T. 2009 Protein Crystallization: Second Edition. International University Line, La Jolla, California, 500 pp.

Berne, B.J.; Pecora, R., 1976, Dynamic Light Scattering; Wiley: New York,

Bond, J. S. and Butler, P.E. (1987). Annu. Rev. Biochem., 56, 333

Cork, C., O'Neill, J., Taylor, J. & Earnest, T. (2006). Acta Cryst. D62, 852-858.

Dalling, M.J. (1986) Plant proteolytic enzymes, 2 Vol. CRC Press, Boca Raton, FL

Doudna, J. A., Grosshans, C., Gooding, A., and Kundrot, C.E. (1993).Proc. Natl.Acad. Sci. USA, 90, 7829

Giege, R., Dock, A.C., Kern, D., Loeber, B., Thierrry, J. and Moras, D. (1986). J. Cryst. Growth, 76, 554

Giege, R. (1987). In Crystallography in molecular biology , NATO ASI Series, Vol. 126, pp. 15-26. Plenum Press, New York and London

Judge, R. A., Swift, K. & Gonzalez, C. (2005). Acta Cryst. D61, 60-66.

Karger, B. L. and Hancock, W. S. (1996). Methods in enzymology, Vol. 270, pp. 1-611 and Vol. 271, pp. 1-543. Academic press London

Konev S. V. (1967), Fluorescence and Phosphorescence of Proteins and Nucleic Acids, Plenum Press, New York

le Maire, A., Gelin, M., Pochet, S., Hoh, F., Pirocchi, M., Guichou, J.-F., Ferrer, J.-L. & Labesse, G. (2011). Acta Cryst. D67, 747-755.

Matthews, B.W. (1968), Journal of Molecular Biology, Volume 33, Issue 2, pp. 491-497

Mikol, V., Hirsche E., Giege R. (1989), FEBS letters Vol. 258, Issue 1, pp. 63-68

McPherson , A. (1985). In Methodsin enzymology, Academic Press, London. Vol. 114, p. 112

McPherson, A., Malkin, A. J., Kuznetsov, Y.G., and Koszelak, S. (1996), J.Cryst. growth, 168, 74

McPherson, A., Shlichta, P. (1988), Science, Vol. 239 Issue 4838, pp 385-387.

Nederlof, I., Hosseini, R., Georgieva, D., Juo, J., Li, Dianfan, Abrahams, J.P. (2011) Crystal Growth & Design, 11 (4), pp 1170-1176

Perrin, D. D. and Dempsey, B. (1974). Buffer for pH and metal ion control. Chapman and Hall Ltd., London and New York.

Takehara, M.; Ino, K.; Takakusagi, Y.; Oshikane, H.; Nureki, O.; Ebina, T.; Mizukami, F.; Sakaguchi, K. (2008), Analytical Biochemistry, 373 (2), pp 322-329.

Thygesen., J., Krumholz, S., Levin, L., Zaytsev-Bashan, A., Harms., J., Barrels, H., (1996). J. Cryst. Growth, 168, 308

Robert M.C., Lefaucheux F. 1988, Journal of Crystal Growth, Vol.90, Issues 1-3, pp 358-367,

Salemme, F.R. (1972), Archives of Biochemistry and Biophysics, Vol. 151, pp 533-539

Skarzynski, T. (2009). Acta Cryst. A, 65, s159

Skouri, M., Lorber, B., Giege, R., Munch, J.P. and Candau, S.J. (1996), J. Cryst Growth, 152, 209

Song J, Mathew D, Jacob SA, Corbett L, Moorhead P, Soltis SM (2007) J Synchrotron Radiation 14, 191-195

Stevens, R.C (2000) Curr. Op. Struct. Biol. 10, 558-563

Sugahara, M.; Asada, Y.; Morikawa, Y.; Kageyama, Y.; Kunishima, N. (2008), Acta Cryst. D 64 (6), pp 686-695.

Thakur, A. S.; Newman, J.; Martin, J. L.; Kobe, B. (2008), Structural Proteomics, Vol. 426, pp 403-409.

Van der Laan, J. M., Swarte, M. B. A., Groendijk, H., Hol, W. G. J., and Drenth , J. (1989). Eur. J. Biochem., 179, 715

Vekilov, P.G. and Rosenberger, F. (1996). J. Cryst. Growth, 158, 540

Wierenga, R. K., Lalk, K. H., and Hol, W.G.J (1987)., J.Mol.Biol., 198, 109

Wampler, R.D, Kissick, D.J.,Dehen, C.J., Gualtieri E.J., Grey, J.L., Wang, H.F., Thompson, D.H., Cheng, J.X. and Simpson, D.J. (2008) J. Am. Chem. Soc, 130, 14076-14077

Wandersman, C. (1989). Mol. Microbiol., 3. 1825

Wuthrich, K. (1995). Acta Cryst. D, pp 51, 249

9

Protein Crystal Growth Under High Pressure

Yoshihisa Suzuki

Institute of Technology and Science,
The University of Tokushima,
Japan

1. Introduction

In this chapter, I would like to describe following two main roles of high pressure (up to 250 MPa) on protein crystal growth.

1. High pressure as a tool for enhancing crystallization of a protein
2. High pressure as a tool for modifying a three-dimensional (3D) structure of a protein molecule

For the first role, Visuri et al. reported that the total amount of obtained crystals of glucose isomerase (GI) was drastically increased with increasing pressure, for the first time (Visuri et al., 1990). Such drastic enhancements probably play an important role in increasing the success rate of 3D structure analysis of protein molecules, since crystallization is still the rate limiting step in the structure analysis process. Although they were the pioneers of this field, they did not do further studies on the growth mechanisms of GI crystals. After their pioneer work, many studies have been done on solubility (Section 2), nucleation (Section 3), and growth kinetics (Section 4) under high pressure. Here I would like to review and classify these studies, and present the potential of high pressure as a tool for enhancing protein crystallization.

For the second role, Kundrot & Richards were the pioneers of this field. They analyzed 3D structure of hen egg-white lysozyme under high pressure at the atomic level, for the first time (Kundrot & Richards, 1987). High-pressure protein crystallography is a prerequisite to understanding effects of pressure on an enzymatic activity of a protein at the atomic level (Makimoto et al., 1984). The structural information also plays an important role in the studies of deep-sea organisms (Yayanos, 1986). Pressure probably influences the protein structure through the structure of surrounding water molecules. Thus the protein structure under high pressure has to be solved with water of hydration at ambient temperatures, since a flash cooling method obviously influences the crystal structure (Charron et al., 2002); a freezing process of the method probably influences the structure of the surrounding water molecules, and the process prevents us from an *"in situ"* analysis of the protein structure with water of hydration. In addition, mainly due to the technical difficulties, total number of studies on high-pressure protein crystallography is not so many at this stage. In Section 5, I would like to review the studies on *"in situ"* protein structure analysis under high pressure, and present a new methodology for an ideal *"in situ"* structure analysis.

2. Solubility of protein crystals under high pressure

Solubility is generally important and indispensable to study an equilibrium state between a solution and crystal. Solubility is usually measured at the beginning of the crystallization research to determine the supersaturation σ ($\sigma = \ln(C/C_e)$, C: concentration of the solution, C_e: solubility), because the crystallization phenomena are often well described by using σ. Supersaturation is also named as the *driving force for crystallization*. For the studies at atmospheric pressure, σ has been useful for the discussion on the protein crystallization (Rosenberger et al., 1996). Thus, for a high-pressure study, the high-pressure σ is also useful to discuss the mechanisms of protein crystallization. To determine the high-pressure σ, the high-pressure solubility is indispensable.

2.1 Methodology

Many researchers have measured the solubility under high pressure. However, the solubility varies according to the method, even though the composition of the solution is almost the same (Suzuki et al., 2000b). The variation prevents us from the quantitative discussion. Therefore, the more sophisticated method is expected to measure the more accurate solubility. Here I would like to present several methods for the measurement of solubility under high pressure with their merits and demerits.

2.1.1 Change in the concentration of a supernatant solution with time (*ex situ*)

In general, this method is the most popular one for the solubility of crystals of small molecules. Groß et al. and Lorber et al. reported the solubility of tetragonal lysozyme crystals under high pressure by *ex situ* measurement (Groß & Jaenicke, 1991; Lorber et al., 1996). They incubated the supersaturated solution under high pressure for a certain period, then reduced the pressure and measured the concentration in the supernatant solution. They measured the solubility only from the supersaturated state, where the solubility is always uncertain from 0 mgmL^{-1} to the asymptotic concentration. In addition, Suzuki et al. showed that an asymptotic concentration from a supersaturated state did not correspond to that from an undersaturated state in a realistic time scale (Suzuki et al., 2000b). Although this method provides concentration data directly, it takes very long time to attain an equilibrium condition.

2.1.2 Change in the concentration of a supernatant solution with time (*in situ*)

We have designed an *in situ* precise method of solubility measurement using a Mach-Zehnder interferometer (Suzuki et al., 1998) and measured *in situ* the solubility of lysozyme (Suzuki et al., 2000b). Using the method, the relative change in the concentration with time during equilibration was measured accurately and continuously starting from a supersaturated state (growth relaxation) and an undersaturated state (dissolution relaxation). The asymptotic concentration for the dissolution relaxation was regarded as the solubility. This method is more precise than *ex situ* one (described in 2.1.1). However, a long time period is still required to establish a solubility curve by this method, and this remains as one of the most serious barriers to further high-pressure studies.

2.1.3 Change in the crystal size with pressure

Takano et al. provided a much-improved technique based on in situ observation (Takano et al., 1997). They measured the equilibrium pressure in situ from supersaturation and undersaturation. They gradually increased the pressure of the sample over a period of few days, and continuously recorded the images of the lysozyme crystal. The solubility was determined from changes in both the size of the crystal and the amount of the transmitted light through the crystal. Although they could shorten the period for one plot, a long time period was still required to establish a solubility curve.

We also measured the solubility of triclinic lysozyme crystals by observing the crystals before and after pressurization for two hours with high precision, although the measurements had been conducted *ex situ* (Suzuki et al., 2011).

2.1.4 Change in the concentration distribution around a crystal (*in situ*)

In order to decrease the time necessary for the solubility measurement under high pressure, another interferometric technique has been developed, which can determine the solubility of lysozyme within at most 3 hours (Sazaki et al., 1999). In the interferometric method, an equilibrium temperature of a given concentration is determined by observing the concentration distribution around a crystal. The distribution can be visualized by using a Michelson interferometer (Sazaki et al., 1999).

Under high pressure, the concentration distribution around the crystals was observed in situ with the Michelson interferometer. Figure 1 shows interferograms of the solution around a GI crystal under 100 MPa (Suzuki et al., 2002b). Here, the concentration of glucose isomerase was 35.4 mgmL^{-1}. If the temperature of the sample was set lower than its equilibrium temperature, the crystal grew (24.7 °C), and the fringes were bent in the vicinity of the crystal (Fig. 1(a)), because of the decrease in the concentration around the crystal. On the other hand, when the temperature was raised higher than its equilibrium temperature (44.1 °C), the crystal dissolved and the fringes bent in the opposite direction (Fig. 1(b)). From observation of the fringes around the crystal, we determined the equilibrium temperature of the crystal and solution of a given concentration.

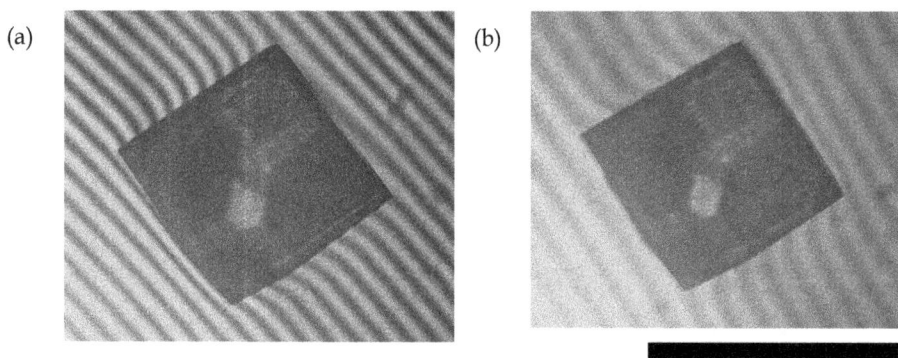

Fig. 1. Interferograms around the glucose isomerase crystal under 100 MPa (Suzuki et al., 2002b). Concentration of glucose isomerase in bulk solution is 35.4 mgmL^{-1}. (a) Growth (24.7 °C), (b) dissolution (44.1 °C). The scale bar represents 1 mm.

Although this technique reduced the measurement time for one data point drastically (within 3 hours), the error of the data points was generally larger than the method of 2.1.3.

2.1.5 Change in the position of steps or the morphology of ledges of crystals (*in situ*)

Among the many studies on protein solubilities so far, *in situ* observation of steps on crystal faces using a laser confocal microscope combined with a differential interference contrast microscope (LCM-DIM (Sazaki et al., 2004)) has been the most powerful method (step-observation method) for measuring the equilibrium temperatures T_e of protein crystals (Van Driessche et al., 2009; Fujiwara et al., 2010). Van Driessche et al. reported that this method yielded the highest precision in measurements of T_e of tetragonal hen egg-white lysozyme crystals (Van Driessche et al., 2009), and we found it produced the fastest results (Fujiwara et al., 2010). For high-pressure solubility, Fujiwara et al. measured that fastest and with highest precision, at this stage.

To tell the truth, we applied this method to measure high-pressure solubility of GI crystals for the first time (Suzuki et al., 2009, 2010a). In these papers, we also use the changes in the morphology of a ledge of a crystal, while the precision was not so high as that of the data measured by Fujiwara et al.

2.2 Solubility data

In addition to the above studies, many studies on the solubility of proteins under high pressure have been reported. The solubility, C_e, of tetragonal (Groß & Jaenicke, 2001; Lorber et al., 1996; Takano et al., 1997; Sazaki et al., 1999; Suzuki et al., 2000a, 2000b, 2002a; Kadri et al., 2002; Fujiwara et al., 2010) and monoclinic (Asai et al., 2004) hen lysozyme, turkey lysozyme (Kadri et al., 2002), and subtilisin (Webb et al., 1999; Waghmare et al., 2000a) crystal increased with increasing pressure, while that of orthorhombic hen lysozyme (Sazaki et al., 1999; Suzuki et al., 2002a), glucose isomerase (Suzuki et al, 2002b, 2005, 2009), and thaumatin (Kadri et al., 2002) crystal decreased with increasing pressure.

Protein crystals	Pressure dependence	Measurement time for one plot	Accuracy
Hen Lysozyme			
Tetragonal	↑ (Positive)	< 70 minutes	< ± 0.7 K in T_e
		(Fujiwara et al., 2010)	
Orthorhombic	↓ (Negative)	< 3 hours	< ± 4.8 K in T_e
		(Sazaki et al., 1999)	
Monoclinic	↑	< 6 hours	< ± 0.4 mgmL^{-1} in C_e
		(Asai et al, 2004)	
Triclinic	↑	< 6 hours	< ± 1.0 mgmL^{-1} in C_e
		(Suzuki et al., 2011)	
Turkey Lysozyme	↑	9 days	-
		(Kadri et al., 2002)	
Purafect Subtilisin	↑	7 days	-
		(Webb et al., 1999)	
Glucose Isomerase	↓	< 90 minutes	< ± 2.5 K in T_e
		(Suzuki et al., 2009)	
Thaumatin	↓	9 days	-
		(Kadri et al., 2002)	

Table 1. Effects of pressure on the solubility of proteins.

The above results are listed in Table 1. In general, the decrease in solubility with pressure results in the increase in nucleation rates and growth rates of crystals. From Table 1, three of eight crystals exhibit the decrease in solubility with pressure. Thus, application of high pressure to a protein solution would be useful for crystallizing previously uncrystallized proteins.

2.3 Thermodynamic analyses

From solubility data, thermodynamic parameters are often calculated using van't Hoff plots. If we assume that the effect of the activity coefficient is negligible, we can estimate the partial molar enthalpy of dissolution, ΔH, from Eq. (1) and the partial molar entropy of dissolution, ΔS, from Eq. (2).

$$\frac{\partial \ln C_e}{\partial(1/T)} = \frac{-\Delta H}{R}, \tag{1}$$

$$\frac{\partial RT \ln C_e}{\partial T} = \Delta S, \tag{2}$$

where C_e: solubility (mg mL^{-1}); R: gas constant.

To estimate ΔH lnC_e is plotted against T^{-1}. If we assume that ΔH does not depend on temperature, ΔH is estimated from the slope by linear fitting of the plot. In this section, weighted fitting was done, because the temperature error was large at lower concentration region. ΔS is estimated from T-RTlnC_e plots.

From the dependence of pressure on solubility, if we assume that the effect of the activity coefficient is negligible, the volume change accompanying the dissolution, ΔV ($\Delta V \equiv \bar{V} - V_c$, $\bar{V}$: the partial molar volume of the solute, V_c: molar volume of the crystal), is expressed as,

$$\Delta V = -RT[\frac{\partial \ln C_e}{\partial P}]_T. \tag{3}$$

If ΔV does not depend on pressure up to P MPa, the molar volume change accompanying the dissolution at 0.1 MPa is expressed as,

$$\Delta V = -RT\frac{\ln C_{e, P} - \ln C_{e, 0.1}}{P-0.1}, \tag{4}$$

where $C_{e, P}$ and $C_{e, 0.1}$ indicate the solubility at P MPa and 0.1 MPa, respectively.

All the above thermodynamic functions (ΔH, ΔS and ΔV) reported so far are listed in Table 2. Negative value of ΔV indicates that the partial molar volume of a protein, $\bar{V}$, is smaller than the molar volume of a crystal, V_c, and *vice versa*. Figure 2 represents a simplified model of the states of the protein in the crystal and in solution.

Consider now the change from crystalline to the solution state. If we neglect any change in volume of the protein molecule, the bulk water, and the waters of hydration 2 (those around parts of the protein exposed in both crystal and solution), then ΔV is given by the volume of the waters of hydration 1 (around the contact surfaces of the protein) minus the volume occupied by these same water molecules as "free" water when the protein is in the crystalline state. For this volume change to be negative, as found for tetragonal lysozyme crystals, the water molecules must be more tightly packed when hydrating the contact regions than when "free" in the bulk water. This, in fact, is expected to be the case for contacts containing a large number of hydrophilic residues. Correspondingly, the positive volume change on dissolution of glucose isomerase crystals implies that the contact surfaces tend to structure the waters of hydration such that they occupy a larger volume than in the bulk. It predicts that the contacts in glucose isomerase crystals should be more hydrophobic than in tetragonal lysozyme crystals.

Protein crystals	ΔV / cm³mol⁻¹		ΔH / kJmol⁻¹		ΔS / Jmol⁻¹K⁻¹	
	0.1 MPa	Authors	0.1 MPa	100 MPa	0.1 MPa	100 MPa
Hen Lysozyme						
Tetragonal	-18±46	S&S	130±10	70±10	460±40	280±40
	-11.6	L				
	-5	We				
	-3.0±0.5	K				
Orthorhombic	5±18	S&S	35±3	35±5	140±10	140±20
Monoclinic		A	102±6	79±2		
Triclinic		S2011	113±4	97±4		
Turkey Lysozyme	-15±1	K				
Purafect Subtilisin	-21±1	We				
	-30±7	Wa				
Glucose Isomerase	54±31	S2002	160±40	210±60	420±100	580±180
Thaumatin	11±1	K				

Table 2. Thermodynamic functions of dissolution obtained by solubility data of protein crystals. Characters listed in Authors column indicate references as follows. S&S: (Sazaki et al., 1999; Suzuki et al., 2002a); L: (Lorber et al., 1996); We: (Webb et al., 1999); K: (Kadrı et al., 2002); A: (Asai et al., 2004); S2011: (Suzuki et al., 2011); Wa: (Waghmare et al., 2000a); S2002: (Suzuki et al., 2002b).

The decrease in ΔS of tetragonal lysozyme crystal with pressure can be explained by a decrease in ΔV with pressure. Since the solution is more compressible than the crystal, the magnitude of ΔV is smaller under high pressure than under atmospheric pressure. Smaller $|\Delta V|$ under high pressure can lead to a smaller change in a degree of freedom. In the case of GI crystals, on the other hand, the increase in ΔS can be explained by the increase in ΔV with pressure. The change in ΔH is still not clear. Further crystallographic study on the hydration of the intermolecular contact regions or the other independent measurements of ΔV, ΔH and ΔS may explain these phenomena.

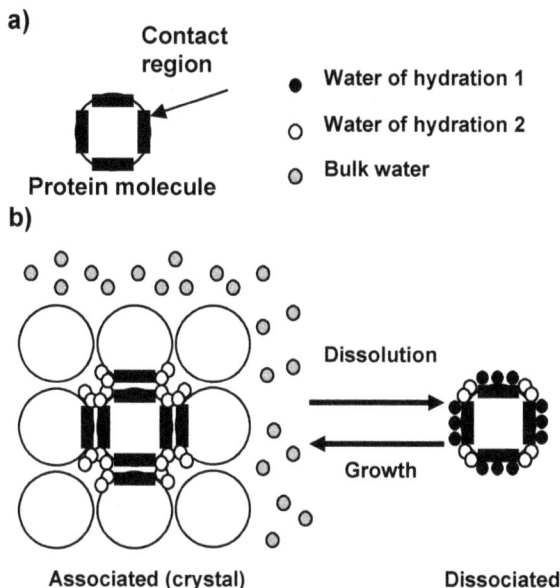

Fig. 2. Schematic diagrams of protein molecules in the crystal and solution (Suzuki et al., 2002a). In the crystal, the molecules are in contact with each other at the regions indicated in black. Hydrated water 1 and 2 represent water molecules hydrating the contact and non-contact regions of the protein, respectively. The volume change on dissolution is given mainly by the difference in volume between hydrated water 1 and bulk water.

3. Nucleation of protein crystals under high pressure

The mechanisms of high-pressure acceleration of 3D nucleation will play the most important role in the improvement of the success rate of crystallization, since the success rate of the 3D nucleation corresponds to that of the crystallization. The precise analyses of the supersaturation dependencies of 3D nucleation rate, J, will clarify the mechanisms.

Except for the data presented in our studies on GI crystals (Maruoka et al., 2010; Suzuki et al., 2010c), the effects of pressure on J have not been reported yet. Although Groß et al. discussed the effects of pressure on the nucleation kinetics using the Oosawa theory of protein self-assembly (Groß et al., 1993), and the group of Glatz discussed the activation volume of the nucleation using the number of crystals (Saikumar et al., 1998; Webb et al., 1999; Waghmare et al., 2000b; Pan & Glatz, 2002), neither group measured J under high pressure directly.

Thus, in this section, I would like to focus on our studies on GI crystals (Suzuki et al., 2009, 2010c; Maruoka et al., 2010).

3.1 Classical nucleation theory

The 3D nucleation rate (Volmer & Weber, 1926), J, is modified and expressed as follows (Suzuki et al., 1994):

$$J = vsn_t \exp(\frac{-\Delta G^*}{kT}), \tag{5}$$

where v, s, n_t, ΔG^*, k, and T represent the collision rate of GI tetramers with critical nuclei, the sticking parameter for the addition of a GI tetramer to a critical nucleus, the number of GI tetramers in the unit volume of a solution, the Gibbs free energy for the formation of a critical nucleus of a GI crystal, the Boltzmann constant, and the absolute temperature, respectively. The variables s and ΔG^* can be expressed as follows (Boistelle & Lopez-Valero, 1990):

$$s = n \exp(\frac{-\varepsilon}{kT}), \tag{6}$$

and

$$\Delta G^* = \frac{f\Omega^2\gamma^3}{(kT\sigma)^2}, \tag{7}$$

where n, ε, f, Ω and γ represent the total number of tetramers adjacent to the surface of a nucleus, the activation energy for the addition of a GI tetramer to a critical nucleus, the shape factor, the average volume occupied by a GI tetramer, and the surface free energy of the GI crystal, respectively. Substituting equation (6) and (7) for (5) and taking the natural log of both sides, we obtain the following expression:

$$\ln J = \ln(vnnt) - \frac{\varepsilon}{kT} - \frac{f\Omega^2\gamma^3}{(kT)^3} \times \frac{1}{\sigma^2}. \tag{8}$$

3.2 Methodology

A high-pressure vessel with transparent sapphire windows was used (Maruoka et al., 2010; Suzuki et al., 2010c). An inner cell (inner volume = $1 \times 6 \times 20$ mm^3) for *in situ* observation was made of glass slides, and equipped with soft silicone tubes for sample loading. The cell was set in the vessel, and crystals in the cell under high pressure were observed through the sapphire windows using a stereoscopic microscope (Nikon, SMZ800, objective: EDPlan×2 (N. A. = 0.2)). The solution and pressure medium were separated by soft silicone tubes of the cell. The solution around the crystals was pressurized *via* the tubes. The pressure in the vessel was well controlled automatically by a feedback system with a pressure sensor (accuracy of pressure: ± 0.5 MPa) and could be kept constant for a long time. The temperature of the cell was directly controlled using a Cu jacket with a Peltier element. We could control the temperature from 15.0 to 35.0 °C with the accuracy of ± 0.2 °C.

A supersaturated solution of a given GI concentration was transferred into an inner cell. The number of the observable crystals per unit volume N was counted with time t using a stereoscopic microscope. The nucleation rate J is defined as the slope of the tangent line of the $t - N$ plots at the point of inflection. In practice, we fit Gompertz function, which is a sigmoid function, to the $t - N$ plots, since Foubert et al. fit the Gompertz function to their data of released crystallization heat of fat crystals, and the fit of the Gompertz model seemed to be better than that of the mostly used Avrami model (Foubert et al., 2003). The Gompertz function we used is expressed as,

$$N = a\exp\{-\exp[-k(t - t_c)]\} \qquad (9)$$

where t represents time, and a, k, and t_c are fitting parameters. We assume that J is defined as the slope of the tangent line of the Gompertz function at the point of inflection, since the slope provides the maximum value. From eq. (9), J is expressed as,

$$J = ak/e \qquad (10)$$

Induction time τ is calculated by substituting $N = 1$ into eq. (9), and expressed as,

$$\tau = t_c - (1/k)\ln(-\ln(1/a)) \qquad (11)$$

3.3 Three-dimensional nucleation rates

N increased with time in a sigmoidal-like fashion (Fig. 3). The Gompertz function fitted well all the $t - N$ plots. From the fitting parameters and eq. (10), J was calculated and plotted against σ (Fig. 4). J increased with increasing pressure at the same σ. We also determined τ^{-1} using eq. (11). τ^{-1} also increased with increasing pressure at the same σ (Fig. 4). The increase in J and τ^{-1} with pressure at the same σ indicates that they are kinetically accelerated under high pressure.

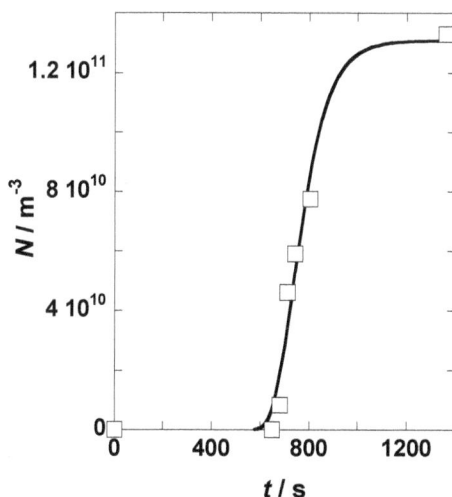

Fig. 3. Time course of the number of observed microcrystals at $T = 20\,°C$, $C = 27.07\ \mathrm{mgmL^{-1}}$, and $P = 100$ MPa (Maruoka et al., 2010). Solid curve indicates a Gompertz function.

Although nucleation of protein crystals under high pressure has been already studied by a few researchers (Suzuki et al., 1994; Waghmare et al., 2000b; Pan & Glatz, 2002), no one has succeeded in measuring J directly and discussing the dependence of J on σ. We previously measured J of tetragonal lysozyme crystals under high pressure by *in situ* observation of the number of crystals using a diamond anvil cell (Suzuki et al., 1994). J decreased with increasing pressure at a constant concentration. However, since the solubility was not measured at that time, we could not separate the effects of solubility change under high pressure. Waghmare et al. and Pan et al. assumed that the final number of subtilisin crystals was proportional to J

(Waghmare et al., 2000b; Pan & Glatz, 2002), and they did not observe the transient number of the crystals. Thus, our results shown in Fig. 4 successfully clarified, for the first time, that the 3D nucleation of GI crystals was kinetically accelerated under high pressure.

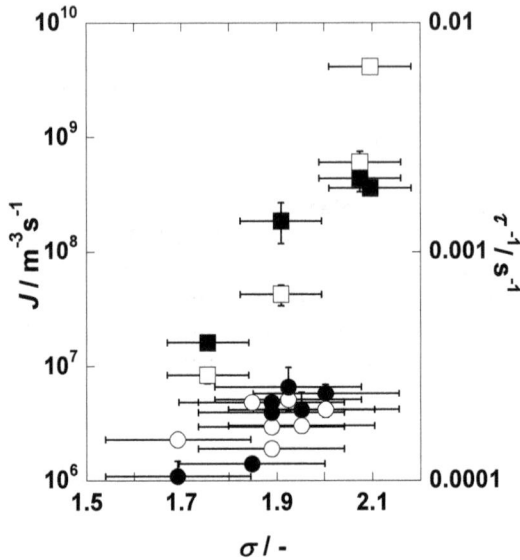

Fig. 4. J and τ^{-1} with supersaturation σ (Maruoka et al., 2010). Open and closed symbols indicate J and τ^{-1}, respectively. Circles and squares indicate the data measured under 0.1 and 100 MPa, respectively.

3.4 Kinetic analyses

We plotted the natural logarithm of J against $1 / \sigma^2$. $\ln J$ increased with increasing pressure at the same value, $1 / \sigma^2$. Using eq. (10) and tentatively assuming that f equals 25 (Boistelle & Lopez-Valero, 1990), the average surface free energies γ at 20°C are calculated to be $(8 \pm 3) \times 10^{-5}$ and $(9 \pm 2) \times 10^{-5}$ Jm^{-2} at 0.10 and 100 MPa, respectively. γ does not change with pressure within experimental errors. This result does not correspond with our previous results, in which γ decreased with increasing pressure (Suzuki et al., 2005). This inconsistency is mainly due to the experimental errors in J. To confirm the pressure dependency of γ in detail, we will need to measure the two-dimensional (2D) nucleation rates with σ under high pressure (Suzuki et al., 2009, 2010a; Van Driessche et al., 2007).

On the other hand, the intercept of the linear fitting function shown in Fig. 2, $\ln(\nu n n_t) - \varepsilon / kT$, at 100 MPa is much larger than that at 0.10 MPa. This indicates that the activation energy for the addition of a GI tetramer to a critical nucleus, ε, decreases drastically with increasing pressure, since ν should not change so much, and n and n_t of 100 MPa are less than those of 0.10 MPa at the same σ.

3.5 Two-dimensional (2D) nucleation rates

We preliminarily measured *in situ* 2D nucleation rates J_s of 2D islands on the {011} face of glucose isomerase crystals at 0.1, 25 and 50 MPa. J_s increased with increasing pressure. For these plots, GI concentration in the bulk solution C (= 5.6 mg mL^{-1}) and temperature (T = 26.4 °C) were constant throughout the measurement. Thus, the increases in J_s are completely due to the increase in pressure.

4. Growth kinetics of protein crystals under high pressure

To understand the mechanisms of crystal growth precisely, growth kinetics should be clarified. In this section, I would like to describe mainly following two topics.
1. Effects of high pressure on growth rates of crystal faces, R
2. Effects of high pressure on step velocities, V

4.1 Effects of high pressure on growth rates of crystal faces, R
Kinetic analyses of R provide useful information about growth mechanisms. Pressure effects on the kinetics of R of protein crystals are listed in Table 3.

Protein crystals		Pressure effects	Authors
Hen Lysozyme			
	Tetragonal	Inhibition	Suzuki et al., 2000a
	Orthorhombic	Inhibition	Nagatoshi et al., 2003
	Monoclinic	Acceleration	Asai et al., 2004
Purafect Subtilisin		Inhibition	Waghmare et al., 2000a
Glucose Isomerase		Acceleration	Suzuki et al., 2005

Table 3. Effects of pressure on the growth kinetics of protein crystals.

4.1.1 Growth theory
How does pressure affects R kinetically? The following three hypotheses are conceivable. (1) An increasing pressure reduces the volume of the system, and thus elevates the protein concentration. (2) The rising pressure brings about changes in the crystals' growth mode. (3) Changes in growth parameters such as an activation energy, surface free energy, etc. occur with elevations in pressure.
(1) Elevation of the protein concentration through a reduction in the system volume
Let us first consider hypothesis (1). How much does the concentration change with increasing pressure? Kundrot & Richards reported that from 0.1 to 100 MPa the volume contraction of a tetragonal hen lysozyme (t-HEWL) crystal, the solvent in the crystal and the bulk solution were 1.1, 3.7 and 3.7 % (Kundrot & Richards, 1987, 1988), respectively. For a t-HEWL crystal, regardless of the increase in the protein concentration resulting from the volume contraction, the growth kinetics decelerates with increasing pressure (Suzuki et al., 2000a). In the case of the other protein crystals listed in Table 3, the volume contraction of the system is probably of the same order as that of the t-HEWL crystal (i. e. several percent). Thus, this hypothesis (1) can not explain the inhibition of the growth kinetics of protein crystals. In addition, it hardly explains the significant acceleration of the growth kinetics of monoclinic hen lysozyme and GI crystals.

(2) Changes in the crystals' growth mode

To evaluate the second hypothesis, we should first confirm the crystal growth mode under all the growth conditions. Since all the crystals described in this review had clear facets, the crystals formed via a layer-by-layer growth mechanism. Therefore, they must have grown in a spiral growth mode with screw dislocations or in a 2D nucleation growth mode (Fig. 5).

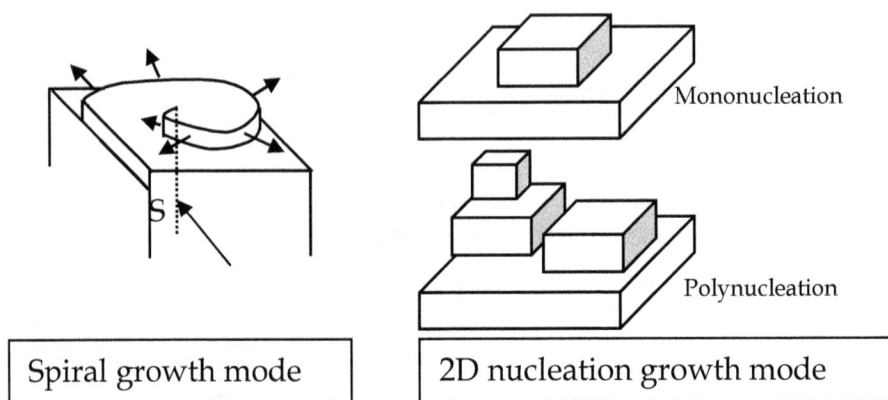

Mononucleation

Polynucleation

| Spiral growth mode | 2D nucleation growth mode |

Fig. 5. Schematic illustrations of typical growth modes of crystals with clear faces.

If the density of the screw dislocations is sufficiently low, the growth rate R of the spiral growth mode is expressed as (Burton et al., 1951; Cabrera & Levine, 1956),

$$R = \frac{K_s h}{19 f_0 \kappa} \sigma^2 . \tag{12}$$

Here, K_s is a step kinetic coefficient, h a step height, f_0 the area which is occupied by a molecule on the crystal face, and κ a ledge free energy. In Table 3, only for t-HEWL and orthorhombic lysozyme (o-HEWL) and GI crystals, R of specific faces of the crystals were precisely measured. Supersaturation dependencies of R of the above three crystals were not fitted well using the equation (12).

Next, we take into account the 2D nucleation growth model. Actually, there are two models that can represent the 2D nucleation growth mode: one operating by mononucleation and the other by polynucleation. Through the following reasoning, we judged that the latter is the growth mode in the present cases. The growth rate in the mononucleation mode is proportional to the surface area of the relevant face (Markov, 1995). However, all R referred in the present review did not depend on the surface area, although different crystals of different size were used. In addition, to confirm the growth mode directly, we observed the surface topography in situ using a reflection type laser confocal microscope combined with a differential interference contrast microscope (LCM-DIM system) (Sazaki et al., 2004) for GI crystals. The 2D nucleation and subsequent lateral growth of the 2D islands were clearly observable. Thus, we concluded that the GI crystal grew in the polynucleation mode. In the case of t-HEWL, we also confirmed polynucleation growth under high pressure.

R in the polynucleation mode is expressed as (Suzuki et al., 2000a, 2005; Nagatoshi et al., 2003),

$$R = k_1 \exp(\frac{2\sigma}{3})(\exp\sigma - 1)^{2/3}\sigma^{1/6}\exp(\frac{-k_2}{\sigma}), \tag{13}$$

where k_1 and k_2 are expressed as,

$$k_1 = (\frac{\pi}{3})^{1/3}a^{13/3}h^{4/3}v\lambda_0^{-2/3}C_e^{4/3}\exp(-\frac{\varepsilon + \varepsilon_{ad} + 2\varepsilon_{kink}}{3kT}), \tag{14}$$

and

$$k_2 = \frac{\pi\gamma^2}{3k^2T^2}. \tag{15}$$

In Eqs. (13), (14) and (15), the following symbols are used: a is the distance between the molecules in the crystal; h is the step height; v is the thermal frequency of a solute; λ_0 is the average distance between the kinks on a step; ε is the activation energy for a solute molecule to be incorporated into a critical nucleus; ε_{ad} is the activation energy for a solute molecule to be adsorbed on the crystal surface; ε_{kink} is the activation energy for a solute molecule to be incorporated into a kink site; γ is the molecular surface energy that represents the excess free energy due to unsatisfied bonds of a molecule at a step edge; and k is the Boltzmann constant. By nonlinear least squares fitting, eq. (13) reproduces the experimental data well. All experimental data were best fitted to the 2D nucleation growth mode of the polynucleation type. Hence we concluded that there was no change in the growth mode with increasing pressure. Thus pressure effects are mainly due to (3) Changes in growth parameters such as an activation energy, surface free energy, etc. with pressure.

4.1.2 Summary of kinetic analysis of R

For t-HEWL, o-HEWL, and GI crystals, R were measured with σ under high pressure. By fitting these data with the equation (13), kinetic constants k_1 and k_2 were calculated as shown in Table 4.

The value of k_1 of the {110} face of t-HEWL crystals at 100 MPA for the best fit ($k_1 = 7.1 \times 10^6$ nms-1) is extraordinarily large and this value has less physical meaning. This is owing to the lack of data and error of R. As in eq. (14), there are too many factors to determine which dominate the increase in k_1 with pressure. Thus, I also showed the result in which I fixed k_1 value. The increase in k_1 and the decrease in k_2 result in the acceleration of growth kinetics. In Table 4, the dependence of k_1 and k_2 on pressure is generally complicated.

First, in the case of t-HEWL crystal, for both faces, k_1 increases with increasing pressure, while k_2 increases. This shows that the effect of the increase in surface free energy dominates the overall inhibition of the growth kinetics. Second, results of o-HEWL crystals show different pressure dependencies. Both k_1 and k_2 decrease with an increase of pressure. The decrease in k_1 dominates the inhibition of the growth kinetics under high pressure. Third, the acceleration of growth kinetics of GI crystals is mainly due to the decrease in surface free energy (k_2).

To study the effects of pressure on each parameter precisely, further accumulation of the data of R is needed.

Protein crystals and constants	Authors	Pressure / MPa		
		0.1	50	100
Hen Lysozyme				
Tetragonal	Suzuki et al., 2000a			
{110} face				
k_1 / nms^{-1}		0.84	3.5	7.1×10^6
k_2 / -		3.6	8.7	45
k_2 (k_1 = 0.84 nms^{-1} fixed) / -		3.6	5.4	9.6
{101} face				
k_1 / nms^{-1}		0.14	0.33	0.86
k_2 / -		1.0	3.8	6.5
k_2 (k_1 = 0.14 nms^{-1} fixed) / -		1.0	2.0	2.6
Orthorhombic	Nagatoshi et al., 2003			
k_1 / nms^{-1}		4.7 ± 1.3		1.7 ± 0.6
k_2 / -		2.0 ± 0.4		1.5 ± 0.5
Glucose Isomerase	Suzuki et al., 2005			
{101} face				
k_1 / nms^{-1}			7 ± 6	0.60 ± 0.08
k_2 / -			18 ± 2	3 ± 1

Table 4. Kinetic constants of 2D polynucleation growth.

4.2 Effects of high pressure on step velocities, V

In the case of a 2D nucleation growth model, since R is proportional to $J_s^{1/3}V^{2/3}$ (J_s: 2D nucleation rate, V: step velocity) (Markov, 1995), the model analyses of R are indirect and prevent further detailed analysis. *In situ* observation of the steps enables us to directly and separately measure J_s and V, which are necessary to elucidate the causes of high-pressure acceleration of the nucleation and growth. Hence, direct observation of individual elementary steps plays a crucial role in studies of crystallization under high pressure.

Direct observations of individual elementary steps on protein crystal surfaces have been carried out mainly by atomic force microscopy (AFM) (Durbin & Carlson, 1992; Durbin et al., 1993; McPherson et al., 2000). However, AFM does not work under high pressure (> 6 atm) at the present stage (Higgins et al., 1998). Besides, the scan of a cantilever would potentially affect the soft surfaces of protein crystals. Advanced optical microscopy is a promising alternative to directly and noninvasively observe individual elementary steps even under high pressure. Among various kinds of advanced optical microscopy, we adopted laser confocal optical microscopy combined with differential interference contrast microscopy (LCM-DIM), by which we have already succeeded in observing the elementary steps of GI crystals under atmospheric pressure (Suzuki et al., 2005). The development of a high-pressure vessel with an optical window thin enough to suppress optical aberration, also played a crucial role in the LCM-DIM system.

At this stage, V of GI crystals under high pressure are only available data (Suzuki et al, 2009, 2010a).

4.2.1 Theory of step velocities, V

Assuming a direct incorporation process, step velocity on a specific face of a GI crystal V is expressed as follows (Suzuki et al., 2009, 2010a):

$$V = \beta_{step}\Omega(C - C_e) \tag{16}$$

where β_{step} and Ω represent the step kinetic coefficient of the incorporation process of GI tetramers, which are the growth units of GI crystals, at kink sites on steps of GI crystals and the average volume occupied by a GI tetramer in the crystal, respectively. We used bulk concentration C instead of the concentration adjacent to a crystal surface, C_{surf}. β_{step} is expressed as follows (Chernov, 1984):

$$\beta_{step} = v\frac{p}{\lambda}a\exp(-\frac{\varepsilon_{kink}}{kT}), \tag{17}$$

where v, p, λ, a, and ε_{kink} represent the vibrational frequency of a GI tetramer, unit cell length parallel to a step, kink spacing, unit cell length perpendicular to the step, and activation energy of the incorporation of a GI tetramer into a kink site on the GI crystal surface, respectively. The variables v, p, and a seldom change with increasing pressure. λ probably does not change with increasing pressure too much, since the shape of a step does not change with increasing pressure (Suzuki et al., 2009, 2010a). Most of the steps on GI crystal surfaces were straight ones and the shape of the steps did not change with increasing pressure. This means that λ did not change significantly; the change in β_{step} was mainly due to the change in ε_{kink}.

Based on the dependence of pressure on β_{step}, the volume change in going to the activation state in the incorporation process of GI molecules at the kink site on the step, $\Delta V^{\ddagger}$ ($\Delta V^{\ddagger} \equiv V^{\ddagger} - \overline{V}$, $V^{\ddagger}$: partial molar volume of the activated GI tetramer in the solution), is expressed as follows (Laidler, 1987):

$$\Delta V^{\ddagger} = -RT[\frac{\partial \ln \beta_{step}}{\partial P}]_T. \tag{18}$$

If $\Delta V^{\ddagger}$ does not depend on pressure up to P, $\Delta V^{\ddagger}$ at 0.10 MPa is expressed as follows:

$$\Delta V^{\ddagger} = -RT\frac{\ln \beta_{step,P} - \ln \beta_{step,0.10}}{P - 0.10}, \tag{19}$$

where $\beta_{step,P}$ and $\beta_{step,0.10}$ indicate the step kinetic coefficients at P and 0.10 MPa, respectively.

4.2.2 Experimental

This study made use of an LCM-DIM system (Olympus Optical Co., Ltd.). To measure V of GI crystals precisely, a super-luminescent diode (SLD) laser (Amonics Ltd., model ASLD68-050-B-FA: λ = 680 nm), whose coherence length is about 10 μm, was adopted as a light source.

Figure 6 shows a schematic illustration of a high-pressure vessel and a GI crystal. A high-pressure vessel with a 1-mm-thick sapphire window (Syn-corporation, Ltd., PC-100-MS) was specially designed and used for the *in situ* observation of crystal surfaces under high pressure. We used an O-ring to provide a seal between the sapphire window and a stainless steel support. The surface of the support attached to the sapphire window was processed by mirror polishing to increase the pressure that the O-ring could withstand. In this study, achieving a balance in the thickness of the sapphire window was particularly important, since a thinner window decreases optical aberration, while a thicker one raises the withstand pressure. To our knowledge, the vessel used in this study provides top performance in *in situ* observation of crystal surfaces. The volume of sample space in this vessel is 8.3 mm^3 (4.3 mm in height and 1.6 mm in diameter).

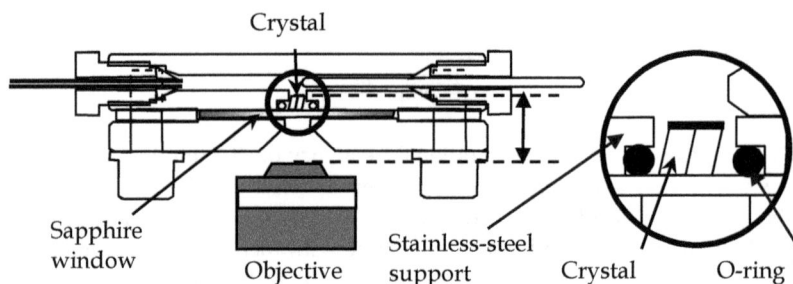

Fig. 6. Schematic illustration of an experimental setup.

To compensate for the optical aberration caused by the light transmitted through the sapphire window, an objective with a compensation ring for a cover glass with a thickness of 0 - 2 mm (Olympus Optical Co., Ltd., SLCPlanFl 40X) also played an important role. Precise adjustments of the compensation ring of the objective and the shear amount of the Nomarski prism were indispensable for obtaining a high contrast level of elementary steps.

For *in situ* observation of elementary steps under high pressure, seed crystals were placed directly on the sapphire window of the high-pressure vessel to minimize the optical aberration. The seed crystals were prepared as follows. A suspension of GI crystals (Hampton Research Co., Ltd., HR7-100), containing 0.91 M ammonium sulfate and 1 mM magnesium sulfate in 6 mM tris hydrochloride buffer (pH = 7.0) (Tris-HCl is known as the most insensitive buffer to pressure) (Neuman et al., 1973), was dissolved (~ 33 mg mL^{-1}) and filtered (Suzuki et al., 2002b). Then the filtrate was transferred onto the sapphire window and sealed with an o-ring and a glass slide.

After a few small crystals appeared on the window at 10 °C, the crystals were grown at room temperature (~ 22 °C) until they reached an appropriate size for the observation (typically ≥ 100 μm). The crystals placed on the window were rinsed with a GI solution of 5.6 mg mL^{-1}, and then the window with the crystals was fitted into the high-pressure vessel filled with a GI solution of 5.6 mg mL^{-1}. In this study we prepared ≤ 10 crystals (size ~ 150 μm) in the 1.6 mmφ O-ring on the sapphire window. Thus, the average separation between the crystals was ~ 300 μm.

4.2.3 Step velocities

Step velocities V on the {011} faces at 0.1 and 50 MPa were measured in the range of protein concentrations $C = 5.3 - 8.9$ mg mL^{-1}. As shown in Figure 7(a), V increased with increasing pressure. The increase in V is attributed to both kinetic and thermodynamic contributions as shown in eq. (16).

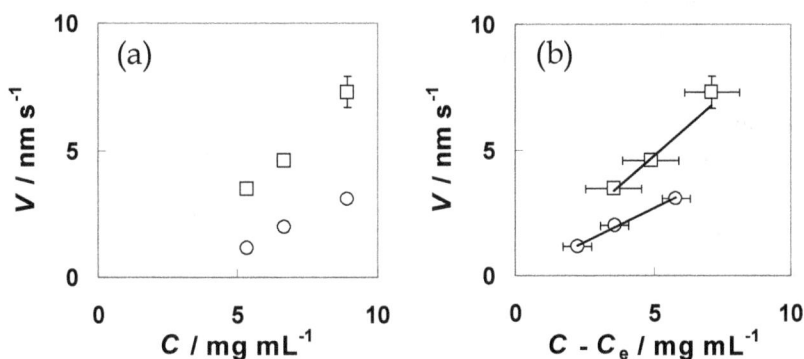

Fig. 7. Step velocities V on the {011} faces of GI crystals as a function of C (a) and $C - C_e$ (b) (Suzuki et al., 2009, 2010a). V was measured at 0.1 MPa (○) and 50 MPa (□). Temperature was 25.0 °C. The lines shown in (b) indicate the results of weighed linear fitting.

To separate the kinetic contribution (β_{step}) from the thermodynamic one ($C - C_e$), we replotted V as a function of $C - C_e$ (Figure 7 (b)). The slopes of the straight lines shown in Figure 7 (b) correspond to $\beta_{step}\Omega$ in eq. (16) at 0.1 and 50 MPa. We have measured Ω under 100 MPa by X-ray crystallography, and found that Ω decreased by only 1.1% with increasing pressure: Ω were $(4.79 \pm 0.03) \times 10^{-25}$ and $(4.74 \pm 0.08) \times 10^{-25}$ m^3 at 0.1 and 100 MPa, respectively (Tsukamoto, 2009). Thus, we concluded that β_{step} increased with increasing pressure kinetically. β_{step} thus obtained were $(3.2 \pm 0.2) \times 10^{-7}$ and $(5.7 \pm 0.9) \times 10^{-7}$ m s^{-1} at 0.1 and 50 MPa (here we assume that Ω at 50 MPa is same as that at 0.1 MPa), respectively. From these data, we calculated $\Delta V^{\ddagger}$ to be -28 ± 8 cm^{3}mol^{-1} using equation (19).

C_e values at 25°C (2.6 ± 1.4) and (0.8 ± 0.4) mgmL^{-1} at 0.10 and 50 MPa, respectively (Suzuki et al., 2009, 2010a). From these data, we calculated ΔV to be -60 ± 40 cm^{3}mol^{-1} using equation (4). The absolute value of $\Delta V^{\ddagger}$ is almost half that of ΔV.

Such volumetric parameters are strongly related to the dehydration process during the incorporation of a growth unit into a kink site. Thus, further data accumulation will be useful for understanding the dehydration process, which should be one of the most important mechanisms of protein crystallization.

5. X-ray crystallography of protein crystals under high pressure

From the viewpoint of "in situ" high-pressure protein crystallography at the atomic level, five reports have been published so far (Kundrot & Richards, 1987; Urayama et al., 2002; Collins et al., 2005, 2007; Colloc'h et al., 2006). Kundrot et al. reported a three-dimensional structure of a protein (lysozyme) molecule under 100 MPa for the first time (Kundrot &

Richards, 1987). They used a Beryllium (Be) vessel with a stainless steel capillary tube. They revealed anisotropic contraction of the molecule and the increase in the number of ordered water in the crystal with increasing pressure. Urayama et al. and Collins et al. also used Be vessels (Urayama et al., 2002; Collins et al., 2005, 2007). Colloc'h et al. used a diamond anvil cell (DAC), and they could collect 2.3 Å resolution data of urate oxidase (Colloc'h et al., 2006).

However, each method has some problems. All the Be vessels equipped capillary tubes, and the tubes were obstacles to the free rotation of the vessels on goniometers during data collection processes. In the case of a DAC, the accuracy of pressure measurements with ruby fluorescence is low, although the DAC can generate much higher pressures than the Be vessels can do. The error of the pressure measurements in a DAC is usually larger than 10 MPa. In addition, there are usually geometrical constraints on data collection processes with a DAC.

A stand-alone type Be vessel solves all the above problems. Without connecting to the capillary tube, the vessel can freely rotate. With a simple free-piston type pressure indicator, we can monitor pressure in the vessel. In this section, I would like to present our most recent work on high-pressure x-ray protein crystallography.

5.1 Methodology
A stand-alone type high-pressure Be vessel (Syn-corporation Ltd.) was constructed for "in situ" high-pressure protein crystallography at the atomic level (Suzuki et al., 2010b). The vessel equips a Be tube, a stainless steel base, a pressure valve, a coupler joint and a free-piston type pressure indicator. The pressure indicator was composed of a free piston and two springs, and pressure was monitored within ± 1 MPa. From calibration plots of the indicator, we obtained the following equation.

$$h = (0.46 \pm 0.05) + (0.0983 \pm 0.0009) \ P. \qquad (20)$$

Here h is the displacement of the piston in mm, and P shows pressure in MPa. The Be tube contains 1% BeO, which reduces X-ray transmittance, and this BeO content is less than that in Urayama's Be tube (2.5%) (Urayama et al., 2002). The thickness of the tube wall is 1.08 mm, and it is also less than that of Kundrot's tube (2.25 mm) (Kundrot & Richards, 1987).

5.2 High pressure X-ray analyses of crystals grown at ambient pressure
Glucose isomerase from *Streptomyces rubiginosus* (Hampton Research, HR7-100) was used without further purification. A GI crystal (~ 0.5 mm) was prepared at atmospheric pressure on the inner wall of a glass capillary (Hampton Research, HR6-164) with its growing solution. The solution contains 0.91 M ammonium sulfate, 1 mM magnesium sulfate, and these are dissolved in 6 mM tris hydrochloride buffer (pH = 7.0). The details of the preparation of the crystal are as follows. First, smaller seed crystals were prepared as described elsewhere (Suzuki et al, 2002b). Second, one of the smaller seed crystals was transferred into a growth solution (the GI concentration of the solution was 28 mgmL^{-1}) in a glass capillary. Third, the seed crystal was incubated for 3 days at 26 °C. Last, the capillary with the crystal and solution was transferred into our high-pressure vessel without replacing the solution. We did not remove the solution, since hydrostatic pressure should be transmitted *via* the solution. A diffraction data set was collected at room temperature on an imaging-plate detector (Rigaku Co., R-AXIS VII) using a rotating copper-anode in-house

generator operating at 40 kV and 20 mA (0.8 kW). Such a relatively mild condition is suitable for "in situ" structure analyses, since a high-intensity X-ray radiation easily increases the temperature of a crystal and water molecules around the crystal; the crystal easily dissolves and deteriorates.

Rotation diffraction spots of a GI crystal and powder diffraction rings of a Be tube are shown in Fig. 8. The data were processed using Crystal Clear (Rigaku Corporation, Tokyo).

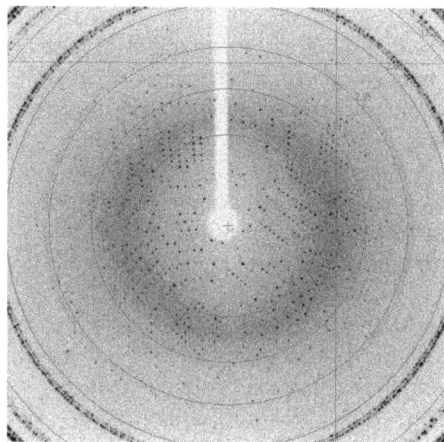

Fig. 8. Rotation diffraction spots of a GI crystal and diffraction rings of a polycrystalline Be tube (Suzuki et al., 2010b).

We successfully collected a 2.0 Å resolution data set of a GI crystal. Pressure could be kept constant at 100 ± 1 MPa for > 24 hours in stand-alone conditions (without connecting to a pressure generator). Although the crystal dissolved a little after the data collection process (> 3 hours irradiation with X-rays), we confirmed that this vessel is sufficiently useful for "in situ" high-pressure protein crystallography.

Strictly speaking, no one has done true "in situ" high-pressure protein crystallography, and a direct result has not been achieved yet. Kundrot et al., Urayama et al., Collins et al., and Colloc'h et al. prepared their crystals at atmospheric pressure, and then pressurized and analyzed the crystals. In such cases, proteins in a crystal shrink with keeping their bonding structure in the crystal during the pressurization. Thus, the pressurized structure in the crystal can be different from that in a solution. In contrast, Charron et al. prepared thaumatin crystals under high pressure, and analyzed the crystals at 100 K after depressurization (Charron et al., 2002). The 3D molecular structures of thaumatin in the depressurized crystals were essentially same as those of unpressurized control crystals. The result suggested that the crystal lattice of thaumatin is elastic. Our group analyzed depressurized, pressurized, and unpressurized GI crystals (Tsukamoto, 2009). A GI structure of the depressurized crystal was essentially same as that of thee unpressurized crystal. Only a GI structure of the pressurized crystal shrinked a little under 100 MPa. This result seems to support Charron's conclusion. However, all the above results are indirect.

To achieve direct results, we should collect a high-resolution, "in situ", and high-pressure data set of a crystal that has nucleated and grown under high pressure. Our setup will achieve the direct results. We can incubate a sufficiently supersaturated protein solution in our vessel under a pressure as long as possible with connecting to a pressure-generating apparatus. After the appropriate nucleation and sufficient growth of crystals from the solution, we can separate the vessel from the capillary tube and directly collect a high-resolution diffraction data set of the crystals with keeping the pressure in the vessel constant.

6. Conclusions

In this chapter, I have presented the great potentialities of high pressure for the promotion of studies on the fundamental growth mechanisms of protein crystals and correlation between the function and 3D structures of protein molecules. Key points in this review are described shortly as follows.

1. As a tool for enhancing the crystal growth, three of eight proteins show the decrease in its solubility under high pressure. Application of high pressure during the screening processes would be useful because of such high probability.
2. Acceleration of growth and nucleation kinetics of glucose isomerase crystals occurred under high pressure.
3. Step velocities under high pressure provided us direct information on activation volume. Activation volume was negative in the case of glucose isomerase crystals. Precise discussion on the activation volume will be useful for understanding dehydration mechanisms during an incorporation process of a protein molecule into a kink site.
4. Usefulness of our standalone-type Be vessel for high-pressure protein crystallography was confirmed. With the vessel, precise high-pressure 3D structure analysis of protein crystals which are also grown under high pressure.

7. Acknowledgments

In this chapter, our studies were supported mainly by two research and education programs and two grants. Studies on solubility, nucleation rates, face growth rates, and step velocities under high pressure were partially supported by "Program for an improvement of education" promoted by the University of Tokushima, the inter-university cooperative research program of the Institute for Materials Research, Tohoku University, and Grants-in Aid (Nos. 16760014 and 19760009 (Y.S.)) for Scientific Research from the Ministry of Education, Culture, Sports, Science and Technology of Japan.

8. References

Asai, T., Suzuki, Y., Sazaki, G., Tamura, K., Sawada, T., & Nakajima, K. (2004). Effects of High Pressure on the Solubility and Growth Kinetics of Monoclinic Lysozyme Crystals. *Cellular and Molecular Biology*, Vol. 50, No. 4, (June 2004), pp. 329-334, ISSN 0145-5680

Boistelle, R., & Lopez-Valero, I. (1990). Growth Units and Nucleation: the Case of Calcium Phosphates. *Journal of Crystal Growth*, Vol. 102, No. 3, (May 1990), pp. 609-617, ISSN 0022-0248

Burton, W.K., Cabrera, N., & Frank, F.C. (1951). The Growth of Crystals and the Equilibrium Structure of Their Surfaces. Philosophical Transactions of the Royal Society of London A., Vol. 243, No. 866, (June 1951), pp. 299-358, ISSN 1471-2962

Cabrera, N., & Levine, M.M. (1956). On the Dislocation Theory of Evaporation of Crystals. Philosophical Magazine, Vol. 1, No. 5, (May 1956), pp. 450-458, ISSN 0031-8086

Charron, C., Robert, M.-C., Capelle, B., Kadri, A., Jenner, G., Giegé, R., & Lorber, B. (2002). X-ray Diffraction Properties of Protein Crystals Prepared in Agarose Gel under Hydrostatic Pressure. Journal of Crystal Growth, Vol. 245, No. 3-4, (November 2002), pp. 321-333, ISSN 0022-0248

Chernov, A.A. (1984). Modern Crystallography III Crystal Growth, Springer-Verlag, ISBN 3-540-11516-1, Berlin Heidelberg New York Tokyo

Collins, M.D., Hummer, G., Quillin, M.L., Matthews, B.W., & Gruner, S.M. (2005). Cooperative Water Filling of a Nonpolar Protein Cavity Observed by High-Pressure Crystallography and Simulation. Proceedings of the National Academy of Sciences of the United States of America, Vol. 102, No. 46, (November 2005), pp. 16668-16671, ISSN 0027-8424

Collins, M.D., Quillin, M.L., Hummer, G., Matthews, B.W., & Gruner, S.M. (2007). Structural Rigidity of a Large Cavity-Containing Protein Revealed by High-Pressure Crystallography. Journal of Molecular Biology, Vol. 367, No.3, (March 2007), pp. 752-763, ISSN 0022-2836

Colloc'h, N., Girard, E., Dhaussy, A.-C., Kahn, R., Ascone, I., Mezouar, M., & Fourme, R. (2006). High Pressure Macromolecular Crystallography: the 140-MPa Crystal Structure at 2.3 Å Resolution of Urate Oxidase, a 135-kDa Tetrameric Assembly. Biochimica et Biophysica Acta, Vol. 1764 (2006), No. 3, (March 2006), pp 391-397, ISSN 1570-9639

Durbin, S.D., & Carlson, W.E. (1992). Lysozyme Crystal Growth Studied by Atomic Force Microscopy. Journal of Crystal Growth, Vol. 122, No. 1-4, (August 1992), pp. 71-79, ISSN 0022-0248

Durbin, S.D., Carlson, W.E., & Saros, M.T. (1993). In Situ Studies of Protein Crystal Growth by Atomic Force Microscopy. Journal of Physics D: Applied Physics, Vol. 26, No. 8B, (August 1993), pp. B128-B132, ISSN 0022-3727

Foubert, I., Dewettinck, K., & Vanrolleghem, P.A. (2003). Modelling of the Crystallization Kinetics of Fats. Trends in Food Science & Technology, Vol. 14, No. 3, (March 2003), pp. 79-92, ISSN 0924-2244

Fujiwara, T., Suzuki, Y., Sazaki, G., & Tamura, K. (2010). Solubility Measurements of Protein Crystals under High Pressure by In Situ Observation of Steps on Crystal Surfaces. Journal of Physics: Conference Series, Vol. 215, (March 2010), pp. 012159-1-5, ISSN 1742-6588

Groß, M., & Jaenicke, R. (1991). Growth Inhibition of Lysozyme Crystals at High Hydrostatic Pressure. FEBS Letters, Vol. 284, No. 1, (June 1991), pp. 87-90, ISSN 0014-5793

Groß, M., & Jaenicke, R. (1993). A Kinetic Model Explaining the Effects of Hydrostatic Pressure on Nucleation and Growth of Lysozyme Crystals. Biophysical Chemistry, Vol. 45, No. 3, (January 1993), pp. 245-252, ISSN 0301-4622

Higgins, S.R., Eggleston, C.M., Knauss, K.G., Boro, C.O. (1998). A Hydrothermal Atomic Force Microscope for Imaging in Aqueous Solution up to 150 °C. *Review of Scientific Instruments*, Vol. 69, No. 8, (August 1998), pp. 2994-2998, ISSN 0034-6748

Kadri, A., Lorber, B., Jenner, G., & Giegé, R. (2002). Effects of Pressure on the Crystallization and the Solubility of Proteins in Agarose Gel. *Journal of Crystal Growth*, Vol. 245, No. 1-2, (November 2002), pp. 109-120, ISSN 0022-0248

Kundrot, C.E., & Richards, F.M. (1987). Crystal Structure of Hen Egg-White Lysozyme at a Hydrostatic Pressure of 1000 Atmospheres. *Journal of Molecular Biology*, Vol. 193, No. 1, (January 1987), pp. 157-170, ISSN 0022-2836

Kundrot, C.E., & Richards, F.M. (1988). Effect of Hydrostatic Pressure on the Solvent in Crystals of Hen Egg-White Lysozyme. *Journal of Molecular Biology*, Vol. 200, No. 2, (March 1988), pp. 401-410, ISSN 0022-2836

Laidler, K.J. (1987). *Chemical Kinetics, 3rd ed.*, Harper & Row, ISBN 0060438622, New York, USA

Lorber, B., Jenner, G., & Giegé, R. (1996). Effect of High Hydrostatic Pressure on Nucleation and Growth of Protein Crystals. *Journal of Crystal Growth*, Vol. 158, No. 1-2, (January 1996), pp. 103-117, ISSN 0022-0248

Makimoto, S., Suzuki, K., & Taniguchi, Y. (1984). Pressure Dependence of the a-Chymotrypsion-Catalyzed Hydrolysis of Amide and Anilides. Evidence for the single-proton-transfer mechanism. *The Journal of Physical Chemistry*, Vol. 88, No. 24, (November 1984), pp. 6021-6024, ISSN 0022-3654

Markov, I.V. (1995). *Crystal Growth for Beginners : Fundamentals of Nucleation, Crystal Growth, and Epitaxy*, World Scientific, ISBN 9810215312, Singapore, Singapore

Maruoka, T., Suzuki, Y., & Tamura, K. (2010). Effects of High Pressure on the Three-Dimensional Nucleation Rates of Glucose Isomerase Crystals. *Journal of Physics: Conference Series*, Vol. 215, (March 2010), pp. 012158-1-5, ISSN 1742-6588

McPherson, A., Malkin, A.J., Kuznetsov, Y.G. (2000). Atomic Force Microscopy in the Study of Macromolecular Crystal Growth. Annual Review of Biophysics and Biomolecular Structure, Vol. 29, pp. 361-410, ISSN 1056-8700

Nagatoshi, Y., Sazaki, G., Suzuki, Y., Miyashita, S., Matsui, T., Ujihara, T., Fujiwara, K., Usami, N., & Nakajima, K. (2003). Effects of High Pressure on the Growth Kinetics of Orthorhombic Lysozyme Crystals. *Journal of Crystal Growth*, Vol. 254, No. 1-2, (June 2003), pp 188-195, ISSN 0022-0248

Neuman, R.C., Jr., Kauzmann, W., & Zipp, A. (1973). *The Journal of Physical Chemistry*, Vol. 77, No. 22, (October 1973), pp. 2687-2691, ISSN 0022-3654

Pan, X., & Glatz, C.E. (2002). Solvent Role in Protein Crystallization as Determined by Pressure Dependence of Nucleation Rate and Solubility. *Crystal Growth & Design*, Vol. 2, No. 1, (January 2002), pp. 45-50, ISSN 1528-7483

Rosenberger, F., Vekilov, P.G., Muschol, M., & Thomas, B.R. (1996). Nucleation and Crystallization of Globular Proteins – What We Know and What is Missing. *Journal of Crystal Growth*, Vol. 168, No. 1-4, (October 1996), pp. 1-27, ISSN 0022-0248

Saikumar, M.V., Glatz, C.E., & Larson, M.A. (1998). Lysozyme Crystal Growth and Nucleation Kinetics. *Journal of Crystal Growth*, Vol. 187, No. 2, (May 1998), pp. 277-288, ISSN 0022-0248

Sazaki, G., Matsui, T., Tsukamoto, K., Usami, N., Ujihara, T., Fujiwara, K., & Nakajima, K. (2004). In Situ Observation of Elementary Growth Steps on the Surface of Protein Crystals by Laser Confocal Microscopy. *Journal of Crystal Growth*, Vol. 262, No. 1-4, (February 2004), pp. 536-542, ISSN 0022-0248

Sazaki, G., Nagatoshi, Y., Suzuki, Y., Durbin, S.D., Miyashita, S., Nakada, T., & Komatsu, H. (1999). Solubility of Tetragonal and Orthorhombic Lysozyme Crystals under High Pressure. *Journal of Crystal Growth*, Vol. 196, No. 2-4, (January 1999), pp. 204-209, ISSN 0022-0248

Suzuki, Y., Miyashita, S., Komatsu, H., Sato, K., & Yagi, T. (1994). Crystal Growth of Hen Egg White Lysozyme under High Pressure. Japanese Journal of Applied Physics, Vol. 33, No. 11A, (November 1994), pp. L1568-L1570, ISSN 0021-4922

Suzuki, Y., Sawada, T., Miyashita, S., Komatsu, H. (1998). *In Situ* Measurements of the Solubility of Crystals under High Pressure by an Interferometric Method. *Review of Scientific Instruments*, Vol. 69, No. 7, (July 1998), pp. 2720-2724, ISSN 0034-6748

Suzuki, Y., Miyashita, S., Sazaki, G., Nakada, T., Sawada, T., & Komatsu, H. (2000a). Effects of Pressure on Growth Kinetics of Tetragonal Lysozyme Crystals. *Journal of Crystal Growth*, Vol. 208, No. 1-4, (January 2000), pp. 638-644, ISSN 0022-0248

Suzuki, Y., Sawada, T., Miyashita, S., Komatsu, H., Sazaki, G., & Nakada, T. (2000b). An Interferometric Study of the Solubility of Lysozyme Crystals under High Pressure. *Journal of Crystal Growth*, Vol. 209, No. 4, (February 2000), pp. 1018-1022, ISSN 0022-0248

Suzuki, Y., Sazaki, G., Miyashita, S., Sawada, T., Tamura, K., & Komatsu, H. (2002a). Protein Crystallization under High Pressure. *Biochimica et Biophysica Acta*, Vol. 1595, No. 1-2, (March 2002), pp. 345-356, ISSN 0167-4838

Suzuki, Y., Sazaki, G., Visuri, K., Tamura, K., Nakajima, K., & Yanagiya, S. (2002b). Significant Decrease in the Solubility of Glucose Isomerase Crystals under High Pressure. *Crystal Growth & Design*, Vol. 2, No. 5, (September 2002), pp 321-324, ISSN 1528-7483

Suzuki, Y., Sazaki, G., Matsui, T., Nakajima, K., & Tamura, K. (2005). High-Pressure Acceleration of the Growth Kinetics of Glucose Isomerase Crystals. *The Journal of Physical Chemistry B*, Vol. 109, No. 8, (March 2005), pp. 3222-3226, ISSN 1089-5647

Suzuki, Y., Sazaki, G., Matsumoto, M., Nagasawa, M., Nakajima, K., & Tamura, K. (2009). First Direct Observation of Elementary Steps on the Surfaces of Glucose Isomerase Crystals under High Pressure. *Crystal Growth & Design*, Vol. 9, No. 10, (October 2009), pp. 4289-4295, ISSN 1528-7483

Suzuki, Y., Sazaki, G., Matsumoto, M., Nagasawa, M., Nakajima, K., & Tamura, K. (2010a). First Direct Observation of Elementary Steps on the Surfaces of Glucose Isomerase Crystals under High Pressure (Additions and Corrections). *Crystal Growth & Design*, Vol. 10, No. 4, (April 2010), pp. 2020-2020, ISSN 1528-7483

Suzuki, Y., Tsukamoto, M., Sakuraba, H., Matsumoto, M., Nagasawa, M., & Tamura, K. (2010b). Design of a standalone-type beryllium vessel for high-pressure protein crystallography. *Review of Scientific Instruments*, Vol. 81, No. 8, (August 2010), pp. 084302-1-3, ISSN 0034-6748

Suzuki, Y., Maruoka, T., & Tamura, K. (2010c). Activation Volume of Crystallization and Effects of Pressure on the Three-Dimensional Nucleation Rate of Glucose Isomerase. *High Pressure Research*, Vol. 30, No. 4, (December 2010), pp. 483-489, ISSN 0895-7959

Suzuki, Y., Konda, E., Hondoh, H., & Tamura, K. (2011). Effects of Temperature, Pressure, and pH on the Solubility of Triclinic Lysozyme Crystals. *Journal of Crystal Growth*, Vol. 318, No. 1, (March 2011), pp. 1085-1088, ISSN 0022-0248

Takano, K.J., Harigae, H., Kawamura, Y. & Ataka, M. (1997). Effect of Hydrostatic Pressure on the Crystallization of Lysozyme Based on In Situ Observations. *Journal of Crystal Growth*, Vol. 171, No. 3-4, (February 1997), pp. 554-558, ISSN 0022-0248

Tsukamoto, M. (2009). *Master Thesis*, Graduate School of Advanced Technology and Science, The University of Tokushima

Urayama, P., Phillips, Jr., G.N., & Gruner, S.M. (2002). Probing Substrates in Sperm Whale Myoglobin Using High-Pressure Crystallography. *Structure*, Vol. 10, No. 1, (January 2002), pp. 51-60, ISSN 0969-2126

Van Driessche, A.E.S., Sazaki, G., Otalora, F., Gonzalez-Rico, F.M., Dold, P., Tsukamoto, K., & Nakajima, K. (2007). Direct and Noninvasive Observation of Two-Dimensional Nucleation Behavior of Protein Crystals by Advanced Optical Microscopy. *Crystal Growth & Design*, Vol. 7, No. 7, (September 2007), pp. 1980-1987, ISSN 1528-7483

Van Driessche, A.E.S., Gavira, J.A., Patiño Lopez, L.D., & Otalora, F. (2009). Precise Protein Solubility Determination by Laser Confocal Differential Interference Contrast Microscopy. *Journal of Crystal Growth*, Vol. 311, No. 13, (June 2009), pp. 3479-3484, ISSN 0022-0248

Visuri, K., Kaipainen, E., Kivimaki, J., Niemi, H., Leisola, M., & Palosaari, S. (1990). A New Method for Protein Crystallization Using High Pressure. *Bio/Technology*, Vol. 8, No. 6, (June 1990), pp. 547-549, ISSN 0733-222X

Volmer, M., & Weber, A. (1926). Keimbildung in Übersättigten Gebilden. *Zeitschrift für Physikalische Chemie*, Vol. 199, pp. 277-301, ISSN 0044-3336

Waghmare, R.Y., Webb, J.N., Randolph, T.W., Larson, M.A., & Glatz, C.E. (2000a). Pressure Dependence of Subtilisin Crystallization Kinetics. *Journal of Crystal Growth*, Vol. 208, No. 1-4, (January 2000), pp. 678-686, ISSN 0022-0248

Waghmare, R.Y., Pan, X.J., & Glatz, C.E. (2000b). Pressure and Concentration Dependence of Nucleation Kinetics for Crystallization of Subtilisin. *Journal of Crystal Growth*, Vol. 210, No. 4, (March 2000), pp. 746-752, ISSN 0022-0248

Webb, J.N., Waghmare, R.Y., Carpenter, J.F., Glatz, C.E., & Randolph, T.W. (1999). Pressure Effect on Subtilisin Crystallization and Solubility, *Journal of Crystal Growth*, Vol. 205, No. 4, (September 1999), pp. 563-574, ISSN 0022-0248

Yayanos, A.A. (1986). Evolutional and Ecological Implications of the Properties of Deep-Sea Barophilic Bacteria. *Proceedings of the National Academy of Sciences of the United States of America*, Vol. 83, No. 24, (December 1986), pp. 9542-9546, ISSN 0027-8424

10

Rational and Irrational Approaches to Convince a Protein to Crystallize

André Abts, Christian K. W. Schwarz, Britta Tschapek,
Sander H. J. Smits and Lutz Schmitt
*Institute of Biochemistry, Heinrich-Heine University, Düsseldorf,
Germany*

1. Introduction

The importance of structural biology has been highlighted in the past few years not only as part of drug discovery programs in the pharmaceutical industry but also by structural genomics programs. Mutations of human proteins have been long recognized as the source of severe diseases and a structural knowledge of the consequences of a mutation might open up new approaches of drugs and cure. Although the function of a protein can be studied by several biochemical and/or biophysical techniques, a detailed molecular understanding of the protein of interest can only be obtained by combining functional data with the knowledge of the three-dimensional structure. In principle three techniques exist to determine a protein structure, namely X-ray crystallography, nuclear magnetic resonance spectroscopy (NMR) and electron microscopy (EM). According to the protein data bank (pdb; http://www.rcsb.org) that provides a general and open-access platform for structures of biomolecules, X-ray crystallography contributes more than 90% of all structures in the pdb, a clear emphasis of the importance of this technique.

To perform X-ray crystallography it is essential to have large amounts of pure and homogenous protein to perform an even today still "trail and error"-based screening matrix to obtain well diffracting protein crystals. Therefore, successful protein crystallization requires three major and crucial steps, all of them associate with specific problems and challenges that need to be overcome and solved. These steps are (I) protein expression, (II) protein purification and (III) the empirical search for crystallization conditions. As summarized in Figure 1, every single step needs to be optimized along the long and stoney road to obtain protein crystals suitable for structure determination of your "most-beloved" protein via X-ray crystallography. This chapter will focus on these three steps and suggests strategies how to perform and optimize each of these three steps on the road of protein structure determination.

2. Protein expression (I)

To crystallize a protein, the first requirement is the expression of your protein in high amounts and most importantly on a regular basis. This implies that it is possible to obtain a freshly purified protein at least weekly. In general, it is possible to express a protein either homologously or heterologously (see Figure 1 – (I) expression). Especially for large proteins,

I) PROTEIN EXPRESSION

prokaryotic   eukaryotic

gram (+)   gram (-)   yeast   cell culture
e.g. *L. lactis*   e.g. *E. coli*   e.g. *P. pastoris*   e.g *insect cells*

- change expression
  system / conditions
- ...

(high) expressed protein

II) PROTEIN PURIFICATION

affinity chromatography

- change metal ion
- change eluent
- second affinity tag
- ion exchange
- ...

pure protein

activity ?

- buffer screen
- flourescence-based
  stability screen
- ...

size exclusion chromatography

pure homogenous protein

III) PROTEIN CRYSTALLIZATION

screening conditions

- check homogeneity
  with SEC / LS
- mutants
- fix conformation
- ...

initial crystals

optimization

crystals

- screening
  other buffers
- ...

diffracting ?

- add ligands or
  additives
- change
  temperature
- ...

collect dataset

protein structure

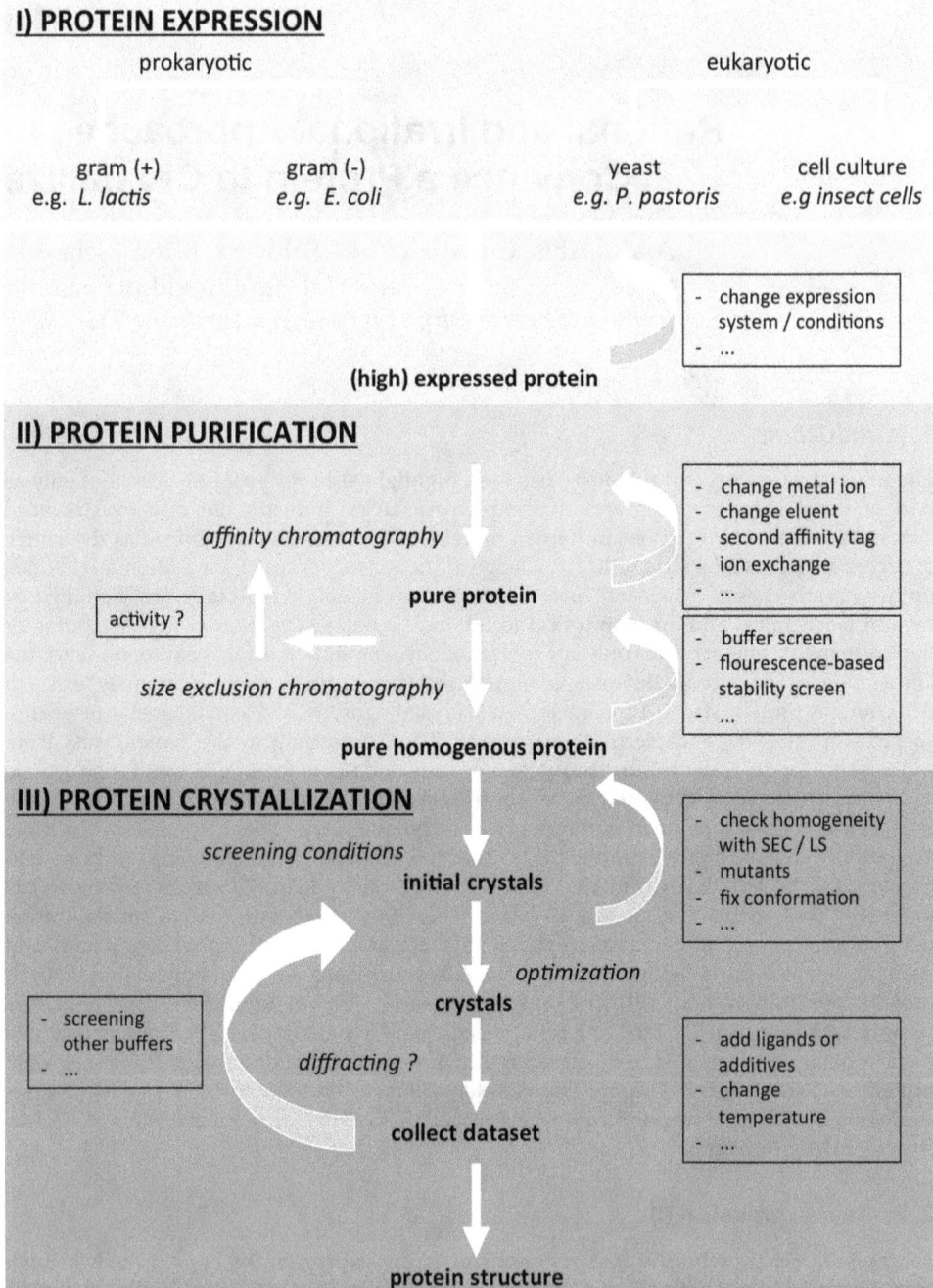

Fig. 1. Schema highlighting the three steps towards a protein crystal. (I) Expression (II) Purification (III) Protein crystallization.

proteins containing a co-factor or a ligand, the natural habitat is likely the best choice to express the protein. However, often the natural host, for example humans, produce only low amounts of protein and suitable overexpression protocolls are not available. To circumvent this problem, several expression strains, cell lines as well as a large number of expression vectors have been developed to allow expression of any protein in a different host (heterologous expression). In general, the used organisms for protein expression can be divided into two different groups: prokaryotic and eukaryotic expression hosts. The natural organism of the protein of interest mainly dictates the choice, which expression system to use. If working with a bacterial protein, it is very likely that also a prokaryotic host is able to express the protein in high amounts. The same holds true for proteins originating from a eukaryotic host, which is likely best overexpressed in a eukaryotic host. These proteins often require posttranslational modification such as glycosylation or disulfide bond formation, which are possible in eukaryotic expression hosts. The most common used heterologous expression host is the gram negative bacteria *Escherichia coli* since it is commercially available and a large number of expression cassettes have been developed. Thus, it is the most widely used expression system with expression rates of several mg/L of culture. The best characterised and understood expression hosts are described in more detail below and the commecially available systems are listed in table 1.

2.1 Expression hosts – Prokaryotic
2.1.1 Gram negative - *E. coli*
As mentioned above, *E. coli* is the most common used expression system (Figure 1 – (I) expression, left side). This is further highlighted by the fact that 80% of all protein structures deposited in the protein data bank were overexpressed in *E. coli* (Sorensen and Mortensen 2005). There are several advantages promoting *E. coli* as expression host: (A) Cultivation of *E. coli* is simple and a doubling time of 30 minutes is rather quick allowing the fast generation of biomass, (B) genetics are well understood and any genetical manipulation is well established, (C) expression levels of up to 60 % of the total protein mass within the cell make the next step, protein purification rather straight forward and finally (D) the cultivation does require only standard equipment normally present in every biochemical laboratory and therefore expression using *E.coli* is relatively cheap. In the last decades, many different plasmid based expression systems have been developed such as the pET vector systems, which contain several different expression plasmids with a choice for the affinity tag on either termini of the protein as well as the possiblity to use a dual cassette when expressing two or more proteins at once. The selection pressure derived from different antibiotics, and the resistance genes encoded on these plasmids further simplify laboratory practice. Only cells harbouring the right plasmids are able to grow and therefore expenditure on sterility is low.

The typical *E. coli* expression system is plasmid-based, which can be transferred to different *E. coli* strains ((Sorensen and Mortensen 2005), Novagen pET vector table). An *E. coli* expression vector consists mainly of five important parts: the replicon, a resistance marker, a promotor and a so-called multiple cloning site (MCS) (Baneyx 1999; Jonasson, Liljeqvist et al. 2002). The replicon is the crucial part of a plasmid to maintain it inside a cell. It is recognized and duplicated by the replication machinery (Baneyx 1999). The selection marker allows the identification of cells carrying a plasmid as it encodes for a resistance, e.g. against antibiotics (see above)(Sorensen and Mortensen 2005). The promotor sequence is the recognition site for the RNA polymerase, however it is inactive under initial cultivation conditions. The addition of an inducer (sometimes also a temperature change) switches the promotor from 'off' to 'on' whereby the expression is initiated (Jana and Deb 2005). Common inducers are isopropyl-β-D-thiogalactopyranosid (IPTG) in the pET system or arabinose for pBAD vectors (Invitrogen™). The multiple cloning site (MCS) is a short DNA segment combining many (up to 20) restriction sites. This feature simplified the insertion of genes into the plasmid enormously and made cloning procedures very convenient. However, new cloning strategies, which are independent of restriction enzymes and ligases, are emerging and will replace the standard approaches some day (see for example Li and Elledge 2007).

2.1.2 Gram positive - *L. lactis*
Within the prokaryotic expression system also some gram (+) bacteria are used for protein overexpression (Figure 1 – (I) expression left side). Here, lactic acid bacteria play a privotal role. They are used in food industry and are known since 1873 when Joseph Lister isolated the first strain (Teuber 1995; Mierau and Kleerebezem 2005). For the overexpression of recombinant proteins, there are many lactic acid bacteria around, *lactococci, lactobacilli, streptococci* and *leuconostocs*. (Gasson 1983; van de Guchte, Kok et al. 1992; de Vos and Vaughan 1994). The best characterized and most widely used host is *Lactococcus lactis*, which is famous for its usage within food fermentation and like for *E. coli* the genome, metabolisms and molecular modifications are well known and established (Bolotin, Wincker et al. 2001; Guillot, Gitton et al. 2003). Thus, it has been called the 'bug of the next millennium'(Konings, Kok et al. 2000). All established gram (+) bacteria expression hosts are able to overexpress proteins homologously or heterologously. Since only one, namely the cyctoplasmic membrane is present (Kunji, Slotboom et al. 2003), this host is in comparison to the two-membrane system of gram (-) bacteria, a good choice to express eukaryotic as well as prokaryotic membrane proteins or proteins with membrane anchors (Kunji, Slotboom et al. 2003). The promoter used for expression in *L. lactis* is induced by the external addition of nisin. Nisin is an antimicrobial active peptide, which interacts with lipid II in the cytoplasmic membrane of gram (+) bacteria and causes cell lysis. Interestingly, nisin is produced by *L. lactis* itself. The expression strain is deleted of the nisin producing genes and therefore external nisin can be used as inductor. Nisin binds to NisK, which as part of a two-component system, phosphorylates NisR, which in turn binds to the promotor P_{nisA} thereby allowing synthesis of the protein located downstream on the plasmid. Since nisin is also active against *L. lactis* itself, the concentration range of nisin used in such expression studies is relatively narrow to circumvent killing of the *L. lactis* expression strain. This is clearly a draw back of this expression system since expression can basically only turned on with a certain nisin concentration. Between the inducer concentration and the expressed protein a linear behaviour is observed. Unfortunately the nisin concentration range between the minimal and maximum nisin concentration is very small, nisin concentration higher than 25 ng per liter of cells, cause cell death.

However, *L. lactis* has been proven to be a very efficient expression system. Kuipers et al. created many expression hosts and plasmids to produce any protein of interest by cloning it downstream of the P_{nisA} promotor. With this nisin inducible (NICE) expression system, it is now possible to induce the protein production with minimal concentration (0.1 – 5 ng) of nisin (de Ruyter, Kuipers et al. 1996; Kuipers, de Ruyter et al. 1998). The amount of produced recombinant protein can reach up to 50 % of the total intracellular proteins (Kuipers, Beerthuyzen et al. 1995; de Ruyter, Kuipers et al. 1996). Following a few examples for expressed proteins in *L. lactis* whose structures have been solved: an ECF-type ABC transporter (PDB:3RLB)(Erkens, Berntsson et al. 2011), a peptide binding protein OppA

(PDB:3RYA)(Berntsson, Doeven et al. 2009) and the multidrug binding transcriptional regulator LmrR (PDB:3F8B)(Madoori, Agustiandari et al. 2009).

2.2 Eukaryotic expression hosts

The great benefit of choosing a eukaryotic host for overexpression of a protein of interest are the availability of a posttranslational modification system as well as the frequently enhanced protein folding (Midgett and Madden 2007). Eukaryotic proteins tend to misfold or lack biological activity when expressed in prokaryotic expression systems such as *E. coli* (Cregg, Cereghino et al. 2000; Midgett and Madden 2007). To overexpress these proteins, different yeast strains, insect cells or even mammalian cell lines have been developed as expression hosts (Figure 1 - (I) expression, right side). Eukaryotic expression systems are often more expensive, provide low expression levels and are sometimes hard to handle, when compared to bacterial systems. However, the genetic and cellular contexts are more similar to the original protein-expressing organism (Midgett and Madden 2007). In the following sections, some of the commonly employed eukaryotic expression systems will be described.

2.2.1 Yeast expression systems - *Saccharomyces cerevisiae* and *Pichia pastoris*

The most widely used yeast strains to express protein are *Saccharomyces cerevisiae* and *Pichia pastoris*, which offer the major advantage of a posttranslational modification system for glycosylation, proteolytic processing as well as disulfide bond formation, which for some proteins are essential for the function and/or correct folding (Cregg, Cereghino et al. 2000; Midgett and Madden 2007). The handling of yeast expression systems is similar to prokaryotic systems with respect to the genetic background and cultivation. Similar to the bacterial vector systems, expression in yeast starts with a plasmid-based cloning part which can be performed in *E. coli* (Cregg 2007). Afterwards the expression cassette gets integrated into the genome by simple homologous recombination in the yeast. One major advantage in *P. pastoris* is the insertion of multiple copies of the protein DNA-sequence into genomic DNA, which increases expression yield.

The biggest advantage of yeast as expression system is that well established protocols for fermentation are available. Optimal fermentation of *P. pastoris* can end up with more than 130 gram of cells per liter of culture. Even if expression levels in the cell are not that high the mass of cells easily compensates for this disadvantage (Wegner 1990; Cregg, Cereghino et al. 2000; Hunt 2005; Cregg 2007; Midgett and Madden 2007). Examples of crystal structures from proteins expressed in *P. pastoris* are a human monoamine oxidase B (PDB:3PO7) (Binda, Aldeco et al. 2010) and a protein involved in cell adhesion NCAM2 IG3-4 (PDB:2XY1)(Kulahin, Kristensen et al. 2011).

2.2.2 Insect cells

The expression system in insect cells is beside yeast a well-characterised alternative to express eukaryotic proteins (Midgett and Madden 2007). As insect cells are higher eukaryotic systems their posttranslational modification machinery can carry out more complex alterations than yeast strains. They also have a machinery for the folding of mammalian proteins. The most commonly used vector system for recombinant protein expression in insect cells is baculovirus, which can also be used for gene transfer and expression in mammalian cells (Smith, Summers et al. 1983; D., L.K. et al. 1992; Altmann, Staudacher et al. 1999). A few examples of proteins expressed in insect cells that resulted in crystal structures are the transferase Ack1 (PDB:3EQP)(Kopecky, Hao et al. 2008), a human hydrolase (PDB: 2PMS)(Senkovich, Cook et al. 2007) and myosin VI (PDB:2BKI)(Menetrey, Bahloul et al. 2005).

2.2.3 Mammalian cell lines

The expression of proteins in mammalian cell lines is the most expensive and complex alternative. Especially for human membrane proteins this expression system has been proven to express the most active protein (Tate, Haase et al. 2003; Lundstrom 2006; Lundstrom, Wagner et al. 2006; Eifler, Duckely et al. 2007). The resulting protein amount, however, obtained from mammalian cell lines is mostly only sufficient for functional studies. Using mammalian cells lines is the most challenging variant of protein overexpression and therefore only choosen if any of the other expression system described failed. Some examples of protein structures expressed in mammalian cell lines are the hydrolase PCSK9 (PDB:2QTW)(Hampton, Knuth et al. 2007) and the acetylcholine receptor AChBP (PDB:2BYQ)(Hansen, Sulzenbacher et al. 2006).

Table 1 sums up advantages and disadvantages of the above mentioned overexpression systems used for protein crystallography.

Expression system	Pros	Cons
Prokaryotic		
Gram negative E. coli	- costs - simplicity (genetic/culture) - yield	- inclusion bodies - protein folding - posttranslational modifications - protein secretion
Gram positive L.lactis	- costs - protein secretion - one membrane	- posttranslational modifications
Eukaryotic		
Yeast	- high cell densities - costs - simple cultivation - posttranslational modifications	- hyper glycosylation, non-native - lipid composition is different to mammalian cells
Insect cells	- more native lipid environment - good track record of functional proteins	- costs - non-native lipid environment - glycosylation pattern different to mammalian cells - protein amount
Mammalian cells	- native conditions for human membrane proteins to investigate diseases - posttranslational modifications - lipid environment - good track record of functional protein	- costs - difficult to establish - protein amount

Table 1. Overview of expression systems. Summarized are the advantages and disadvantages.

3. Purification

After having expressed your protein of interest, the race for crystals is by no means finished. The next step on the long road to structure determination is to isolate the protein or - phrasing it differently - to remove all other proteins present in the cell (Figure 1 – (II) purification). An elegant method to do so is the genetic attachment of an affinity tag on either site of the protein or in some cases on both sides (Waugh 2005). This affinity tag has the possibility to bind high affine to a immobilized ligand on a matrix, while all other proteins have a much more reduced binding affinity and therefore flow through the matrix (Figure 1 – (II) purification 1st step). This allows a one-step purification, which in almost all cases is relatively harmless for the protein and likely does not interfere with folding and/or overall structure of the protein. There are a lot of affinity tags available as well as matrix materials (Terpe 2003). The well known and most often used affinity tag is the poly-histidine tag (Porath, Carlsson et al. 1975; Gaberc-Porekar and Menart 2001), which can vary in length as well as in position but the overall purification strategy is the same. From all the structures solved nowadays, almost 60 % of the proteins are purified via a histidine tag; mainly due to the great purification efficiency, which can be as large as 90% after a single purification step (Gaberc-Porekar and Menart 2001; Arnau, Lauritzen et al. 2006). Therefore, most commercially available expression systems and methods contain a his-tag encoded on the plasmid. Besides the his-tag, there are other tags avaible and used for protein purification, of which the Strep-, CBP-, GST-, MBP-tag are described below.

3.1 Choice of the right tag
3.1.1 Polyhistidine-tag (his-tag)

As mentioned above the polyhistidine-tag is the most common affinity tag and the required affinity resins and chemicals are relatively inexpensive. The purification step is a so-called immobilized metal ion affinity chromatography (IMAC) (Porath, Carlsson et al. 1975). Here, a matrix is able to bind bivalent metal ions. For example nitrilotriactetic acid (NTA), which is a chelator and binds metal ions like Ni^{2+}, Zn^{2+}, Co^{2+} or Cu^{2+} (Hochuli, Dobeli et al. 1987). These metal ions have a high affinity to the imidazole group of the amino acid histidine. A stretch of histindines in a row with for example an E. coli protein is very unusual. Thus, the genetical introduction of several, in most cases 6- 10 histidines in a row selects for specific binding of this protein. As eluant very elegantly imidazole can be used, which competes with the histidine tag and elutes the protein of interest. When used in low concentrations, imidazole can also be used to remove unspecifically bound proteins, which bind with low affinity to the matrix (Hefti, Van Vugt-Van der Toorn et al. 2001). Normally, a protein with a 6-10 histidine tag should be bound to the matrix relative strongly and 100-250 mM imidazole in the buffer is required to elute the protein from the resin. In contrast, proteins with a low affinity to the matrix can already be eluted with 10-50 mM imidazole (the "impurities" of E. coli). Therefore, a linear imidazol gradient, for example, separates the protein of interest and impurities (Hochuli, Dobeli et al. 1987; Gaberc-Porekar and Menart 2001). Although the polyhistidine-tag is the most common and mostly an efficient variant, there are a few applications where the his-tag can cause problems. Metalloproteins can interact either directly with the his-tag or with the ions immobilized on the matrix. In comparison to some other affinity-tags, the specificity of the his-tag is not that high and in some cases this results in the co-purification of other proteins (Waugh 2005).

3.1.2 Strep-tag

In comparison to the his-tag, which binds to immobilized metal ions, the strep-tag II constists of a small octapeptide (WSHPQFEK), which binds to the protein streptavidin (Schmidt, Koepke et al. 1996). The commercial available matrix is a streptavidin variant and is called Strep-Tactin. This variant is able to bind the Strep-tag II octapeptide under mild buffer conditions and can be gently eluted with biotin derivates such as desthiobiotin (Schmidt, Koepke et al. 1996; Voss and Skerra 1997). Especially for metal-ion containing enzymes it is a promising alternative to the his-tag (Groß, Pisa et al. 2002). However, as chemicals are more expensive and the matrix has a lower binding capacity, compared to NTA resins, it is often not the first option choosen. Moreover, it cannot be used under denaturing conditions since Strep-Tactin denatures and will not bind the tag anymore (Terpe 2003; Waugh 2005). Examples of proteins crystallized after a Strep-tag purification are OpuBC (PDB:3R6U)(Pittelkow, Tschapek et al. 2011) and *Af*ProX (PDB:3MAM)(Tschapek, Pittelkow et al. 2011) as well as the sodium dependent glycine betain transporter BetP from *Corynebacterium glutamicum* (PDB:2WIT)(Ressl, Terwisscha van Scheltinga et al. 2009).

3.1.3 CBP-tag

Another peptide tag, is the calmodulin binding peptide, first described in 1992 (Stofko-Hahn, Carr et al. 1992). This peptide is prolonged compared to the Strep-tag II, consisting of 26 amino acids and binds with nanomolar affinity to calmodulin in the presence of Ca^{2+} (Blumenthal, Takio et al. 1985). It is derived from the C-terminus of the skeletal-muscle myosin light-chain kinase, which makes the system an excellent choice for proteins expressed using a prokaryotic expression system, since in prokaryotic systems nearly no protein interacts with calmodulin. This allows extensive washing to remove impurities and elution with EGTA, which complexes specifically Ca^{2+}, and a protein recovery around 90 % can be achieved (Terpe 2003). A drawback of this tag however is that the CBP tag can only be fused to the C-terminus of the protein since it has been shown that CBP on the N-terminus negatively influences the translation and thereby the expression rate (Zheng, Simcox et al. 1997).

3.1.4 GST-tag

With respect to the length of the tags, the his-tag contains only a few amino acids, the Strep-tag II and the CBP-tag already contain 8 – 26 amino acids, but it is possible to fuse whole proteins with 26 – 40 kDa to a recombinant protein. Here, the high affinity binding of the protein to their substrate is used to purify the protein of interest (Smith and Johnson 1988). In the case of the glutathione S-transferase (GST, 26 kDa) the protein specifically binds to immobilized glutathione. To elute the fusion protein from the resin, non-denaturating buffer conditions employing reduced glutathione are used (Terpe 2003). The tag can help to protect the recombinant protein from degradation by cellular proteases. It is recommended to cleave off the GST-tag after purification with a specific protease like thrombin or TEV (Tobacco Etch Virus) protease (Terpe 2003).

3.1.5 MBP-tag

Another affnitiy tag, which can be fused to the protein of interest, is the maltose binding protein (MBP) from *E. coli*. This protein has a molecular weight of 40 kDa and has the ability to bind to a cross-linked amylose matrix. The binding affinity is in the micro molar range and the tag can be used in a pH range from 7.0 – 8.0, however, denaturating buffer conditions are not possible (di Guan, Li et al. 1988). The elution of the recombinant protein is recommended with 10 mM maltose. A great opportunity of the MBP-tag is the increasing solubility effect of the recombinant protein in prokaryotic expression systems and even more pronounced in eukaryotic systems (Sachdev and Chirgwin 1999). Like the CBP-tag, a fusion at the N-terminal side might influence translation and expression rates (Sachdev and Chirgwin 1999).

3.1.6 Tag position and double tags

As described above, the position of the tag either at the N- or C-terminus has a considerable influence on translation and expression rate as well as on the biological function (Arnau, Lauritzen et al. 2006). If information regarding activity of the protein is already available especially about the location of interaction sites, this should be included in the protein design, meaning tag position etc. In general, the tag should be placed at the position of the protein, which is less important for interactions and/or expression. To minimize the influence of the tag on folding and/or activity in some cases it helps to create a linker region of a few amino acids between the tag and the protein (Gingras, Aebersold et al. 2005). A very efficient and sophisticated solution is, the addition of amino acids between tag and protein of interest, which functions not only as an accessibility increasing factor, but, also encodes for a recognition site for proteases like thrombin or TEV. Due to this arrangement the tag – protein interaction is minimized and the tag can be cleaved off if necessary (Arnau, Lauritzen et al. 2006). In some special cases a combination of two affinity tags results in enhanced solubility and more efficient purification. To enhance the purity of a protein, often a construct of two different short affinity tags like his-tag and Strep-tag or CBP-tag can be engineered (Rubio, Shen et al. 2005). Also a combination of two his-tag or two strep-tag kept apart by a linker region enhances the binding affinity extremely. This allows more stringent washing steps prior to elution of the protein (Fischer, Leech et al. 2011).

3.2 Size exclusion chromatography and ion exchange chromatography

Despite the usage of affinity tags a second purification step is sometimes required (Figure 1 – (II) purification). Which kind of purification procedure is required depends on the nature of impurities. If these impurities differ in molar mass compared to the protein of interest, a method based on size separation can be applied. Size exclusion chromatography (SEC) also separates different oligomeric species of the protein from each other, which otherwise would strongly inhibit crystallization and also allows analysis of stability and monodispersity of the protein (Regnier 1983a; Regnier 1983b).

However, in many cases, SEC is not sufficient to remove all impurities. Then separation by overall charge of the protein might be an option. Depending on the isoelectric point of the protein either anion or cation exchange chromatography can be performed. The protein binds to a matrix under very low ionic strength and is eluted afterwards either by increasing the ionic strength or by pH variation. Similar results can be achieved by hydrophobic interaction chromatography. Here, proteins with different surface properties show differences in their binding strength and binding of the protein is done inversely as during ion exchange chromatography. High ionic strength favors protein binding to a hydrophobic matrix and elution takes place when reducing the ionic strength. Although there are many other possibilities to increase the purity of a protein, the above mentioned techniques are without any doubt the most widely used and general applicable methods.

3.3 How to get a homogenous protein solution

In some cases isolated proteins are stable and homogenous at high concentrations after the purification and can be directly used for crystallization experiments. Often, however, the protein does not behave ideal and precipitates at high concentrations or forms aggregates or inhomogenous, oligomeric species; all of them prohibit crystal growth. SEC is a very elegant method to visualize the stability and oligomeric state of a protein. If the stability or the homogeneity of a protein sample is critical, you need to adapt your purification protocol and search for an optimized procedure. Different approaches are summarized below, for example a buffer screen to enhance protein solubility, multi-angle light scattering experiments to determine the absolute mass and the oligomeric state of the protein sample or fluorescence-based experiments to investigate the stability of the protein of interest.

3.3.1 Purified proteins – An *in vitro* system

After a protein is expressed in a soluble form, the subsequent purification procedure changes the environment of the protein dramatically. The cytoplasm of the cells, where the overexpression takes place, is packed with macromolecules. In *E. coli*, for example, the concentrations of proteins, RNAs and DNAs are about 320 mg/mL, 120 mg/mL, and 18 mg/mL, respectively (Cayley, Lewis et al. 1991; Zimmerman and Trach 1991; Elowitz, Surette et al. 1999) resulting in an overall concentration of macromolecules of above 450 mg/mL. During cell lysis and the first purification step, likely an IMAC (see above), the protein is separated from almost all other cell components. This rigorous procedure is accompanied with a severe change of the environment into an *in vitro* system. As a result proteins often tend to aggregate, precipitate or form inhomogeneous oligomeric states that prevent the formation of crystals in further experiments. Therefore one of the biggest challenges in structural studies is the preparation of protein solutions with high concentrations (as a rule of thumb 10-20 mg/mL) in a homogenous state. To fulfill these requirements, the *in vitro* system needs to be optimized with respect to different parameters as highlighted in *Figure 1 – (II) purification*. If a sufficient protein sample cannot be obtained, different strategies are available to increase the important characteristics of the protein: purity and homogeneity. As mentioned above, the usage of different metal ions during IMAC, ion exchange, a second affinity chromatography etc. can be sufficient to enhance purity. This might also lead to an increased stability. However, if the stability and/or homogeneity of a protein is still a problem, screening for a new buffer composition is essential to succeed during crystallization trials.

3.3.2 Buffer composition

Many examples illustrate the importance of an adequate buffer composition for protein stability, homogeneity, conformation, and activity (Urh, York et al. 1995; Holm and Hansen 2001; Jancarik, Pufan et al. 2004; Collins, Stevens et al. 2005). Some buffers are very frequently used and recommended by manufactures (see for example Qiagen, Roche, New England BioLabs, Fermentas, etc.). All of them contain a buffer reagent that keeps the pH constant in a well-defined range. Well-known examples are phosphate, tris (hydroxymethyl) aminomethane (Tris), or HEPES (4-(2-hydroxyethyl)-1-piperazineethanesulfonic acid) that buffer at the physiological relevant pH range of 6- 9 (Durst and Staples 1972; Chagnon and Corbeil 1973; Tornquist, Paallysaho et al. 1995). In recent years, the development of other buffer systems has been quite successful (Taha 2005) (for a list of buffers and corresponding pH ranges, see for example: http://delloyd.50megs.com/moreinfo/buffers2.html). Next to the well-defined pH, the stability and homogeneity of proteins depend on many other parameters, for example ionic strength, the presence of ligands and/or co-factors, divalent ions, glycerol, etc. The appropriate buffer composition cannot be predicted so far and needs to be identified by trial-and-error approaches.

3.4 Protein purification – How to overcome problems
In this part we would like to present some pitfalls that might occur during protein purification and provide some ‚rationales‘ to overcome these problems. As usual, the crucial step of solving a problem is its identification. Here, we are trying to sensitize the reader to indications, which might point towards problems related to instability and/or inhomogeneity of the protein sample. Moreover, such problems cannot always be recognized without the adequate technique(s). Therefore, we are introducing techniques that are capable to visualize the state of proteins.

3.4.1 Visible protein precipitations during IMAC
A very obvious stability problem is the formation of precipitations in the elution fractions of a chromatography step (see Figure 2). In this example, the his-tagged protein was eluted with a linear imidazole gradient from 10 to 500 mM imidazole and eluted at about 250 mM imidazole. Protein precipitation occured immediately after elution (Figure 2A and B) and continued (Figure 2C) resulting in a low amount of soluble protein. This aggregation can be reduced by dilution with a IMAC buffer (typically lacking imidazole) immediatly after the elution. Thereby, dilution hinders the concentration-dependent aggregation. In many cases, this rational is not sufficient to prevent precipitation. After applying, for example, a *buffer screen* (see Figure 1 – (II) purification) the new defined buffer is used for the chromatography or the eluting protein is diluted into the new buffer (see Figure 2D).
Other elution strategies of his-tagged proteins from an IMAC column are available. As described before, competing the poly-histidine from the IMAC column by imidazole is the most common elution strategy, however, for some proteins other strategies are superior, for example, replacing imidazole by histidine. Imidazole is only a mimic for histidine. If one uses histidine instead of imidazole aggregation can be avoided as concentration of the eluent can be reduced by a factor of ten. An example for a protein sensitive to imidazole concentration is shown in Figure 3B. Here a comparative SEC chromatogram is shown. After elution from the IMAC column with imidazole only a very small amount of the protein elutes at the volume corresponding to the size of a monomer or the dimer, respectively (Figure 3B, continuous line). Most of the protein passes the column very fast and elutes at the void volume indicating large radii meaning aggregated protein. Yields of dimeric (at about 150 mL) and monomeric (at about 180 mL) proteins are strongly increased after an elution with histidine (dashed line) compared to an elution with imidazole (continous line) and only the monomeric species could be crystallized (data not shown). The choice of the eluent in IMAC might therefore be an important step in a purification protocol. Another elution strategy of his-tagged proteins is a pH change from 8 to 4. In an acidic environment, histidines become positively charged and are therefore released from the column matrix. This strategy results in a sharp elution from the matrix and the protein is eluted highly concentrated. Although this strategy is recommended by the manufacturers (see GE Healthcare, Qiagen, etc.) the desired protein needs to retain activity at acidic pHs. The

bivalent metal ions (Ni^{2+}, Co^{2+}, Zn^{2+},..., see above) that complex the his-tag can be removed from the matrix by chelating reagents as ethylenediaminetetraacetic acid (EDTA) as another elution strategy (Muller, Arndt et al. 1998)

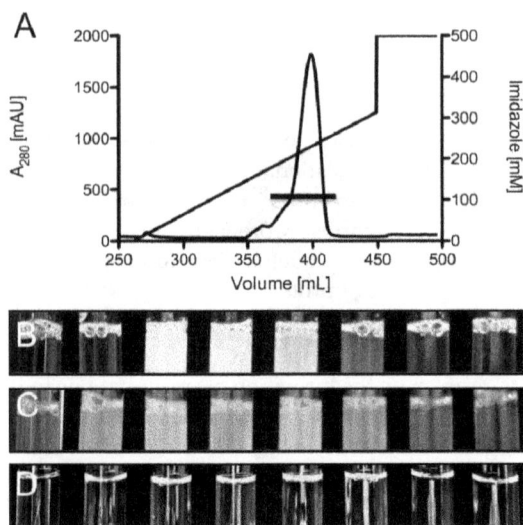

Fig. 2. Elution fractions of an IMAC. The protein was eluted via a linear imidazole gradient from 10 to 500 mM and the absorption at 280 nm was recorded. The elution fractions were collected and photographed. A: IMAC chromatogram of the his-tagged protein. Elution fractions containing the desired protein (indicated by a bar) are collected and shown in B – D. B and C: Elution fractions of the protein in 50 mM Tris-HCl, 150 mM NaCl, pH 8.0 immediately and 10 min after the elution, respectively. D: The elution fractions were immediately mixed in a 1 to 1 ratio with a buffer that enhances the protein stability (50 mM citrate, 50 mM LiCl, pH 6.00) evaluated during a solubility screen.

3.4.2 Invisible aggregations

Sometimes aggregation of proteins in solution can not be detected directly by eye. This inhomogeneity of protein samples can be visualized SEC, a method that separates proteins by their hydrodynamical radius (see above). Protein aggregates are eluting at the void volume, since they are clumbed together resulting in a big hydrodynamical radius (see Figure 3A and B). If invisible aggregation is detected the buffer composition needs to be adjusted. In one case we applied this technique to visualize the state of a protein after an IMAC, and the resulting elution profile is shown in Figure 3A (continous line). Comparable to the imidazole-induced precipitation described above, the protein aggregated and elutes within the void volume of the column (about 40 mL). Moreover, several other protein species elute from 55 to 80 mL indicating a highly inhomogeneous protein sample. The running buffer of the SEC was 50 mM Tris-HCl, pH 8.0 and 150 mM NaCl. Remarkably, a simple change to a new buffer (20 mM HEPES, 150 mM NaCl, pH 7.0) Resulted in a stable and homogenous protein sample (Figure 3A, dotted line), which was suitable for crystallization trials. Next to the rigorous change in the homogeneity of the protein, the biological activity of the protein could only be determined in the new buffer system. The influence of the buffer composition for the protein activity is a well-known phenomenon (Urh, York et al. 1995; Holm and Hansen 2001; Zaitseva, Jenewein et al. 2005) and in many cases the activity goes hand in hand with an optimal buffer for the purification. Mentionable, the new buffer was not found by trial-and-error approaches. We searched for literature dealing with homologous proteins, especially for established purification protocols. This literature search revealed the new buffer, illustrating that not every step towards a protein structure determination must be a trial-and-error process.

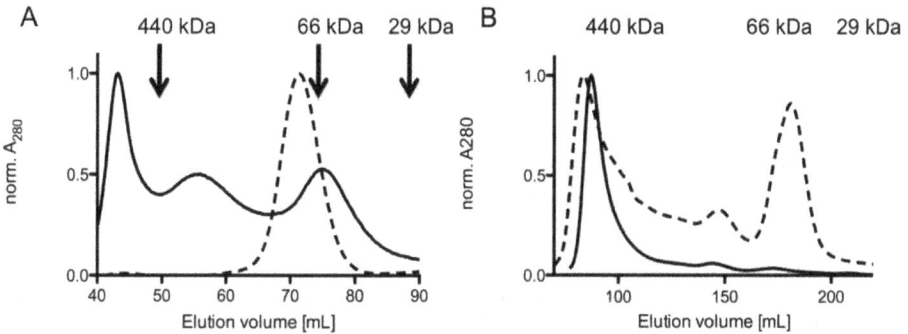

Fig. 3. Size exclusion chromatograms (UV 280nm) of proteins in different buffers. A: The homogeneity of a protein was analyzed in two different buffers; continuous line: 50 mM Tris-HCl, 150 mM NaCl, pH 8.00; dotted line: 20 mM Hepes, 150 mM NaCl, pH 7.00. B: The protein was eluted of the IMAC column either with imidazole (continous line) or with histidine (dotted line), concentrated and applied to the SEC.

Another example for the influence of the buffer composition was published bei Mavaro et al. (Mavaro, Abts et al. 2011). Instead of the buffer agent, the ionic strength of the buffer was the crucial determinant. Purification of the protein in low-salt buffer resulted in an inhomogenous protein sample containing a mixture of aggregates, dimers and monomers without biological activity. However, a simple change to high-salt buffer allowed the purification of homogenous dimeric protein, that was able to bind its substrate.

3.4.3 Overcoming protein instability
In the previous sections different strategies were mentioned to enhance the stability and the homogeneity of purified proteins and in all cases the buffer composition was the solution. Still, the essential question how to determine the optimal buffer to make a protein feel happy *in solution* is not answered? Some rationales and experiences are listed above: different elution strategies for IMAC purifications, the usage of frequently used buffer agents and a literature research for established purification protocols of related proteins. However, in many cases these approaches do not solve the problems occuring during the purification. But, is there a general methodology to overcome the problems? Unfortunately, the answer is as frustrating as challenging - there is not a general panacea around for the right buffer composition of a protein. If a new buffer needs to be found, trial-and-error

approaches have to be applied. A lot of different parameters are influencing the state of a protein, i. e. the buffer agent, the salt concentration, presence of metal ions with different valences, the hydrophobicity, and even the temperature of the buffer. The analysis of the protein in different buffers can be done by SEC and/or light scattering experiments. However, screening of all the different variables is very labor- and cost-intensive, and time-consuming, moreover only combinations of two or more additives might be sufficient to enhance the solubility and homogeneity of the protein. Therefore high-throughput methods are needed that handle a lot of different conditions simultaneously using as few as possible protein sample.

3.4.4 Buffer screen – Enhancing the solubility

Many publications are available suggesting methods for a solubility screening to allow the crystallization of initially inhomogeneous, aggregating protein samples (Jancarik, Pufan et al. 2004; Zaitseva, Holland et al. 2004; Collins, Stevens et al. 2005; Sala and de Marco 2010; Schwarz, Tschapek et al. 2011). In all of these methods aggregating protein samples are mixed with commercially available crystallization screens incubated for a period of time, and analyzed for precipitation visually using a light microscope. Screening conditions resulting in no precipitations are analyzed upon their composition, and protein samples are further examined with respect to their solubility and homogeneity under these conditions by SEC or light scattering experiments. This technique allows high-throughput screening in a 96-well format, where an automated pipetting system mixes only 50-200 nL of protein solution with 50-200 nL of buffer solutions to minimize the needed protein sample and increase the screening efficiency. Several buffer screens are commercially available that cover many different buffer agents, salt concentrations and other buffer parameters (i. e. from Hampton Research, Molecular Dimensions, Sigma, Jena Bioscience, Qiagen). After a solubility screening was applied, we were able to stabilize a previously unstable protein at concentrations above 3 mg/mL (see above "Protein precipitations during IMAC" and Figure 2D) at concentrations of up to 100 mg/mL for weeks (Schwarz, Tschapek et al. 2011). Typically, the new buffer (50 mM citrate, 50 mM $LiCl_2$, pH 6,00) should be used during the entire purification procedure starting with cell lysis. In the described case, the new buffer contains citrate, which is incompatible with an IMAC purification. Therefore the protein was immediately mixed with the new buffer after the elution of the IMAC column.

3.4.5 Size-exclusion chromatography versus light scattering experiments

Size-exclusion chromatography (SEC) and light scattering experiments (LS) are very helpful tools to analyze the homogeneity (Collins, Stevens et al. 2005) and the molecular mass of proteins; however both of them have advantages and disadvantages compared to each other. In SEC experiments proteins are separated based on their hydrodynamic radius by partitioning between a mobile phase and a stationary liquid within the pores of a matrix. All SEC columns are characterized by the volumes V_0, the liquid volume in the interstitial space between particles, V_i, the volume contained in the matrix pores and V_T, the total diffusion volume ($V_0 + V_i$) (Regnier 1983a; Regnier 1983b). In dependency of the hydrodynamic radius molecules are eluting at specific retention volumes in between V_0 and V_T with big molecules eluting first. After a calibration of a SEC column with proteins of known molecular weight (i. e. Sigma-Aldrich, "Kit for Molecular Weights") the molecular mass of the protein of interest can be roughly estimated; the elution volume is correlated to the $\log_{10}$ of the molecular weight (therefore, the hydrodynamic radius is considered to be proportional to the molecular weight). However, many extraneous mechanisms such as adsorptive, hydrophobic and ionic effects are further limiting the correlation between the retention volume and the molecular mass giving sometimes rise to wrong estimations.

Light scattering (LS) experiments can be applied to overcome these disadvantages and investigate the exact molecular weight of the protein sample. The rayleigh scattering of particles of monochromatic light depends directly on the molar mass of the particle. If you know the exact number of particles you can calculate the average molar mass of these particles. This technique is very powerful when used online after separation of the protein depending on their hydrodynamic radius, meaning SEC. This technique is always superior to normal SEC but requires special equipment and especially more time. However, if the protein fold is not really globular or other effects occur (see above: ionic, hydrophobic, etc.) assumption on size and oligomeric state based on SEC is not possible at all. For protein crystallization information about monodispersity, which can be provided by such an experiment, is an additional benefit.

3.4.6 Analysis of the homogeneity – High-throughput methods

Despite the development of various sophisticated methods, a bottleneck of homogeneity screening is high-throughput analysis. As mentioned above, proteins need to be analyzed by SEC and/or LS experiments after visual read-out of the protein-buffer droplets. Therefore, fluorescence-based solubility screens were developed that allow the high-throughput analyzes of many samples in a 96-well format (Ericsson, Hallberg et al. 2006; Alexandrov, Mileni et al. 2008; Kean, Cleverley et al. 2008). All these assays use fluorophores as reporters of the protein state. A suitable fluorophore is, for example, Sypro Orange, which exhibits different fluorescence properties as a function of its environment. This dye is almost dark in hydrophilic environment, however, after binding to hydrophobic molecules, it emits light at 570 nm. In inhomogenous and unfolded protein samples hydrophobic amino acids are exposed to the surface of proteins (Murphy, Privalov et al. 1990). An increase in the fluorescence signal of Sypro Orange correlates therefore with unfolding events of proteins. The homogeneity screening can be performed in basically two ways: temperature- or time-dependent. For the first setup the protein sample is heated gradually in distinct steps (i. e. 1 °C) and the emission is monitored at 570 nm. Hereby, a "melting" temperature is determined, which is characterized by 50% fluorescence of the maximal fluorescence at the highest temperature; the higher the melting temperature, the higher the stability of the protein (Ericsson, Hallberg et al. 2006). Secondly, the protein sample is incubated at a specific temperature (i. e. 40°C) and the fluorescence is measured for a period of time. The "half-life" time, at which 50 % fluorescence of the maximum fluorescence in one sample is detected, can be compared to all buffer conditions. In Figure 4 an example of the time-dependent approach is shown. Here, the protein is incubated in different buffers with various salt concentrations. The emission of Sypro Orange is recorded each minute at 570 nm. An analysis of all time-dependent fluorescence plots indicates that the protein is most stable in buffers containing 125 mM NaCl but unfolds fast in 1 M ammonium sulfate. These assays result in qualitative indications about a favourable environment of proteins that enhance the stability. Ericsson et al. proved the concept of this method by applying it to different proteins (Ericsson, Hallberg et al. 2006). The stability optimization yielded a twofold increase in initial crystallization leads. Moreover these assays enable the search for putative ligands of the protein. Upon binding of a substrate in the binding pocket or an inhibitor, the stability of the protein increases, which can be detected experimentally.

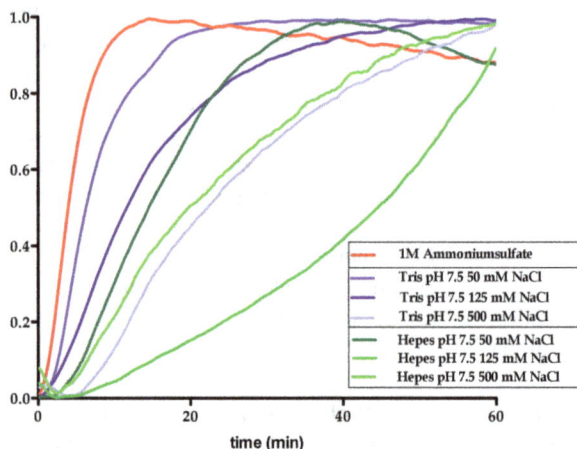

Fig. 4. Time-dependent stability optimization screen using Sypro Orange as reporter. The protein is diluted 1:50 into each test buffer containing Sypro Orange, excited with 490 nm and the fluorescence at 570 nm is measured for 60 minutes automatically with a PLATE READER (Fluorostar, BMG Labtech). Normalized fluorescence is plotted against the time.

4. Protein crystallization: Introduction

Protein crystals suitable for X-ray diffraction experiments and usable for subsequent structure determination are normally relatively large with a size of at least 10 to 100 μm. In contrast to crystals of mineral compounds, protein crystals are rather soft and sensitive to mechanical stress and temperature fluctuations. These properties are due to weak interactions between single proteins within the crystal, their high flexibility as well as the size of the macromolecules. The periodic network of building blocks is held together by dipole-dipole interactions, hydrogen bonds, salt bridges, van der Waals contacts or hydrophobic interactions. All of them have binding energies in the low kcal/mol range.

Especially the limited number of crystal contacts and their directionality are the largest difference to the high interactions generally observed in salt crystals. An example of the interactions within a protein crystal is shown in Figure 5. This picture highlights the main pitfalls in protein crystallization. A protein is a highly irregular shaped and flexible macromolecule which allows weak and stinted interactions at very specific locations of its surface. All vacuity is filled with buffer, in general not contributing to any kind of interactions between the protein molecules. Figure 5A shows a protein of around 30 kDa, which crystallizes in a rather small unit cell (shown in black). Only one protein monomer is located in the asymmetric unit of the unit cell, the other shown monomers represent symmetry related proteins. Figure 5B highlights the three-dimensional packing of protein molecules within a crystal.

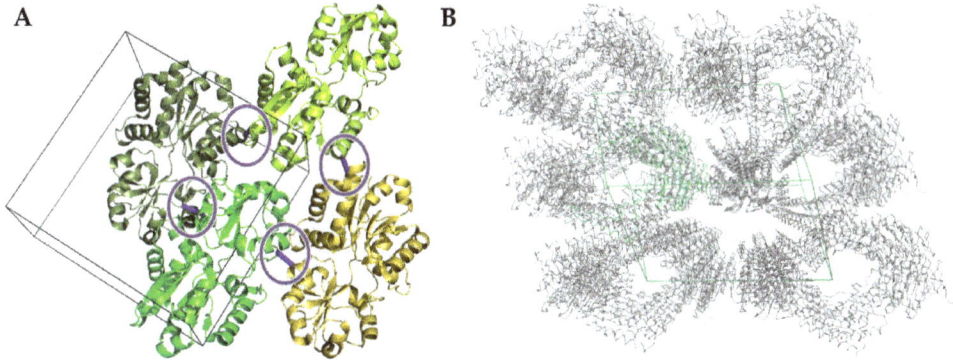

Fig. 5. Example of the packing within a crystal. A: The unit cell is shown in black, crystal contacts are highlighted with purple circles and lines. B: Three-dimensional crystal packing of a different protein. The unit cell as well as one protein monomer are depicted in green.

The flexibility as well as the other mentioned characteristics of proteins are responsible for the problems occuring during crystallization trials and despite extensive efforts not every protein is suitable for crystallization. If one cannot generate crystals one has to move back several steps and change the properties of the protein, e. g. surface properties by mutation of single amino acids, truncation of the protein or sometimes only changing buffer compositions that result in a more suitable protein for crystallization (see Figure 1 and also below). There are several prediction servers available that help choosing the 'right' protein and modification (Linding, Jensen et al. 2003; Goldschmidt, Cooper et al. 2007). However, protein crystallization still remains an empirical approach, sometimes called voodoo, while crystallography is science.

4.1 Phase diagram

The conditions or protocols for obtaining good crystals are still poorly understood and despite all progress and efforts protein crystallization is a trail-and-error approach. However, a step towards a better understanding of crystal growth can be achieved by analyzing the phase diagram of a protein-water mixture. The phase diagram is a simple illustration to help understanding how protein crystals are formed. Mostly, it is shown as a function of two ambient conditions that can be manipulated, i. e. the temperature and the concentration. Three-dimensional diagrams (two dependent parameters) have also been reported (Sauter, Lorber et al. 1999) and even a few more complex ones have been determined as well (Ewing, Forsythe et al. 1994). Figure 6 shows a schematic phase diagram for a protein solution as a function of protein concentration and precipitant concentration. The phase diagram is broken down into four distinct zones (Rosenbaum and Zukoski 1996; Haas and Drenth 1999; Asherie 2004):

1. Undersaturated zone: Under these condition the protein will stay in solution as neither the concentration of the protein nor of the precipitant is high enough to reach supersaturation.
2. Precipitation zone: Is the protein concentration or the precipitant concentration too high, the protein precipitates out of solution; this kind of solid material is not useful for crystallographic studies.

3. Labil zone: This is the most important configuration of the two parameters, as nucleation and initial crystal growth take place under these conditions.
4. Metastable zone: After initial crystals are formed and start growing in the labil zone, protein concentration decreases in the drop and the metastable zone will be reached. Here the crystal can grow further to its final maximum size.

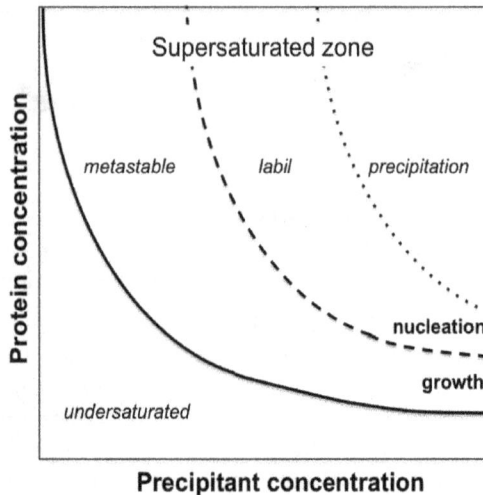

Precipitant concentration

Fig. 6. A basic solubility phase diagram for a given temperature (adapted from (Rupp 2007).

The curve separating the undersaturated zone from the supersaturated one is called *solubility curve*. If conditions are chosen below the solubility curve, the protein will stay in solution and never crystallize. This means when a protein crystal is placed in a solvent, which is free of protein, it will start to dissolve. If the volume of the droplet is small enough it will not dissolve completely: it will stop dissolving when the concentration of the protein in the droplet reaches a certain level. At this concentration the crystal loses protein molecules at the same rate at which protein associate to the crystal – the system is at equilibrium. Determination of the solubility of the protein of interest might be a helpful information at the beginning of crystallization experiments. This can be done in a two-dimensional screen varying for example ammonium sulfate concentrations as well as the protein concentration.

4.2 Crystallization techniques
Crystallization is a phase transition phenomenon. Protein crystals grow from a supersaturated aqueous protein solution. Varying the concentration of precipitant, protein and additives, pH, temperature and other parameters induce the supersaturation. However, as mentioned before, prediction of this kind of phase diagrams is *a priori* impossible.
Protein crystallization can be divided into two main steps:
1. Generating initial crystals: 'Searching the needle in a haystack'
2. Empirical optimization of these crystallization condition
The first step is mostly based on experiences from other crystallization trials with different proteins. Nowadays several supplier offer crystallization screens that contain solutions for

initial experiments that were used successfully in the past for crystallization trials (Jancarik and Kim 1991), so-called "sparse matrix screens". There are also some trials around to use more systematic approaches (Brzozowski and Walton 2000) to get more information about solubility prior and simultaneous to crystallization (incomplete factorials, solubility assays). Both kinds of screens can be applied to different crystallization techniques.

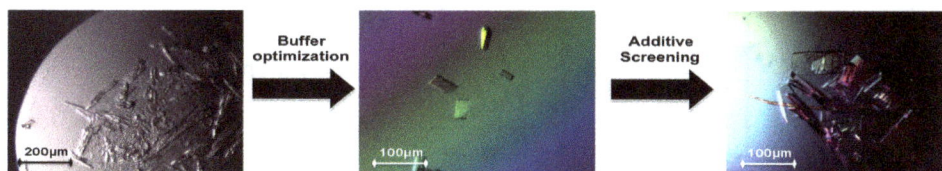

Fig. 7. Crystal optimization. First steps in crystal optimization are shown. Initial protein crystals look weak and fragile, after screening around this initial buffer composition crystal evaluation by eye results in less fragile, homogeneous looking crystals. However, diffraction quality was poor. Therefore an additive screening was performed that resulted in a different crystal form. These crystals finally were able to diffract X-rays to a reasonable resolution.

A lead/hit in that initial step might not be a 'real' crystal rather than a crystalline precipitate or just phase separation. In the next step, fine-tuning the buffer composition further optimizes this hit. Varying pH, salt concentration, type and concentration of precipitant and protein concentration are expected to yield larger and hopefully also better-diffracting crystals. In this step, the chemicals used are much more defined and therefore it is a more systematic than empirical screening (see Figure 7).

4.2.1 Vapor diffusion
The most popular and simplest technique to obtain protein crystals is the vapor diffusion method either in the sitting or hanging drop variant (see Figure 8). For both a defined volume (mostly < 1µl) of protein solution is mixed with an equivalent volume of screening solution and then equilibrated against the original precipitant/screening concentration. During this equilibration, the vapor pressure of the solution rises as the protein crystallizes (protein in solution lowers water activity) while the water evaporates to maintain equilibrium, which causes the precipitant concentration to rise. Therefore, if the crystal growth is sensitive to the precipitant concentration, vapor diffusion can rapidly force the mixture to unstable conditions where growth and nucleation are too rapid. This is the main disadvantage of vapor diffusion: Growing large crystals might be problematic!

4.2.2 Micro batch method
In this set-up the protein solution is mixed with screening solution at concentrations required for supersaturation right at the beginning of the experiment. Typical drop sizes of micro batch experiments ranges from 1-2 µl. The drop is then covered with oil, which acts as an inert sealing to protect the drops during incubation from evaporation (see Figure 8).

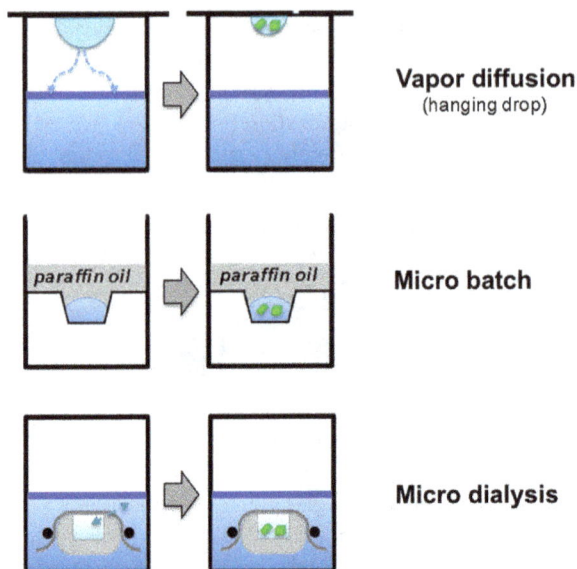

Fig. 8. Protein crystallization techniques. Schematic representation of a) vapor diffusion, b) micro batch and c) micro dialysis crystallization techniques widely used for crystal growth (adapted from (Drenth 2006)).

4.2.3 Micro dialysis
Dialysis is another way to change the buffer composition and increase its concentration in the crystallization experiment gradually (see Figure 8). Micro-dialysis buttons are exposed to different screening buffers. This method requires rather high amounts of protein but might yield large crystals.

After obtaining initial crystal hits in a commercial screen the tough part of crystal optimization starts. By varying pH, salt concentration, temperature, precipitant concentration or protein concentration these initial crystals should be reproduced and become larger, more regular shaped or are simply growing faster. A further improvement of crystal quality might be achieved by the addition of small amounts of so called 'additives'. At this point basically each chemical compound might be sufficient to improve the crystal quality. Luckily, there are some preferable working additives, which have been proven to produce better crystal in more than one case. Especially compounds that are known to reduce undirected interactions in proteins like organic solvents, i. e. DMSO or phenol, or detergents and reducing agents are very often used at this stage and helpful to force more homogeneous well diffracting crystals.

4.3 Crystal nucleation
There are two fundamental steps during protein crystallization: Nucleation and crystal growth. If one cannot obtain single crystals of adequate quality for analysis, this is generally a consequence of problems associated with the growth phase (see above). But failure to obtain any crystals at all or failure to obtain single, supportable nuclei reflects difficulties in the nucleation step. Therefore control of nucleation is a powerful tool to optimize protein crystals

or sometimes it is the only way to get crystals at all. Nucleation can take place either homogeneous meaning in the bulk of the solution, when the supersaturation is high enough for the free-energy barrier to nucleus formation to be overcome or heterogeneous mostly by solid material in the crystallization solution. This can also occur even when the supersaturation is not achieved. Therefore in order to control nucleation one has to work with highly clean solutions to avoid nucleation by the second mentioned possibility. The nucleation zone can be bypassed by insertion of crystals, crystal seeds or other nucleants to the protein/precipitant mixture. Addition of crystals or tiny fragments of crystals is called seeding. This method is then subdivided into macro- and micro-seeding dependent on the size of the nucleant added. In macro-seeding experiments one single, already well-formed but small crystal is placed into a new crystallization solution at lower saturation. Microseeding in contrast requires small fragments of a crystal or almost invisible microcrystalline precipitate. These 'seeds' are then transferred into a fresh crystallization solution either by a seeding wand which is dipped into the microseed mixture to pick up seeds and then touched across the surface of the new drop or by a animal whisker or hair that is stroked over the surface of the parent crystal to trap the nuclei and then is drawn through the new drop. As this method also enhances the speed of crystal growth it can be used with sensitive substrate that undergo decomposition over time. Oswald et al. proved this in 2008 by solving the structure of ChoX from *Sinorhizobium meliloti* in complex with a highly hydrolyzing substrate, acetylcholine (Oswald, Smits et al. 2008). In classical vapor diffusion experiments crystals appear after four weeks but data showed only little electron density in the ligand-binding site and turned out to result from a choline bound instead of acetylcholine. Hydroxylation was favored due to the relatively long time for crystal growth but also because of an acetic pH in the crystallization set-up. To circumvent these problems accelerated crystal growth was required. In this case micro-seeding results in crystals suitable for data collection in less than 24 hours.

Recent years more effort in nucleation control yielded in fancy materials that can be used as nuclei for crystals. These methods use the second way of nuclei formation, as a solid material is introduce into the crystallization solution as an 'universal' nucleant (Chayen, Saridakis et al. 2006). There have been several substances that have been tried more or less successful. Some have been useful for individual proteins, but mostly they were not applicable in general (McPherson and Shlichta 1988; Chayen, Radcliffe et al. 1993; Blow, Chayen et al. 1994). In 2001, Chayen et al. proposed the idea of using porous silicon whose pore size is comparable with the size of a protein molecule. In theory such pores may confine and concentrate the protein molecules at the surface of the silicone and thereby encourage them to form crystal nuclei (Chayen, Saridakis et al. 2001). These nucleants have made it to commercial availability (www.moleculardimensions.com) and have proven to be suitable for different kinds of proteins and even membrane proteins that have not been possible to crystallize before formed nice crystals in the presence of these nucleants.

4.4 Cryoprotection

Exposure of a protein crystal at room temperature results in dramatic radiation damage due to radicals formed by the ionizing X-ray photons. To reduce that harmful disintegration of the protein crystal the crystal is cooled to ≈100K with the help of liquid nitrogen (Low, Chen et al. 1966; Hope 1988; Rodgers 1994; Garman 1999). However, it is common for the cooling process to disrupt the crystal order and decrease diffraction quality. Thus, the crystal must be cooled fast so that the water in the solvent channels is in the vitreous rather than in the crystalline state at the end of this procedure. As for pure water this cooling has to take place very quick (10^{-5}s, (Johari, Hallbrucker et al. 1987), some water molecules can be replaced by a cryoprotective solution prior to cooling (Juers and Matthews 2004). This exceeds the time window for freezing up to 1-2s (Garman and Owen 2006) however, finding a good 'cryoprotectant' for a special protein crystal again involves substantial screening. Once flash frozen in liquid nitrogen, the crystal must be kept below the glass transition temperature of the cryobuffer at or below 155K at all times (Weik, Kryger et al. 2001).

4.5 What can you do when all efforts did not succeed in crystals?
4.5.1 Buffer composition – Again!
The choice of the right buffer used for crystallization experiments is very crucial. As shown above, every protein needs its own buffer composition to feel kind of happy in this aqueous artificial environment. Especially as high protein concentrations (>10mg/ml) are required for crystallization, one might has to test several buffer compositions again (see also Figure 1). As a rule of thumb you should obtain around 50% of clear drops immediately after mixing protein and buffer solution. If you detect drastically more precipitation in your drops you should think first about lower protein concentration but of course secondly about changing your buffer system again.

4.5.2 How to obtain a rigid protein suitable for crystallization?
To overcome the problem of flexibility of some regions in the protein addition of ligands is often a very powerful tool to fix the protein in a single conformation that is more favorable for crystallization. A good example for this strategy is the crystallization of so-called substrate binding proteins (for a recent review see (Berntsson, Smits et al. 2010)). These proteins catch their substrate in the periplasm of bacteria or on the outer membrane of archaea and then deliver it to their cognate transport system located in the membrane. The mechanism of substrate binding is quite well understood. These binding proteins all consist of two domains, which rotate towards each other during the binding event. In solution without substrate they are quite flexible and NMR-studies proved a equilibrium between open and closed conformation (Tang, Schwieters et al. 2007). Analysis of all available structures for this class of proteins showed that more than 95% were crystallized with a ligand bound (Berntsson, Smits et al. 2010). Thus, a stabilization of the two domains seems to simplify crystal contact formation dramatically. Although people always want to obtain a functional conformation of the protein in their crystal structure, it is sometimes helpful to think about how to stop the protein from doing its job. A non-functional protein is in general less flexible and fixed in one conformation. One example for successful implementation of this strategy is the crystal structure of NhaA from *Escherichia coli* solved in 2005 (Hunte, Screpanti et al. 2005). Here, Hunte et al., downregulated the protein activity by working at an acidic pH of 4. Although the protein shows almost no activity at this pH the structure reveals the basis for mechanism of Na+/H+ exchange and also its regulation by pH could be understood.

4.5.3 Rational protein design for crystallization: Surface engineering
The first example of rational protein design that yielded a good diffracting protein crystal is given by Lawson et al. in 1991 (Lawson, Artymiuk et al. 1991). They compared amino acids

involved in crystal contact formation of the rat ferritin protein L. (which is highly homologous to human ferritin H, the target protein) with the amino acids present at that position in human ferritin H. A replacement of Lys86, found in the human sequence, with Glu, which occurs in rat, recreated a Ca^{2+} binding bridge that mediates crystal contacts in the rat ortholog. As this method was successful for several other proteins (McElroy, Sissom et al. 1992; Braig, Otwinowski et al. 1994; Horwich 2000), a general protocol was required. The concept Derewenda et al. proposed in 2004 is based on the general equation for the free energy that drives protein crystallization:

$$\Delta G = \Delta H - T(\Delta S_{protein} + \Delta S_{solvent})$$

As the enthalpy values of intermolecular interactions in a crystal lattice are rather small (see above), crystallization is very sensitive to entropy changes of both protein and solvent. The formation of ordered protein aggregates carries a negative entropy term. This can only be overcome by positive entropy from the release of water bound to the protein. However, large hydrophilic residues (e.g. lysines, arginies, glutamates, glutamines) exposed on the protein surface need to be ordered. Since they are rather flexible this can cause problems. This can be overcome by mutating large amino acids into smaller ones, for example alanines. Among these large amino acids lysines and glutamates play a particular role, as they are always (with only very few exeptions) located on the protein surface (Baud and Karlin 1999). Both lysines and glutamates are typically disfavored at interfaces in protein protein complexes (Lo Conte, Chothia et al. 1999), therefore it is rather straight forward to assume that lysine and glutamate to alanine mutants are good targets for protein crystallization if wildtype protein hardly forms crystals. However this also means that you have to go several steps backwards on road to a protein structure determination (see Figure 1).

4.5.4 Affinity tag removal: Philosophic question???
Another variant in protein crystallization nowadays is the affinity tag used for purification of the desired protein. The decision about position and choice of the affinity tag are mostly made at the beginning of the long way to a crystal structure (see Figure 1). However, it becomes crucial again at the crystallization step. In general most people like to remove the tag before crystallization to prove a physiological conformation. But, there are examples where the tag played a pivotal role in crystallization (Smits, Mueller et al. 2008a). The crystal structure of the octopine dehydrogenase from *Pecten maximus* is shown in Figure 9 (Smits, Mueller et al. 2008b), with the interactions sides/crystal contacts highlighted in green.

In Figure 9A contacts look quite similar to that presented in Figure 5. However when having a closer look on the his-tag, you recognize that it is located in a cavity formed by another monomer of that protein. In that cavity it can perform several hydrogen bonds with amino acids from the other monomer resulting in a very strong interaction which yields good quality crystals.

4.5.5 Crystallization using antibody fragments
A number of ways to stabilize proteins for crystallography have been developed, for example genetic engineering, co-crystallization with natural ligands and reducing surface

entropy (see above). Recently, crystallization mediated by antibody fragments has moved into the focus of crystallographers especially to obtain crystals of membrane proteins (Ostermeier, Iwata et al. 1995; Hunte and Michel 2002). Membrane protein crystallization is even tougher compared to soluble proteins, because of the amphipathic surface of the molecules. As they are located in the lipid bilayer most of their surface is hydrophobic and must be covered to keep them in solution. This is maintained by detergents. The detergent micelles cover the hydrophobic surface and therefore this area is no longer available to form crystal contacts. Crystal contacts can only be formed by the polar surfaces of these proteins. As many membrane proteins contain only relatively small hydrophilic domains, a strategy to increase the probability of getting well-ordered crystals is required. Antibody fragments can play this role. They can be designed for binding at specific regions in the protein and then function as additional polar domain in the membrane protein complex (for example see (Ostermeier, Iwata et al. 1995; Huber, Steiner et al. 2007).

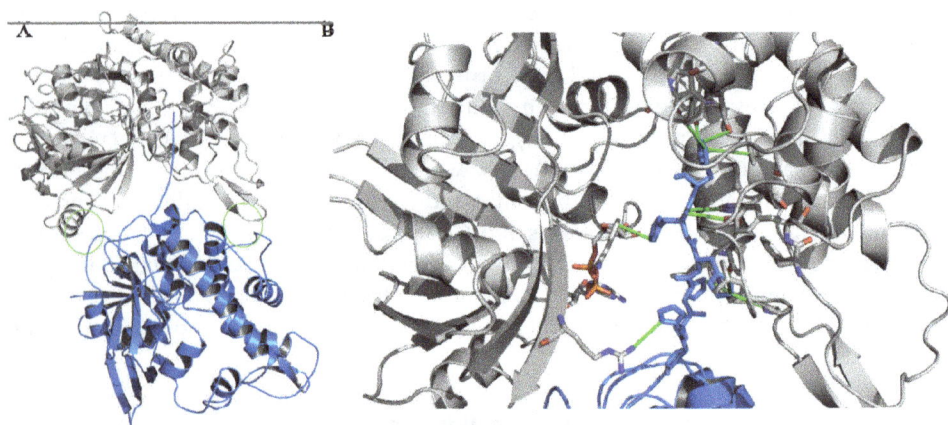

Fig. 9. Crystal contacts in OcDH protein. A: Overall view on two monomers. Surface Crystal contacts are highlighted in the green circles. B: Zoom in on the His-tag of one monomer. The his-tag of one monomer in the crystal structure is located near the binding site in a deep cavity formed by the other monomer. Therefore it is able to form several hydrogen bonds (highlighted in green) with side and main chains of the other protein but also with the ligand bound in this binding site (orange).

5. Conclusion

For what reason do we effort so much work on good quality crystals?
Single good quality crystals constitute an essential prerequisite for structural investigations of biological macromolecules using X-ray diffraction. The harder one works on crystal quality the easier the determination of a reasonable atomic model of the molecule of interest becomes. The vast majority of problems encountered in crystal structure determination can typically be traced back to data-quality issues caused by crystal imperfections. Consequently, although primary focus of structural biology is on the macromolecule that makes up a crystal, there is also considerable interest in the physical properties, nucleation

and growth of the crystals themselves. Statistics of various Structural Genomics Centers proved that protein crystallization is still despite all the progress in the technology of crystallization robotics is still a rather tough field in biological science. Success rate ranges for small prokaryotic proteins from 10-30% and decreases dramatically to a few percent for human proteins. The struggle obtaining crystals for protein structure determination is justified. After all efforst looking at electron density and subsequent the protein structure is still one of the most intriguing as well auspicious parts in structural biology

6. References

Alexandrov, A. I., M. Mileni, et al. (2008). "Microscale fluorescent thermal stability assay for membrane proteins." Structure 16(3): 351-359.

Altmann, F., E. Staudacher, et al. (1999). "Insect cells as hosts for the expression of recombinant glycoproteins." Glycoconj J 16(2): 109-123.

Arnau, J., C. Lauritzen, et al. (2006). "Current strategies for the use of affinity tags and tag removal for the purification of recombinant proteins." Protein expression and purification 48(1): 1-13.

Asherie, N. (2004). "Protein crystallization and phase diagrams." Methods 34(3): 266-272.

Baneyx, F. (1999). "Recombinant protein expression in Escherichia coli." Current opinion in biotechnology 10(5): 411-421.

Baud, F. and S. Karlin (1999). "Measures of residue density in protein structures." Proc Natl Acad Sci U S A 96(22): 12494-12499.

Berntsson, R. P., M. K. Doeven, et al. (2009). "The structural basis for peptide selection by the transport receptor OppA." EMBO J 28(9): 1332-1340.

Berntsson, R. P., S. H. Smits, et al. (2010). "A structural classification of substrate-binding proteins." FEBS Lett 584(12): 2606-2617.

Binda, C., M. Aldeco, et al. (2010). "Interactions of Monoamine Oxidases with the Antiepileptic Drug Zonisamide: Specificity of Inhibition and Structure of the Human Monoamine Oxidase B Complex." J Med Chem.

Blow, D. M., N. E. Chayen, et al. (1994). "Control of nucleation of protein crystals." Protein Sci 3(10): 1638-1643.

Blumenthal, D. K., K. Takio, et al. (1985). "Identification of the calmodulin-binding domain of skeletal muscle myosin light chain kinase." Proc Natl Acad Sci U S A 82(10): 3187-3191.

Bolotin, A., P. Wincker, et al. (2001). "The complete genome sequence of the lactic acid bacterium Lactococcus lactis ssp. lactis IL1403." Genome research 11(5): 731-753.

Braig, K., Z. Otwinowski, et al. (1994). "The crystal structure of the bacterial chaperonin GroEL at 2.8 A." Nature 371(6498): 578-586.

Brzozowski, A. M. and J. Walton (2000). "Clear strategy screens for macromolecular crystallization." J Appl Cryst 34: 97-101.

Cayley, S., B. A. Lewis, et al. (1991). "Characterization of the cytoplasm of Escherichia coli K-12 as a function of external osmolarity. Implications for protein-DNA interactions in vivo." J Mol Biol 222(2): 281-300.

Chagnon, A. and M. Corbeil (1973). "Use of an organic buffer (HEPES) in human lymphocytoid cell line cultures." In Vitro 8(4): 283-287.

Chayen, N. E., J. W. Radcliffe, et al. (1993). "Control of nucleation in the crystallization of lysozyme." Protein Sci 2(1): 113-118.

Chayen, N. E., E. Saridakis, et al. (2001). "Porous silicon: an effective nucleation-inducing material for protein crystallization." J Mol Biol 312(4): 591-595.

Chayen, N. E., E. Saridakis, et al. (2006). "Experiment and theory for heterogeneous nucleation of protein crystals in a porous medium." Proc Natl Acad Sci U S A 103(3): 597-601.

Collins, B., R. C. Stevens, et al. (2005). "Crystallization Optimum Solubility Screening: using crystallization results to identify the optimal buffer for protein crystal formation." Acta Crystallogr Sect F Struct Biol Cryst Commun 61(Pt 12): 1035-1038.

Cregg, J. M. (2007). Pichia Protocols, Humana Press.

Cregg, J. M., J. L. Cereghino, et al. (2000). "Recombinant protein expression in Pichia pastoris." Molecular biotechnology 16(1): 23-52.

D., O. R., M. L.K., et al. (1992). Baculovirus Expression Vectors A Laboratory Manual. New York.

de Ruyter, P. G., O. P. Kuipers, et al. (1996). "Controlled gene expression systems for Lactococcus lactis with the food-grade inducer nisin." Applied and environmental microbiology 62(10): 3662-3667.

de Vos, W. M. and E. E. Vaughan (1994). "Genetics of lactose utilization in lactic acid bacteria." FEMS microbiology reviews 15(2-3): 217-237.

di Guan, C., P. Li, et al. (1988). "Vectors that facilitate the expression and purification of foreign peptides in Escherichia coli by fusion to maltose-binding protein." Gene 67(1): 21-30.

Drenth, J. (2006). "Principles of Protein X-Ray Crystallography." Springer 3.

Durst, R. A. and B. R. Staples (1972). "Tris-tris-HCl: a standard buffer for use in the physiologic pH range." Clin Chem 18(3): 206-208.

Eifler, N., M. Duckely, et al. (2007). "Functional expression of mammalian receptors and membrane channels in different cells." Journal of structural biology 159(2): 179-193.

Elowitz, M. B., M. G. Surette, et al. (1999). "Protein mobility in the cytoplasm of Escherichia coli." J Bacteriol 181(1): 197-203.

Ericsson, U. B., B. M. Hallberg, et al. (2006). "Thermofluor-based high-throughput stability optimization of proteins for structural studies." Anal Biochem 357(2): 289-298.

Erkens, G. B., R. P. Berntsson, et al. (2011). "The structural basis of modularity in ECF-type ABC transporters." Nat Struct Mol Biol 18(7): 755-760.

Ewing, F., E. Forsythe, et al. (1994). "Orthorhombic lysozyme solubility." Acta Crystallogr D Biol Crystallogr 50(Pt 4): 424-428.

Fischer, M., A. P. Leech, et al. (2011). "Comparative Assessment of Different Histidine-Tags for Immobilization of Protein onto Surface Plasmon Resonance Sensorchips." Analytical chemistry.

Gaberc-Porekar, V. and V. Menart (2001). "Perspectives of immobilized-metal affinity chromatography." Journal of biochemical and biophysical methods 49(1-3): 335-360.

Garman, E. (1999). "Cool data: quantity AND quality." Acta Crystallogr D Biol Crystallogr 55(Pt 10): 1641-1653.

Garman, E. F. and R. L. Owen (2006). "Cryocooling and radiation damage in macromolecular crystallography." Acta Crystallogr D Biol Crystallogr 62(Pt 1): 32-47.

Gasson, M. J. (1983). "Plasmid complements of *Streptococcus lactis* NCDO 712 and other lactic streptococci after protoplast-induced curing." Journal of bacteriology 154(1): 1-9.

Gingras, A. C., R. Aebersold, et al. (2005). "Advances in protein complex analysis using mass spectrometry." The Journal of physiology 563(Pt 1): 11-21.

Goldschmidt, L., D. R. Cooper, et al. (2007). "Toward rational protein crystallization: A Web server for the design of crystallizable protein variants." Protein Sci 16(8): 1569-1576.

Groß, R., R. Pisa, et al. (2002). "Isolierung der trimeren Hydrgenase aus Wollinella succinogenes durch StrepTactin-Affinitätschromatographie." Biospektrum 1: 101.

Guillot, A., C. Gitton, et al. (2003). "Proteomic analysis of *Lactococcus lactis*, a lactic acid bacterium." Proteomics 3(3): 337-354.

Haas, C. and J. Drenth (1999). "Understanding protein crystallization on the basis of the phase diagram." Journal of Crystal Growth 196(2-4): 388-394.

Hampton, E. N., M. W. Knuth, et al. (2007). "The self-inhibited structure of full-length PCSK9 at 1.9 A reveals structural homology with resistin within the C-terminal domain." Proc Natl Acad Sci U S A 104(37): 14604-14609.

Hansen, S. B., G. Sulzenbacher, et al. (2006). "Structural characterization of agonist and antagonist-bound acetylcholine-binding protein from *Aplysia californica*." J Mol Neurosci 30(1-2): 101-102.

Hefti, M. H., C. J. Van Vugt-Van der Toorn, et al. (2001). "A novel purification method for histidine-tagged proteins containing a thrombin cleavage site." Analytical biochemistry 295(2): 180-185.

Hochuli, E., H. Dobeli, et al. (1987). "New metal chelate adsorbent selective for proteins and peptides containing neighbouring histidine residues." Journal of chromatography 411: 177-184.

Holm, J. and S. I. Hansen (2001). "Effect of hydrogen ion concentration and buffer composition on ligand binding characteristics and polymerization of cow's milk folate binding protein." Biosci Rep 21(6): 745-753.

Hope, H. (1988). "Cryocrystallography of biological macromolecules: a generally applicable method." Acta Crystallogr B 44 (Pt 1): 22-26.

Horwich, A. (2000). "Working with Paul Sigler." Nat Struct Biol 7(4): 269-270.

Huber, T., D. Steiner, et al. (2007). "In vitro selection and characterization of DARPins and Fab fragments for the co-crystallization of membrane proteins: The Na(+)-citrate symporter CitS as an example." J Struct Biol 159(2): 206-221.

Hunt, I. (2005). "From gene to protein: a review of new and enabling technologies for multi-parallel protein expression." Protein expression and purification 40(1): 1-22.

Hunte, C. and H. Michel (2002). "Crystallisation of membrane proteins mediated by antibody fragments." Curr Opin Struct Biol 12(4): 503-508.

Hunte, C., E. Screpanti, et al. (2005). "Structure of a Na+/H+ antiporter and insights into mechanism of action and regulation by pH." Nature 435(7046): 1197-1202.

Jana, S. and J. K. Deb (2005). "Strategies for efficient production of heterologous proteins in Escherichia coli." Applied microbiology and biotechnology 67(3): 289-298.

Jancarik, J. and S. H. Kim (1991). "Sparse matrix sampling: a screening method for crystallization of proteins." J Appl Cryst 24: 409-411.

Jancarik, J., R. Pufan, et al. (2004). "Optimum solubility (OS) screening: an efficient method to optimize buffer conditions for homogeneity and crystallization of proteins." Acta Crystallogr D Biol Crystallogr 60(Pt 9): 1670-1673.

Johari, G. P., A. Hallbrucker, et al. (1987). "The glass-liquid transition of hyperquenched water." Nature 330: 552-553.

Jonasson, P., S. Liljeqvist, et al. (2002). "Genetic design for facilitated production and recovery of recombinant proteins in Escherichia coli." Biotechnology and applied biochemistry 35(Pt 2): 91-105.

Juers, D. H. and B. W. Matthews (2004). "Cryo-cooling in macromolecular crystallography: advantages, disadvantages and optimization." Q Rev Biophys 37(2): 105-119.

Kean, J., R. M. Cleverley, et al. (2008). "Characterization of a CorA Mg2+ transport channel from Methanococcus jannaschii using a Thermofluor-based stability assay." Mol Membr Biol 25(8): 653-663.

Konings, W. N., J. Kok, et al. (2000). "Lactic acid bacteria: the bugs of the new millennium." Curr Opin Microbiol 3(3): 276-282.

Kopecky, D. J., X. Hao, et al. (2008). "Identification and optimization of N3,N6-diaryl-1H-pyrazolo[3,4-d]pyrimidine-3,6-diamines as a novel class of ACK1 inhibitors." Bioorg Med Chem Lett 18(24): 6352-6356.

Kuipers, O. P., M. M. Beerthuyzen, et al. (1995). "Autoregulation of nisin biosynthesis in Lactococcus lactis by signal transduction." The Journal of biological chemistry 270(45): 27299-27304.

Kuipers, O. P., P. G. de Ruyter, et al. (1998). "Quorum sensing-controlled gene expression in lactic acid bacteria." J Biotechnol 64: 15-21.

Kulahin, N., O. Kristensen, et al. (2011). "Structural model and trans-interaction of the entire ectodomain of the olfactory cell adhesion molecule." Structure 19(2): 203-211.

Kunji, E. R., D. J. Slotboom, et al. (2003). "Lactococcus lactis as host for overproduction of functional membrane proteins." Biochimica et biophysica acta 1610(1): 97-108.

Lawson, D. M., P. J. Artymiuk, et al. (1991). "Solving the structure of human H ferritin by genetically engineering intermolecular crystal contacts." Nature 349(6309): 541-544.

Li, M. Z. and S. J. Elledge (2007). "Harnessing homologous recombination in vitro to generate recombinant DNA via SLIC." Nat Methods 4(3): 251-256.

Linding, R., L. J. Jensen, et al. (2003). "Protein disorder prediction: implications for structural proteomics." Structure 11(11): 1453-1459.

Lo Conte, L., C. Chothia, et al. (1999). "The atomic structure of protein-protein recognition sites." J Mol Biol 285(5): 2177-2198.

Low, B. W., C. C. Chen, et al. (1966). "Studies of insulin crystals at low temperatures: effects on mosaic character and radiation sensitivity." Proc Natl Acad Sci U S A 56(6): 1746-1750.

Lundstrom, K. (2006a). "Structural genomics for membrane proteins." Cellular and molecular life sciences : CMLS 63(22): 2597-2607.

Lundstrom, K., R. Wagner, et al. (2006b). "Structural genomics on membrane proteins: comparison of more than 100 GPCRs in 3 expression systems." Journal of structural and functional genomics 7(2): 77-91.

Madoori, P. K., H. Agustiandari, et al. (2009). "Structure of the transcriptional regulator LmrR and its mechanism of multidrug recognition." EMBO J 28(2): 156-166.

Mavaro, A., A. Abts, et al. (2011). "Substrate recognition and specificity of NISB, the lantibiotic dehydratase involved in nisin biosynthesis." J Biol Chem.

McElroy, H. H., G. W. Sissom, et al. (1992). "Studies on engineering crystallizability by muttion of surface residues of human thymidylate synthase." J. Cryst. Growth 122: 265-272.

McPherson, A. and P. Shlichta (1988). "Heterogeneous and epitaxial nucleation of protein crystals on mineral surfaces." Science 239(4838): 385-387.

Menetrey, J., A. Bahloul, et al. (2005). "The structure of the myosin VI motor reveals the mechanism of directionality reversal." Nature 435(7043): 779-785.

Midgett, C. R. and D. R. Madden (2007). "Breaking the bottleneck: eukaryotic membrane protein expression for high-resolution structural studies." Journal of structural biology 160(3): 265-274.

Mierau, I. and M. Kleerebezem (2005). "10 years of the nisin-controlled gene expression system (NICE) in Lactococcus lactis." Applied microbiology and biotechnology 68(6): 705-717.

Muller, K. M., K. M. Arndt, et al. (1998). "Tandem immobilized metal-ion affinity chromatography/immunoaffinity purification of His-tagged proteins--evaluation of two anti-His-tag monoclonal antibodies." Anal Biochem 259(1): 54-61.

Murphy, K. P., P. L. Privalov, et al. (1990). "Common features of protein unfolding and dissolution of hydrophobic compounds." Science 247(4942): 559-561.

Ostermeier, C., S. Iwata, et al. (1995). "Fv fragment-mediated crystallization of the membrane protein bacterial cytochrome c oxidase." Nat Struct Biol 2(10): 842-846.

Oswald, C., S. H. Smits, et al. (2008). "Microseeding - a powerful tool for crystallizing proteins complexed with hydrolyzable substrates." Int J Mol Sci 9(7): 1131-1141.

Pittelkow, M., B. Tschapek, et al. (2011). "The Crystal Structure of the Substrate-Binding Protein OpuBC from Bacillus subtilis in Complex with Choline." Journal of molecular biology 411(1): 53-67.

Porath, J., J. Carlsson, et al. (1975). "Metal chelate affinity chromatography, a new approach to protein fractionation." Nature 258(5536): 598-599.

Regnier, F. E. (1983a). "High-performance liquid chromatography of biopolymers." Science 222(4621): 245-252.

Regnier, F. E. (1983b). "High-performance liquid chromatography of proteins." Methods Enzymol 91: 137-190.

Ressl, S., A. C. Terwisscha van Scheltinga, et al. (2009). "Molecular basis of transport and regulation in the Na(+)/betaine symporter BetP." Nature 458(7234): 47-52.

Rodgers, D. W. (1994). "Cryocrystallography." Structure 2(12): 1135-1140.

Rosenbaum, D. F. and C. F. Zukoski (1996). "Protein interactions and crystallization." Journal of Crystal Growth 169(4): 752-758.

Rubio, V., Y. Shen, et al. (2005). "An alternative tandem affinity purification strategy applied to Arabidopsis protein complex isolation." The Plant journal : for cell and molecular biology 41(5): 767-778.

Rupp, B. (2007). "Biomolecular Crystallography: Principles, Practice, and Application to Structural Biology." Taylor & Francis Ltd 1.

Sachdev, D. and J. M. Chirgwin (1999). "Properties of soluble fusions between mammalian aspartic proteinases and bacterial maltose-binding protein." Journal of protein chemistry 18(1): 127-136.

Sala, E. and A. de Marco (2010). "Screening optimized protein purification protocols by coupling small-scale expression and mini-size exclusion chromatography." Protein Expr Purif 74(2): 231-235.

Sauter, C., B. Lorber, et al. (1999). "Crystallogenesis studies on yeast aspartyl-tRNA synthetase: use of phase diagram to improve crystal quality." Acta Crystallogr D Biol Crystallogr 55(Pt 1): 149-156.

Schmidt, T. G., J. Koepke, et al. (1996). "Molecular interaction between the Strep-tag affinity peptide and its cognate target, streptavidin." Journal of molecular biology 255(5): 753-766.

Schwarz, C. K., B. Tschapek, et al. (2011). "Crystallization and preliminary X-ray crystallographic studies of an oligomeric species of a refolded C39 peptidase-like domain of the *Escherichia coli* ABC transporter haemolysin B." Acta Crystallogr Sect F Struct Biol Cryst Commun 67(Pt 5): 630-633.

Senkovich, O., W. J. Cook, et al. (2007). "Structure of a complex of human lactoferrin N-lobe with pneumococcal surface protein a provides insight into microbial defense mechanism." Journal of molecular biology 370(4): 701-713.

Smith, D. B. and K. S. Johnson (1988). "Single-step purification of polypeptides expressed in *Escherichia coli* as fusions with glutathione S-transferase." Gene 67(1): 31-40.

Smith, G. E., M. D. Summers, et al. (1983). "Production of human beta interferon in insect cells infected with a baculovirus expression vector." Molecular and cellular biology 3(12): 2156-2165.

Smits, S. H., A. Mueller, et al. (2008a). "Coenzyme- and His-tag-induced crystallization of octopine dehydrogenase." Acta Crystallogr Sect F Struct Biol Cryst Commun 64(Pt 9): 836-839.

Smits, S. H., A. Mueller, et al. (2008b). "A structural basis for substrate selectivity and stereoselectivity in octopine dehydrogenase from *Pecten maximus*." J Mol Biol 381(1): 200-211.

Sorensen, H. P. and K. K. Mortensen (2005). "Advanced genetic strategies for recombinant protein expression in *Escherichia coli*." Journal of biotechnology 115(2): 113-128.

Stofko-Hahn, R. E., D. W. Carr, et al. (1992). "A single step purification for recombinant proteins. Characterization of a microtubule associated protein (MAP 2) fragment which associates with the type II cAMP-dependent protein kinase." FEBS letters 302(3): 274-278.

Taha, M. (2005). "Buffers for the physiological pH range: acidic dissociation constants of zwitterionic compounds in various hydroorganic media." Ann Chim 95(1-2): 105-109.

Tang, C., C. D. Schwieters, et al. (2007). "Open-to-closed transition in apo maltose-binding protein observed by paramagnetic NMR." Nature 449(7165): 1078-1082.

Tate, C. G., J. Haase, et al. (2003). "Comparison of seven different heterologous protein expression systems for the production of the serotonin transporter." Biochimica et biophysica acta 1610(1): 141-153.

Terpe, K. (2003). "Overview of tag protein fusions: from molecular and biochemical fundamentals to commercial systems." Applied microbiology and biotechnology 60(5): 523-533.

Teuber, M. (1995). "The genus Lactococcus." The genera of lactic acid bacteria 2.

Tornquist, K., J. Paallysaho, et al. (1995). "Influence of Hepes- and CO2/HCO(3-)-buffer on Ca2+ transients induced by TRH and elevated K+ in rat pituitary GH4C1 cells." Mol Cell Endocrinol 112(1): 77-82.

Tschapek, B., M. Pittelkow, et al. (2011). "Arg149 Is Involved in Switching the Low Affinity, Open State of the Binding Protein AfProX into Its High Affinity, Closed State." Journal of molecular biology 411(1): 36-52.

Urh, M., D. York, et al. (1995). "Buffer composition mediates a switch between cooperative and independent binding of an initiator protein to DNA." Gene 164(1): 1-7.

van de Guchte, M., J. Kok, et al. (1992). "Gene expression in *Lactococcus lactis*." FEMS microbiology reviews 8(2): 73-92.

Voss, S. and A. Skerra (1997). "Mutagenesis of a flexible loop in streptavidin leads to higher affinity for the Strep-tag II peptide and improved performance in recombinant protein purification." Protein engineering 10(8): 975-982.

Waugh, D. S. (2005). "Making the most of affinity tags." Trends in biotechnology 23(6): 316-320.

Wegner, G. H. (1990). "Emerging applications of the methylotrophic yeasts." FEMS microbiology reviews 7(3-4): 279-283.

Weik, M., G. Kryger, et al. (2001). "Solvent behaviour in flash-cooled protein crystals at cryogenic temperatures." Acta Crystallogr D Biol Crystallogr 57(Pt 4): 566-573.

Zaitseva, J., I. B. Holland, et al. (2004). "The role of CAPS buffer in expanding the crystallization space of the nucleotide-binding domain of the ABC transporter haemolysin B from *Escherichia coli*." Acta Crystallogr D Biol Crystallogr 60(Pt 6): 1076-1084.

Zaitseva, J., S. Jenewein, et al. (2005). "Functional characterization and ATP-induced dimerization of the isolated ABC-domain of the haemolysin B transporter." Biochemistry 44(28): 9680-9690.

Zheng, C. F., T. Simcox, et al. (1997). "A new expression vector for high level protein production, one step purification and direct isotopic labeling of calmodulin-binding peptide fusion proteins." Gene 186(1): 55-60.

Zimmerman, S. B. and S. O. Trach (1991). "Estimation of macromolecule concentrations and excluded volume effects for the cytoplasm of *Escherichia coli*." J Mol Biol 222(3): 599-620.

11

Crystallization of Membrane Proteins: Merohedral Twinning of Crystals

V. Borshchevskiy[2,3] and V. Gordeliy[1,2,3]
1Laboratoire des Protéines Membranaires,
Institut de Biology Structurale J.-P.,
2Research-educational Centre "Bionanophysics",
Moscow Institute of Physics and Technology,
3Institute of Complex Systems (ICS),
ICS-5: Molecular Biophysics, Research Centre Juelich,
1France
2Russia
3Germany

1. Introduction

Membrane proteins are the main functional units of biological membranes. They represent roughly one-third of the proteins encoded in the genome and about 70% of drugs are targeted to membrane proteins. X-ray protein crystallography is one of the most powerful tools to determine protein structure and to provide a basis for understanding molecular mechanisms of protein function. Despite an obvious importance of membrane protein only about 1% of structures in the Protein Data Bank (PDB) are of this type. Moreover, although the number of membrane protein structures deposited to PDB since 1985, date of the first membrane protein structure [1], is increasing it is not yet comparable with the rate achieved for soluble proteins [2]. Currently, the PDB contains more than 70,000 structures, and the structures of membrane proteins do not exceed 500 [3]. Considerable effort made in several laboratories in the last years towards extension of high-throughput crystallography to membrane proteins open a hope of correcting this imbalance. Nevertheless significant challenges must be overcome to achieve this goal. Two major problems toward the determination of membrane proteins structures are: the production of pure, stable and functional protein solubilized in detergents, and the growth of crystals suitable for X-ray crystallography. The latter is often defined as major bottleneck of structural biology of membrane proteins. For a long time, the vapor diffusion method has been the only method which was used to crystallize membrane proteins. This method, which is based on a well-developed approach of crystallization of water soluble proteins, led to relative success, however, it failed to produce crystals of some important membrane proteins. Quite recently new methods were introduced. One of the most promising new method to overcome this problem is the so called *in meso* crystallization approach where lipid systems (e.g. the lipid cubic phase (LCP)) are used as a crystallization matrix. It has been demonstrated that these methods are applicable to different membrane proteins including G-protein-coupled receptors (GPCR), membrane protein complexes and others. One of the first important breakthroughs was bacteriorhodopsin (bR) which for a long time failed to be crystallized by the *in surfo* methods and was solved to resolution about 1.55 Å from the crystals obtained by LCP crystallization. Thanks to the *in meso* method crystallographic structures of almost all functional states of bR are now available with atomic resolution (see [4] for review). Despite this fact the detailed mechanism of bR proton pumping is still to be elucidated. It appeared that a severe problem originates from the tendency of the best (in the sense of resolution) bR crystals to be perfectly twinned. Being a general problem of protein crystallography, twinning may result in controversial structural models of intermediate states in the case of bR. The chapter presented here is aimed to summarize the present knowledge on twinning formation during *in meso* crystallization and the methods to overcome it.

2. *In meso* crystallization

2.1 Crystallization from lipidic cubic phase

A principally new crystallization method – crystallization of membrane proteins in lipidic cubic phases was developed by Rosenbusch and Landau in 1996 [5]. A fundamental difference between methods of standard crystallization and crystallization in the LCP is that in the latter, the solubilized protein is reconstituted back in the native lipid bilayer and after that the crystallization is induced by the addition of a precipitant. Liquid crystalline systems formed by lipids in aqueous media can form infinite bicontinuous periodic minimal surfaces, which have a zero mean curvature and a periodicity in all the three dimensions characterized by a cubic lattice [6-8]. The system consists of two compartments: a continuous curved lipid bilayer forming a three-dimensional well-ordered structure, interwoven with continuous aqueous channels. Macroscopically the phase is very viscous, isotropic, and optically transparent. Membrane cubic phases are found in the cells [9], and they are used in food industry [10] as well as for drug delivery [11]. Practical aspects of crystallization in the lipidic cubic phase look very simple and an example – crystallization of bR – can be described as the following procedure [12]:

1. Weigh into the PCR tube (200 mL) approximately 5 mg of dry MO, incubate tubes with monooleoyl (MO) at 40°C, and spin the lipid down for 10 min at 13,000 × g at room temperature.

2. Keep MO at 40°C during an additional 20 min to gain the isotropic fluid lipidic phase and then let the lipid phase cool to room temperature.

3. Mix 1 mL of prepared 10 mg/mL BR solution comprising about 1.2 w/w% of *n*-octyl-β-D-glucopyranoside (OG) with 1 mg of MO. To gain the cubic phase, centrifuge the PCR tubes with the sample at 10,000 rpm for at least 1 h at 22°C (rotating tubes within the rotor every 15 min by 90°). Incubate the samples during 1-2days in the dark at 20-22°C. An alternative way to prepare the cubic phase is described in [13].

4. Add a precipitant to induce crystallization– a ground powder of KH2PO4 mixed with Na2HPO4 (95/5 w/w) with a final concentration of the salt mixture 1–2.5 M (pH 5.6). Repeat homogenizing centrifugation of samples as described in the previous item. Leave the crystallization batch in the dark at 22°C. bR microcrystals (10–20 mm in diameter) usually appear within 1 weak after induction of crystallization (Fig. 1). This protocol of crystallization is close to the original one provided by Rosenbusch and Landau. An alternative way to do such crystallization (it is used in nanovolume high throughput approach) is to add liquid precipitant to the top of the lipidic phase [13].

5. To separate the crystals from the lipidic phase directly from LCP use mechanical manipulation with microtools or, alternatively, add lipase or detergent to the lipidic phase to destroy the lipid phase at room temperature during several hours or days [14].

LCP approach remains most efficient among all other *in meso* approaches introduced later. Nevertheless, it is not yet clear whether other new methods were properly optimized. In other words it is not yet clear what is the real potential of these methods. Therefore, we will describe briefly three more new approaches

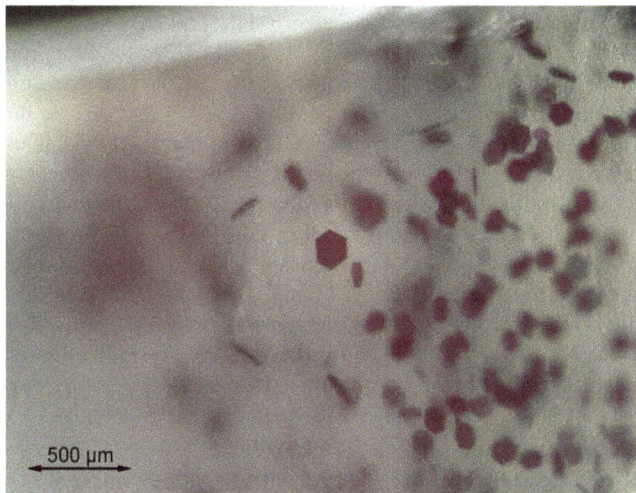

Fig. 1. A crystallization well (a PCR tube) with bR crystals.

2.2 Crystallization from vesicles
An interesting and unusual approach to membrane protein crystallization was proposed in 1998 [15,16]. The authors observed that purple membranes (two-dimensional hexagonal native crystals of bR) treated with the neutral detergent under certain conditions lead to the creation of spherical protein clusters (~50 nm in diameter). Using a standard vapor diffusion method for crystallization from bR vesicles with a high protein/lipid ratio, well diffracting hexagonal crystals were obtained [15-17]. This new crystal belongs to the space group *P*622 with unit cell dimensions of $a = b = 104.7$ Å and $c = 114.1$ Å. The highest announced structural resolution achieved by this method is 2.0 Å. It is not compared to the LCP results obtained with the same protein. Until now there is no evidence that a specific case of bR crystallization from vesicles can be extended to other membrane proteins. However, it is not yet clear whether this approach is limited to some specific cases, like bR, or has a more general application.

2.3 Crystallization from bicelles
Just after the second *in meso* method was published another approach - crystallization from bicelles - was proposed. This method was first applied to obtain well diffracting bR crystals [18,19]. Bicelles, known for quite a long time, are a liquid crystal phase consisting of disc-shaped lipid-rich bilayer particles formed from mixtures of dimyristoyl phosphatidylcholine (DMPC) with certain detergents. The detergents mostly used for such a type of crystallization are either dihexanoyl phosphatidylcholine (DHPC) or zwitterionic bile salt derivative, CHAPSO. The bicelle sizes at a 1:3 DMPC/DHPC molar ratio are: the bilayer thickness – 40 Å and the diameter – 400 Å. The lipid detergent ratios present in the bicellar systems are relatively high compared to standard micellar systems [20,21].

The procedure of crystallization of membrane proteins from bicelles is as follows. The first step is preparation of bicelles. Then, solubilized protein is mixed with bicelles. It is considered, but not directly proven, that at this stage, the protein molecules are reconstituted into bicelles. After that the protein is crystallized by a standard vapor diffusion method. bR crystals grown at room temperature are identical to the previously obtained at 37°C twinned crystals: space group $P2_1$ (2.0 Å resolution) with unit cell dimensions of a = 44.7 Å, b = 108.7 Å, c = 55.8 Å, β = 113.6°. The other room-temperature crystals were not-twinned and belong to space group $C222_1$ (2.2 Å resolution) with the following unit cell dimensions: a = 44.7 Å, b = 102.5 Å, c = 128.2 Å. It is important to note that the crystals of the human β_2-adrenergic GPCR were obtained by this method [22]. The structure was solved to 3.5/3.7 Å resolution. It is considerably lower than what was obtained by protein crystallization in the cubic phase [23]. Taking into account the long and dramatic attempts to crystallize a ligand binding GPCR, there is no doubt it was a new considerable success of the method under discussion. The 2.3 Å resolution structure of the murine voltage dependent anion channel (mVDAC) that reveals a high-resolution presentation of membrane protein architecture was also obtained due to bicelles method [24]. Very recent success of the bicelle-like approach is the crystallization of the membrane part of the respiratory complex I [25].

2.4 Crystallization from sponge phases (L_3-phase)

It is interesting that historically crystallization from the sponge phase was described about 10 years after discovering the LCP approach. This is despite the fact that the sponge phase (L_3-phase) is the liquid analogue of the lipidic cubic phase with the reduced bending rigidity of membranes and without a long-range order. When the bending rigidity of the membrane becomes comparable with a thermal energy the ordered cubic phase structure is perturbed by thermally excited collective out-of-plane fluctuations of membranes. The transformation of the cubic to the sponge phase can be induced by different factors, for instance, via adding a solvent such as polyethyleneglycol (Mw = 400), dimethyl sulfoxide, 2-methyl-2,4-pentanediol (MPD), propylene glycol, or Jeffamine M600 to a lipid/ water system [26]. The diameter of aqueous pores in the MO cubic phase is relatively narrow (ca. 3-6 nm) compared to that of the sponge phase (10–15 nm and more) [27]. Evidently the size of the pores of the L_3-phase is compatible with membrane proteins with large hydrophilic parts and lets them diffuse freely within the plane of the membrane surface [26]. Well diffracting crystals of the reaction center from *Rhodobactersphaeroides* were grown in the L_3 by a conventional hanging-drop scheme of the experiment, and were harvested directly without the addition of lipase or cryoprotectant, and the structure was refined to 2.2 Å resolution. The authors of the work claimed that in contrast to the earlier LCP reaction center structure [28], the mobile ubiquinone could be built and refined. In these experiments, the only additional component (relative to the components of the cubic phase crystallization – the MO/membrane protein/detergent/buffer) was a small amphiphilic molecule 1,2,3-heptanetriol or Jeffamine M600. The structure was solved to resolution 2.35 Å [28]. In another work [29], crystals of the light harvesting II complex suitable for X-ray crystallography were obtained with structural 2.45 Å resolution. In this study, the additives used were KSCN, butanediol, pentaerythritolpropoxylate (PPO), *t*-butanol, Jeffamine, and 2-methyl-2,4-pentanediol (MPD). An advantage of the L_3 approach is that the liquid properties of the sponge phase at room temperature can be used directly in hanging- or sitting-drop vapor-diffusion crystallization by commercially available robots. Recently, a sponge phase sparse matrix crystallization screen consisting of different conditions became available [30]. However, unlike the LCP method, this one has not led to a breakthrough in structural biology of membrane protein. There was no structure of a new membrane protein or a principal improvement in structural resolution achieved by this method. Does it mean that the sponge phase approach does not have the same (or higher) power as the LCP method? We would speculate that this approach can be at least considered as a complementary one to the LCP.

3. Overcoming twinning formation

3.1 Introduction to the merohedral twinning of bR P6$_3$ crystals

Although bR can be crystallized by many methods and in different types of symmetries [5,16,18,31], only P6$_3$ crystal grown by *in meso* crystallization diffracts to the highest resolution. At the same time, these crystals often suffer from perfect merohedral twinning [32].

Twinning is one of the most common crystalline defects. A twin crystal consists of several domains oriented in such a way that their reciprocal lattices are superimposed at least in one dimension [33]. There are two forms of twinning: merohedral and non-merohedral. Only part of reflections of individual crystal domains superimpose in non-merohedral twinning, whereas all reflections are superimposed in three space dimensions in the merohedral form [34]. If only two orientations of twin domains are present the merohedral twinning is called hemihedral. It is the most widespread type of merohedral twinning [33]. The hemihedral twinning is intrinsic for hexagonal P6$_3$ crystals of bR grown in the cubic phase of MO [32,35].

Twinning of bR crystals implies the imposition of reflections with Miller indexes *hkl* and *kh-l*, so that the observed crystal reflections is a weighted sum of two different crystallographic reflections:

$$I^{obs}_{hkl} = (1-\alpha)I_{hkl} + \alpha I_{kh-l}$$
$$I^{obs}_{kh-l} = (1-\alpha)I_{kh-l} + \alpha I_{hkl} \tag{1}$$

Where I^{obs}_{hkl} are crystallographic intensities observed in the X-ray experiment, I_{hkl} are crystallographic intensities of the twin domains and α is the twinning ratio, i.e. the volume fraction of equally oriented domains. Twinning is called perfect when α is close to 50 %. The shape and optical properties of twinned crystals are identical to those without twinning. The presence of twinning and estimation of the twinning ration are only possible by using special analysis methods of the diffraction data [36].

Twinning of the crystals complicates the obtaining of a crystallographic structure of the protein. If the twinning ration is $\alpha \neq 50\%$, then the system (1) can be solved:

$$I_{hkl} = \frac{(1-\alpha)I^{obs}_{hkl} - \alpha I^{obs}_{kh-l}}{1-2\alpha}$$
$$I_{kh-l} = \frac{(1-\alpha)I^{obs}_{kh-l} - \alpha I^{obs}_{hkl}}{1-2\alpha} \tag{2}$$

After that, the usual tools can be applied for crystallographic analysis. However, as follows from (2), the error in intensity calculation increases and tends to infinity as a tends to 50 %[37]. For this reason the presence of crystal twinning worsens the electron density maps and reduces the reliability of protein models.

The perfect hemihedral twinning of bR crystals shows up in the presence of additional two-fold symmetry since $I_{hkl}^{obs} = I_{kh-l}^{obs}$ (see (1) when $\alpha = 50\%$). In this case, the number of independent observations (crystallographic intensities) is two times fewer. The equation system (1) is confluent and the crystallographic intensities cannot be extracted from the X-ray data. In this case, the intensities calculated from the protein model are used to obtain the desired crystallographic intensities according to the equation:

$$I_{hkl} = \frac{I_{hkl}^{obs} + I_{hkl}^{cal} - I_{kh-l}^{cal}}{2}$$
$$I_{kh-l} = \frac{I_{hkl}^{obs} + I_{kh-l}^{cal} - I_{hkl}^{cal}}{2}$$

(3)

where I_{hkl}^{cal} are intensities calculated from the protein model. R-factors of protein models obtained from the perfect twinned data overestimate the model reliability, since the difference between the observed and calculated structural factors is undervalued due to the averaging over the reflections related by the twinning law. Hence, the refined crystallographic R-factors from perfectly twinned data are typically a factor of $1/\sqrt{2}$ lower than for low (or un-)twinned data [36,38,39]. In addition, the use for refinement of the intensities calculated according to (3) introduces additional model bias due to the explicit dependence of the detwinned data on the model itself.

An additional problem for X-ray analysis caused by perfect twinning is the inability to use the experimental difference Fourier map. Basing on the mathematical consideration it was shown about 40 years ago that the difference Fourier electron density maps are most sensitive, accurate and less susceptible to model bias method for observing limited structural changes [40]. The difference map is simply the Fourier transform of the amplitudes $(F_{exc} - F_{gr})$ (where F_{gr} and F_{exc} are the structural factors of the ground and excited state of the protein) and phases are taken from the model of the ground state. This type of maps visualizes the changes in the electron density between the first and second crystallographic datasets. If structural changes are visible at a reasonable significance level within a difference Fourier map, then it is a plausible feature of the experimental data. On the opposite side, if changes arise during crystallographic refinement and are not confirmed by the difference Fourier map, then they are probably artifacts. For this reason, the difference Fourier maps are the main criterion for detecting small structural changes in the macromolecular systems and were used in many studies, for instance: myoglobin-CO complex [41-44], photoactive yellow protein [45-48], sensory rhodopsin II [49] and bR [50-57]. In the case of perfect twinning of protein crystals, structural factors F_{gr} and F_{exc} cannot be obtained, and Fourier difference maps cannot be constructed.

Despite the fact that twinning creates problems for protein crystallography, currently there are no rational effective methods of obtaining untwinned crystals. Similarly there are only a few works published on the systematic study of interrelation between twinning formation and crystallization conditions. Description of the phenomenon of twinning is even poorer for the crystals of membrane proteins and particularly for those obtained by *in meso* crystallization.

However, the twinning problem is of particular importance for the case of bR. Among 28 bR structures obtained from P6$_3$ crystals, 19 are from crystals with perfect twinning [32]. The best resolution of bR crystallographic model is 1.43 Å [58]. However, all the structures with the resolution better than 1.9 Å were obtained from crystals with perfect twinning. The only exception is the structure with a resolution 1.55 Å from the crystal with a twinning ratio of 25 %. All the currently published crystallographic studies devoted to the K, L and M bR intermediate states either have a relatively low resolution (> 2.1 Å) [50-56] or were obtained from perfectly twinned crystals [58-63]. The intermediate state structures built using these data are not consistent with each other [53,56,64]. One of the most probable reasons for this is the twinning problem.

It is well known that the changes in bR structure during the transition from the ground state to intermediates are relatively small [50-55,58,60,62]. Thus, X-ray data of very high quality are required to obtain the structures of intermediate states. In particular, crystals should be untwinned as twinning reduces the quality of the electron density maps and the reliability of protein models, as well as suppresses the utilization of the Fourier difference maps. To elucidate the molecular mechanism of bR proton transport, it is crucial to obtain highly ordered crystals without twinning.

3.2 Physical detwinning of bR crystals

In 2004 [35] it was shown that the twinned crystals of bR consist of large scale domains. Each of the domains is a hexagonal plate with the size in the hexagonal plane equal to that of the whole crystal and the thickness comparable to that of the crystall (as it is shown in Fig.2). In most cases the crystals were split in two plates with no twinning. However in some cases the crystals were split in three and more plates. Thus it may be supposed that most of bR P6$_3$ crystals consist of only two twinning domains. However the presence of three and even more domains is also possible. But the size of these domains is always comparable to the size of the twinned crystal. The attempts to mechanically separate the twin domains had no effect. However it was noted that the slow decrease of mother liquid molarity may result in crystal slicing. Basing on this idea the approach for physical detwinning of bR crystals was proposed. According to the procedure the molarity of salt in mother liquid was slowly reduced from 3 to 1 M which induces splitting of agglutinated plates. Some of the split crystals diffracted well enough to determine the twin ratio which in all cases was equal to zero within the experimental error.

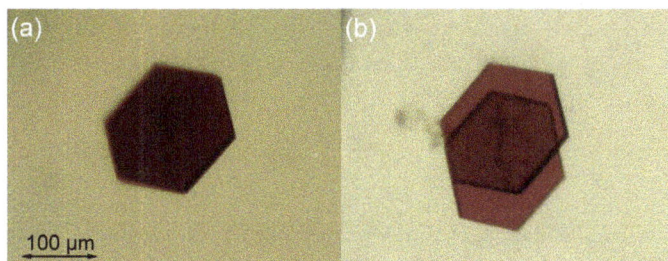

Fig. 2. bR crystal splits into two parts: (a) initial crystal, (b) two parts of the crystal separated by gradual decrease of the salt concentration.

Unfortunately, it turned out that the procedure of physical separation of the crystals often leads to a significant drop in the diffraction quality of the crystals, and therefore is not applicable in practice for obtaining high-resolution X-ray analysis.

3.3 Direct observation of twin domains

As it was mentioned before the twinning fraction of the crystal can only be estimated by the analysis of the statistical distribution of its crystallographic intensities. This implies that to determine the twinning, one has to fulfill the whole procedure of obtaining the crystallographic data, including the dissolution of the crystallographic sample, the separation of crystals from the crystallization matrix and X-ray data collection. Meanwhile, this resource- and time-consuming procedure has a small useful output: nine out of ten crystals have the twinning ratio close to 50 %.

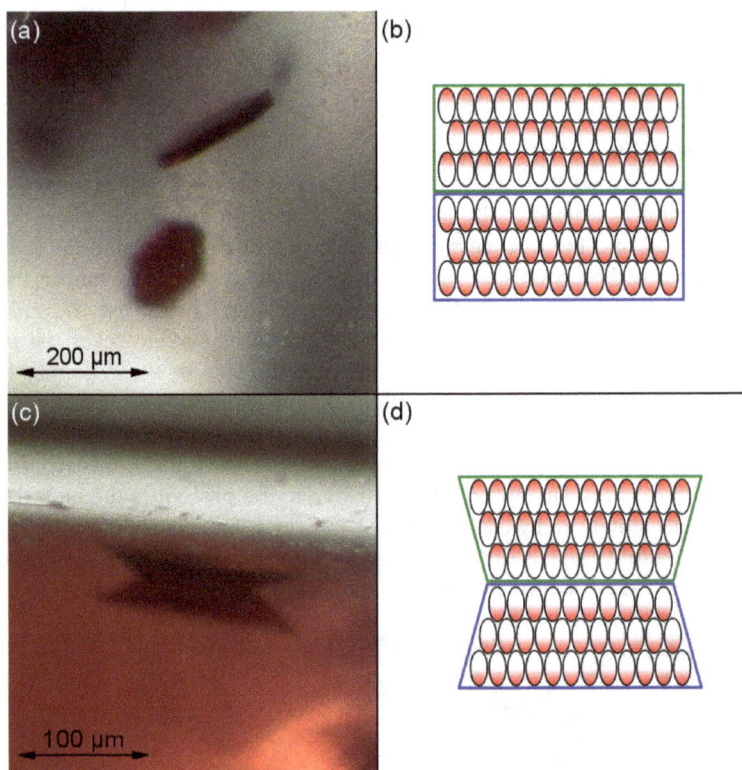

Fig. 3. bR crystals usually obtained by *in meso* crystallization in OG (a) and their schematic representation (b). bR crystals obtained by *in meso* crystallization in CYMAL-5 (c) with their schematic representation (d). Two different twin domains are shown in blue and green color. Red color represents the negative charge of CP side of bR.

One of the ways to simplify this procedure was found during crystallization trials with different detergent types [32]. It was observed that the crystals grown in the presence of 5-cyclohexyl-1-pentyl-β- D-maltoside (CYMAL-5) at concentrations of about 10 % have a shape of two truncated pyramids stuck together along the smaller of the hexagonal sides (Figure 3 c). Crystals in one crystallization probe had all the possible values of relative volumes of domains (from 0 when one domain was missing; to 0.5 when the domains had equal volume). The twinning ratio was surprisingly correlated with the relative domain volumes, which was confirmed by statistical analysis of X-ray intensities. The twinning fraction was close to 0 % when one of the domains was much smaller than the other, and close to 50 % for crystals with approximately equal parts. In addition, some of the crystals were split in two parts during fishing. Each of the domains had no twinning. Thus, it was concluded that the truncated pyramids represent twin domains as shown at Fig.3d. It is possible to select non-twinned crystals by careful inspection of the crystals shape in stereomicroscope, which significantly reduces the time and resources on the procedure for selection of crystals suitable for X-ray diffraction studies and produces additional information about the nature of the twinning formation.

3.4 Interrelation of crystal growth rate and twinning fraction

Additional information on the nature of bR twinning came from the statistical distribution of twinning ratio among several hundreds of crystals [32]. For this purpose bR crystals were grown in a wide range of crystallization conditions: at different concentrations of salt and protein, types and concentrations of detergents. More than 300 crystals were obtained and X-ray data were collected from all of them to determine their twinning ratios.

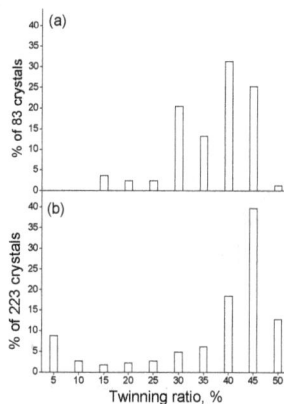

Fig. 4. Distribution of twinning ratios in two groups of crystals with the characteristic growth time less than 1.5 months (empty columns) (a) and more than 1.5 months (hatched columns) (b). The first and second groups consist of 83 and 227 crystals, respectively.

It turns out that regardless of the specific crystallization conditions the crystals with low twinning ratio (< 20 %) were observed with higher probability in samples where the first crystals appeared relatively late (in 2-3 weeks after sample preparation, rather than 2-3 days) and growth proceeds for a longer time period (for ~10 weeks). If the first crystals appeared in the sample relatively early and their growth was rapid then almost all crystals had a high twinning ratio. The distribution of twinning ratio for 83 crystals grown less than 1.5 months and for 227 crystal with growth time of more than 1.5 months is shown at Figure 4. 11 % of the slowly grown crystals had the twinning ratio smaller than 10 %. Meanwhile all the fast grown crystals had the twinning ratio higher than 10 % and only 5 % had the twinning ratios between 10 % and 20 %.

It was suggested before for the soluble protein plastocyanin that slow growth favors the formation of untwinned crystals [39]. Confirmation of this relationship for a membrane protein, probably indicates the general nature of this phenomenon. It is plausible that in all cases when protein crystals suffer from twinning, one should search for the crystallization conditions of slow crystal growth.

3.5 Crystallisation in β-XylOC$_{16+4}$ mesophase

A presumably new approach to obtaining non-twinned bR crystals unexpectedly comes from the *in meso* crystallization in the "exotic" β-XylOC$_{16+4}$ mesophase.

The crystallization trials with this lipid were excited by the inequality of lipid and detergent libraries used for handling membrane proteins. The library of detergents with different hydrophilic and hydrophobic parts used for solubilization, purification and crystallization of membrane proteins is quite large. The fittest detergent may be found in the library for each specific membrane protein. This fact significantly increases the number of crystallized membrane proteins [65]. On the contrary the library of lipids used for the cubic phase creation is discouragingly small. MO is the most common lipid for *in meso* crystallization. Three other monoglycerols are reported to be suitable for this type of crystallization: monopalmitolein [5], monovaccenin [66], 2,3-dihydroxypropyl-(7Z)-hexadec-7-enoate [67] and 2,3-dihydroxypropyl-(7Z)-tetradec-7-enoate [68]. The library of matrix lipids for *in meso* crystallization should be increased for further success of the method.

Recently we presented the results of bR crystallization in the β-XylOC$_{16+4}$ cubic phase used for this purpose for the first time. β-XylOC$_{16+4}$ (Fig.2 in [69]) represents a recently developed isoprenoid-chained lipid family [70,71].

β-XylOC$_{16+4}$ forms a cubic phase almost at the same conditions as MO. It turns to be possible to crystallize bR in the β-XylOC$_{16+4}$ cubic phase using the standard protocol of *in meso* crystallization [69]. Several dozens of crystals were obtained. Three of them diffracted well enough and the X-ray dataset was collected for them. Two crystals diffracted up to 2 Å. The third one was worse and gave diffraction up to 2.7 Å.

The crystals obtained in the cubic phase of β-XylOC$_{16+4}$ and MO have the same P6$_3$ symmetry. The diffraction quality of bR crystals obtained in β-XylOC$_{16+4}$ is better than that of the first bR crystals obtained in MO [72] (the resolution is 2.0 Å and 2.5 Å, correspondingly). A further search for optimal crystallization conditions will possibly improve the diffraction properties as it was done in the case of MO.

It is important to mention that three studied crystals had a low twinning ratio. The twinning ratio was 37 and 34 % in two cases (for the crystals with diffraction resolution of 2.0 Å), and the third crystal (with resolution of 2.7 Å) had no twinning. As follows from §3.4 and [32], only 28 % of crystals obtained in the MO cubic phase have the twinning ratio smaller than 34 %. Thus the probability to find in one crystallization probe three crystals with small twinning ratios is relatively low which is unlikely to be a coincidence. The β-XylOC$_{16+4}$ cubic phase may favor the formation of low-twinned crystals.

3.6 The nature of the twinning phenomenon

Experiments described in 3.2-3.5 gave enough information to produce untwinned bR crystals for the investigation of the proton transport mechanism. On the other hand they gave some hints to understand the nature of the phenomenon of twinning formation in bR crystal.

bR crystals belong to class I in the nomenclature introduced in [73]. The hexagonal plane of bR crystals is perpendicular to the crystallographic axis c which implies that crystal growth occurs trough layer-by-layer two-dimensional nucleation on the ab surfaces of the crystal [74]. This assumption is in accordance with the model of *in meso* crystal growth proposed by M. Caffrey [75] and is confirmed by atomic force microscopy [76]. The contact surface between twinning domains is also perpendicular to c axis as it is demonstrated in Fig.3. Consequently, this surface also emerges as a result of two-dimensional nucleation on the ab-surface.

The contact surface may be formed either by two cytoplasmic (CP) surfaces of bR or two extracellular (EC) ones. The twinning ratio of the majority of crystals is > 30 %, and most of them consist of two domains. This peculiarity may be explained by different energies of interaction for CS-CS and EC-EC contacts in the protein crystal. As follows from the pdb-structure (1C3W [77] for instance) EC surface of bR is almost neutral and CP is negatively charged. On the other hand there is no specific interaction seen in pdb-structures between two adjacent protein layers, they interact by Van-der-Waals contacts between only two amino acids [72]. That means that even a weak electrostatic interaction may play an important role in the total energy of layer interaction.

Thus we can imagine the following process of crystal formation: the first twin domain emerges soon after (or even during) nucleation with two twin domains interacting by their EC surfaces. The crystal itself has two CP surfaces at its external faces. The probability to form a new twin domain on the CP surface is relatively low due to unfavourable electrostatic interaction. Consequently, the crystal continues to grow without formation of new twinning domains.

It may be noted that the distribution of the twinning fraction of slowly growing crystals has a sinuous pattern: there are local maxima with the twinning ratio < 10 % and > 35 %, and a minimum is located in between them. This non-obvious behavior may be explained by computer modelling of the growth of twin crystals.

As it was mentioned before crystal growth occurs through the two-dimensional nucleation at the surface of the crystal (slow step) and a relatively fast growth of the new layer in two dimensions. Thus one can use a one-dimensional model to simulate crystal growth in the direction perpendicular to the ab crystallographic plane. Crystal growth begins from a single layer and proceeds by consecutive addition of new layers to each surface of the crystal alternatively. When a new layer is added three different types of contacts may be formed.

1. CP-EC contacts which corresponds to normal crystal growth. Let us assign to this event a relative probability of 1.
2. EC-EC contacts which corresponds to the formation of the twinning domain. We will assign the probability P_1 to this event.
3. CP-CP contacts which also gives rise to a twinning domain as probability P_2 is assigned to this event.

The usual thickness of P6$_3$ bR crystal is about 20 μm that corresponds to about 4000 protein layers. This number of layers was used in the simulation of the crystal growth.

There are two variables which will dictate the number of formed twinning domains and the twinning ratio of the crystal: the probabilities P_1 and P_2. These probabilities may be varied to fit the experimental dependencies shown at Fig.4.

The first feature noted while exploring this model was that the symmetrical conditions $(P_1 = P_2)$ cannot reproduce the experimental data. Under relatively low probabilities of twinning formation the distribution shows a peak at zero twinning ratio. The height of the peak decreases as the probability of twin formation increases and the distribution over the nonzero range remains quite flat until the peak at zero vanishes (Fig. 5a).

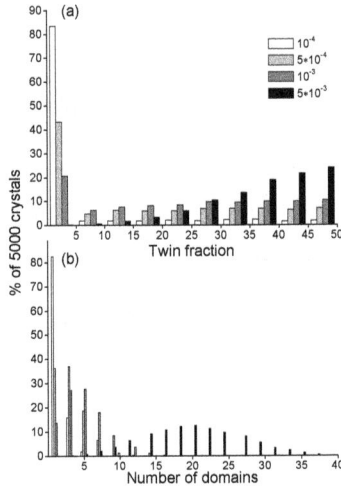

Fig. 5. Distribution of twinning fraction (a) and (b) the number of twinning domains calculated for 5000 crystals under conditions of symmetric domain nucleation $(P_1=P_2)$ and for probabilities in the range 10^{-4} - 5×10^{-3}.

When an asymmetry in the probabilities is introduced to the model with $P_1=10^{-3}$, the peak at zero value changes very little, while the rest of the distribution has low values at low twinning ratios which gradually increase towards higher twinning ratios (Fig. 6a). Two important things are worth noting here. Firstly, when P_2 is smaller than P_1 it has almost no influence on the distribution. P_2 is the probability of forming CP-CP. As it was described above this event is quite improbable because the two negatively charged CP surfaces are pushing apart. Thus P_2 can be fixed at 0 at the following consideration. Second, the introduction of asymmetry leads to a shift in the peak of the number of twin domains distribution (compare Fig. 5b and 6b) from six domains (for $P_1 = P_2 = 10^{-3}$) to two domains $(P_1 = 10^{-3}, P_2 = 0)$ which is in accordance with the experimental results.

Under the asymmetrical conditions the model resembles the experimentally observed distributions. Small changes in P_1 lead to dramatic changes in the fractions of non-twinned and perfectly twinned crystals, while the fraction of crystals with an intermediate twinning ratio changes much more slowly. The best fit of the experimentally observed distributions corresponds to a probability P_1 of 3×10^{-3} for fast crystal growth, where less than 1 % of crystals grow without twinning, and of 1.25×10^{-3} for slow growth, where 10 % of crystals have no twinning (Fig.7).

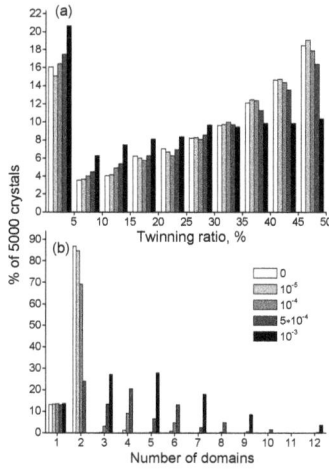

Fig. 6. Distribution of twinning ratio (a) and twinning domains (b) calculated for 5000 crystals under conditions of asymmetric domain nucleation $P_1=10^{-3}$, P_2 in the range between 0 and 10^{-3}.

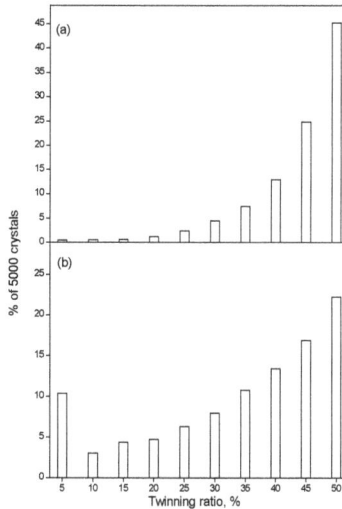

Fig. 7. Modelled distributions of twinning ratios simulating experimental distributions for slow (a) and fast (b) crystal growth. The P_1 probabilities for the models are 1.25×10^{-3} and 3×10^{-3}, correspondingly; $P_2 = 0$.

The model resembles the principal features of the experimentally observed distributions having quite a bad fit at the region of the high twinning ratio. This feature may be explained either by the underestimation of the twinning ratio by computational procedures owing to noise in the diffraction intensities or by the heterogeneity of the crystallization medium, which is responsible for the inevitable differences in the growth rates of different crystal surfaces.

The described model of twinning formation explains how the probabilities P_1 and P_2 determine the type of twinning fraction distribution. However, this model does not explain what is the relation between the rate of crystal growth and probabilities P_1 and P_2. Unfortunately, the theory of *in meso* crystallization is quite poorly understood at the moment and it cannot be used to explain this dependence. However, we can imagine the following thermodynamic explanation:

The limiting step of the crystal growth is the two-dimensional nucleation on the crystal surface. According to the classical two-dimensional theory of crystallization the thermodynamic potential of nucleus formation is [78]:

$$\Delta G = -n(\mu_v - \mu_c) + \sum_i l_i \kappa_i \tag{4}$$

where μ_v and μ_c are chemical potentials of the protein molecule in the volume and on the crystal surface, n is the number of molecules in a nucleus. The second term describes the surface energy, where κ_i is a specific surface energy and l_i is the length of i-*th* edge.

Basing on (4) we can write down the general expression for the free energy:

$$\Delta G^* = -A\Delta\mu + B \tag{5}$$

where A and B are the values which are not dependent on $\Delta\mu = \mu_v - \mu_c$. A depends on the specific surface energy and is virtually equal for normal and twinning nucleation. B depends on the interaction of the molecules in different layers and is significantly different for normal and twinning nucleation.

The rate of two-dimentional nucleation represents as:

$$J \approx \Gamma e^{\frac{-\Delta G^*}{kT}} \tag{6}$$

where Γ poorly depends on $\Delta\mu$. The experimental fact that twinning domains have a macroscopic size results in the condition that the probability of normal layer nucleation is significantly higher than that of the twinning formation. Consequently:

$$J_0 \gg J_1; J_2 \Rightarrow \Delta G_0^* < \Delta G_1^*; \Delta G_2^* \tag{7}$$

where J_0, J_1 and J_2 are the rates for normal and two twinning (CS-CS and EC-EC) nucleations, and ΔG_0^*, ΔG_1^* и ΔG_2^* are the corresponding free energies. The rate of crystal growth is regulated by supersaturation $\Delta\mu$ and at very high values of supersaturation the difference between J_0, J_1 and J_2 vanishes (see (5)). When $\Delta\mu$ decreases, the absolute value of ΔG^* also diminishes and the growth rate drops down. As follows from (5), ΔG_1^* and ΔG_2^* approache 0 faster than ΔG_0^* and under a certain value of $\Delta\mu$ become positive (the formation of twin crystals ceases). Simultaneously, due to exponential dependency (6) the difference between probabilities of normal and twinning nucleations grows.

The described explanation is applicable for any crystals where twinning is formed by two-dimensional nucleation. For this type of crystals the correlation between the growth rate and the probability of twinning formation may be a common feature. Taking into consideration the presence of the lipidic cubic phase may give better understanding of the mechanism of the twinning formation.

The most important feature of the *in meso* crystallization mechanism for this consideration is the presence of lamellar lipid environment around the growing protein crystal (see Fig.1a in [75]). It is obvious that the highly curved transitional lipid phase should be present between the bulky cubic and lamellar phases. The changes in the cubic phase curvature will simultaneously cause the corresponding changes in the curvature of the transitional phase.

It was proposed in [79] that the protein *in meso* crystallization is provoked by excess of elastic energy in the curved lipid bilayer. This type of energy is accumulated by the crystallization system due to hydrophilic-hydrophobic mismatch between the lipid bilayer and the protein molecule and the value of this energy is strongly dependent on the bilayer curvature radius and the length of protein hydrophobic-hydrophilic boarder. The rate of crystal growth is regulated through the changes of the elastic energy caused by variations in the bilayer curvature. The decrease of the curvature radius results in the slowdown of crystal growth.

On the other hand the variations in the length of the hydrophilic-hydrophobic boarder also influence the crystallization rate. There are two substantially different hydrophilic-hydrophobic boarders of the protein molecule (one is at the EC side of the protein and the other is at the CP one). The curved bilayer is also asymmetrical relative to the perpendicular to its surface. Consequently, the elastic energy of deformation is dependent on the orientation of the protein in the curved bilayer.

The protein molecule has to cross the highly curved transitional bilayer during the crystallization and the corresponding energy barrier of this process is different for different orientations of protein molecules. And the character of the elastic energy dependence on the bilayer curvature is also different for the two possible protein orientations.

The decrease of the bilayer curvature during crystallization results in a slowdown of crystal growth and simultaneously reduces the curvature of the transitional region. The energy barriers for two different protein orientations change differently and this fact results in different probabilities of the formation of the normal or twinned protein layer in the crystal.

4. Conclusions

Twinning of protein crystals is an unwelcome phenomenon for crystallographers and may be a barrier, like in the case of bR crystals, on the way to elucidating protein function. For this reason the efforts were applied to understand and overcome it. Nowadays the twinning of bR $P6_3$ crystals is one of the most studied and characterised twinning phenomena of protein crystals.

First of all it was directly shown that the LCP grown twinned crystals of bR consist of large scale domains. Each of the domains is a hexagonal plate with the size equal to that of the whole crystal [35]. It is important the crystals may be split into several non-twinned domains by slow changes of salt concentration in the mother liquid so that the split parts preserved high diffraction quality. Further systematic investigation showed that the rate of crystal growth strongly affects the twinning-ratio distribution of the crystals. Searching for crystallization conditions leading to slow crystal growth, it is possible to select crystallization trials that contained up to 10% non-twinned crystals [32]. In addition, the conditions were found allowing selection of crystals with low twinning by visual inspection of their shape with no need for analysis of the diffraction intensity distribution. This discovery further facilitates the process of selection of non-twinned crystals. The experimental data obtained so far allow the formulation of a theory of twinning formation which in particular sheds some light on the general question of the process of *in meso* crystallization. Most recently some hints were found that the usage of different crystallization matrixes may allow to improve the yield of non-twinned crystals in crystallization [69].

5. Acknowledgements

Authors are grateful to Georg Büldt, Rouslan Efremov and Ekaterina Round for their contribution to the chapter. Authors are supported by the program "Chairesd'excellence" édition 2008 of ANR France, CEA(IBS)-HGF(FZJ) STC 5.1 specific agreement, the German Federal Ministry for Education and Research (PhoNa – Photonic Nanomaterials), the MC grant for training and career development of researchers (Marie Curie, FP7-PEOPLE-2007-1-1-ITN, project SBMPs), an EC FP7 grant for the EDICT consortium (HEALTH-201924), Russian State Contracts No. 02.740.11.0299, 02.740.11.5010, P974 of activity 1.2.2, and No. P211 of activity 1.3.2 of the Federal Target Program "Scientific and Academic Research Cadres of Innovative Russia" for 2009–2013.

6. References

[1] Deisenhofer J. et al., Structure of the protein subunits in the photosynthetic reaction centre of Rhodopseudomonas viridis at 3[angst] resolution, *Nature* 318 (1985) 618-624.

[2] White S.H., The progress of membrane protein structure determination, *Protein Sci.* 13 (2004) 1948-1949.

[3] The Protein Data Bank, http://www.pdb.org/ (2011)

[4] Hirai T. et al., Structural snapshots of conformational changes in a seven-helix membrane protein: lessons from bacteriorhodopsin, *Current Opinion in Structural Biology* 19 (2009) 433-439.

[5] Landau E.M. et Rosenbusch J.P., Lipidic cubic phases: A novel concept for the crystallization of membrane proteins, *Proc.Natl.Acad.Sci.USA* 93 (1996) 14532-14535.

[6] Mariani P. et al., Cubic phases of lipid-containing systems. Structure analysis and biological implications, *J.Mol.Biol.* 204 (1988) 165-189.

[7] Luzzati V. et al., Structure of the cubic phases of lipid-water systems, *Nature* 220 (1968) 485-488.

[8] Scriven L.E., Equilibrium Bicontinuous Structure, *Nature* 263 (1976) 123-125.

[9] Landh T., From entangled membranes to eclectic morphologies: cubic membranes as subcellular space organizers, *FEBS Lett.* 369 (1995) 13-17.

[10] Fontell K., Cubic phases in surfactant and surfactant-like lipid systems, *Colloid & Polymer Science* 268 (1990) 264-285.

[11] B.Ericsson et al., Cubic Phases as Delivery Systems for Peptide Drugs, в: *Polymeric Drugs and Drug Delivery Systems*, American Chemical Society, 1991) 251-265.

[12] V.I.Gordeliy et al., Crystallization in lipidic cubic phases: A case study with Bacteriorhodopsin, *Membrane Protein Protocols: Expression, Purification, and Crystallization*, ed. B.Selinsky, publ.: Humana Press, Totowa NJ, (2003) 305-316.

[13] Caffrey M. et Cherezov V., Crystallizing membrane proteins using lipidic mesophases, *Nature Protocols* 4 (2009) 706-731.

[14] Nollert P. et Landau E.M., Enzymic release of crystals from lipidic cubic phases, *Biochem.Soc.Trans.* 26 (1998) 709-713.

[15] Kouyama T. et al., Polyhedral assembly of a membrane protein in its three-dimensional crystal, *J.Mol.Biol.* 236 (1994) 990-994.

[16] Takeda K. et al., A novel three-dimensional crystal of bacteriorhodopsin obtained by successive fusion of the vesicular assemblies, *J.Mol.Biol.* 283 (1998) 463-474.

[17] Denkov N.D. et al., Electron cryomicroscopy of bacteriorhodopsin vesicles: mechanism of vesicle formation, *Biophys.J.* 74 (1998) 1409-1420.

[18] Faham S. et Bowie J.U., Bicelle crystallization: a new method for crystallizing membrane proteins yields a monomeric bacteriorhodopsin structure, *J.Mol.Biol.* 316 (2002) 1-6.

[19] Faham S. et al., Crystallization of bacteriorhodopsin from bicelle formulations at room temperature, *Protein Sci.* 14 (2005) 836-840.

[20] Sanders C.R. et Schwonek J.P., Characterization of magnetically orientable bilayers in mixtures of dihexanoylphosphatidylcholine and dimyristoylphosphatidylcholine by solid-state NMR, *Biochemistry* 31 (1992) 8898-8905.

[21] Sanders C.R. et Prestegard J.H., Magnetically orientable phospholipid bilayers containing small amounts of a bile salt analogue, CHAPSO, *Biophys.J.* 58 (1990) 447-460.

[22] Rasmussen S.G. et al., Crystal structure of the human beta2 adrenergic G-protein-coupled receptor, *Nature* 450 (2007) 383-387.

[23] Cherezov V. et al., High-resolution crystal structure of an engineered human beta2-adrenergic G protein-coupled receptor, *Science* 318 (2007) 1258-1265.

[24] Ujwal R. et al., The crystal structure of mouse VDAC1 at 2.3 A resolution reveals mechanistic insights into metabolite gating, *Proc.Natl.Acad.Sci.U.S.A* 105 (2008) 17742-17747.

[25] Efremov R.G. et al., The architecture of respiratory complex I, *Nature* 465 (2010) 441-445.

[26] S.Engstrom et al., Solvent-induced sponge (*L3*) phases in the solvent-monoolein-water system, *The Colloid Science of Lipids*, ed. B.Lindman, B.Ninham, publ.: Springer Berlin / Heidelberg,1998) 93-98.

[27] Ridell A. et al., On the water content of the solvent/monoolein/water sponge (L3) phase, *Colloids and Surfaces A: Physicochemical and Engineering Aspects* 228 (2003) 17-24.

[28] Katona G. et al., Lipidic cubic phase crystal structure of the photosynthetic reaction centre from Rhodobacter sphaeroides at 2.35A resolution, *J.Mol.Biol.* 331 (2003) 681-692.

[29] Cherezov V. et al., Room to move: crystallizing membrane proteins in swollen lipidic mesophases, *J.Mol.Biol.* 357 (2006) 1605-1618.

[30] Wohri A.B. et al., A lipidic-sponge phase screen for membrane protein crystallization, *Structure.* 16 (2008) 1003-1009.

[31] Schertler G.F. et al., Orthorhombic crystal form of bacteriorhodopsin nucleated on benzamidine diffracting to 3.6 A resolution, *J.Mol.Biol.* 234 (1993) 156-164.

[32] Borshchevskiy V. et al., Overcoming merohedral twinning in crystals of bacteriorhodopsin grown in lipidic mesophase, *Acta Crystallogr.D.Biol.Crystallogr.* 66 (2010) 26-32.

[33] Parsons S., Introduction to twinning, *Acta Crystallogr.D Biol.Crystallogr.* 59 (2003) 1995-2003.

[34] Dauter Z., Twinned crystals and anomalous phasing, *Acta Cryst. D*59 (2003) 2004-2016.

[35] Efremov R. et al., Physical detwinning of hemihedrally twinned hexagonal crystals of bacteriorhodopsin, *Biophys.J.* 87 (2004) 3608-3613.

[36] Yeates T.O., Detecting and overcoming crystal twinning, *Methods Enzymol.* 276 (1997) 344-358.

[37] Royant A. et al., Detection and characterization of merohedral twinning in two protein crystals: bacteriorhodopsin and p67(phox), *Acta Crystallogr.D Biol.Crystallogr.* 58 (2002) 784-791.

[38] Royant A. et al., Detection and characterization of merohedral twinning in two protein crystals: bacteriorhodopsin and p67(phox), *Acta Crystallogr.D Biol.Crystallogr.* 58 (2002) 784-791.

[39] Redinbo M.R. et al., The 1.5-A crystal structure of plastocyanin from the green alga Chlamydomonas reinhardtii, *Biochemistry* 32 (1993) 10560-10567.

[40] Henderson R. et Moffat J.K., The difference Fourier technique in protein crystallography: errors and their treatment, *Acta Cryst. B*27 (1971) 1414-1420.

[41] Schlichting I. et al., Crystal-Structure of Photolyzed Carbonmonoxy-Myoglobin, *Nature* 371 (1994) 808-812.

[42] Srajer V. et al., Photolysis of the carbon monoxide complex of myoglobin: nanosecond time-resolved crystallography, *Science* 274 (1996) 1726-1729.

[43] Srajer V. et al., Protein conformational relaxation and ligand migration in myoglobin: a nanosecond to millisecond molecular movie from time-resolved Laue X-ray diffraction, *Biochemistry* 40 (2001) 13802-13815.

[44] Schotte F. et al., Watching a protein as it functions with 150-ps time-resolved x-ray crystallography, *Science* 300 (2003) 1944-1947.

[45] Genick U.K. et al., Structure of a protein photocycle intermediate by millisecond time-resolved crystallography, *Science* 275 (1997) 1471-1475.

[46] Genick U.K. et al., Structure at 0.85 A resolution of an early protein photocycle intermediate, *Nature* 392 (1998) 206-209.

[47] Perman B. et al., Energy transduction on the nanosecond time scale: early structural events in a xanthopsin photocycle, *Science* 279 (1998) 1946-1950.

[48] Ren Z. et al., A molecular movie at 1.8 A resolution displays the photocycle of photoactive yellow protein, a eubacterial blue-light receptor, from nanoseconds to seconds, *Biochemistry* 40 (2001) 13788-13801.

[49] Moukhametzianov R. et al., Development of the signal in sensory rhodopsin and its transfer to the cognate transducer, *Nature* 440 (2006) 115-119.

[50] Edman K. et al., High-resolution X-ray structure of an early intermediate in the bacteriorhodopsin photocycle, *Nature* 401 (1999) 822-826.

[51] Sass H.J. et al., Structural alterations for proton translocation in the M state of wild-type bacteriorhodopsin, *Nature* 406 (2000) 649-653.

[52] Royant A. et al., Helix deformation is coupled to vectorial proton transport in the photocycle of bacteriorhodopsin, *Nature* 406 (2000) 645-648.

[53] Matsui Y. et al., Specific damage induced by X-ray radiation and structural changes in the primary photoreaction of bacteriorhodopsin, *J.Mol.Biol.* 324 (2002) 469-481.

[54] Kouyama T. et al., Crystal structure of the L intermediate of bacteriorhodopsin: Evidence for vertical translocation of a water molecule during the proton pumping cycle, *J.Mol.Biol.* 335 (2004) 531-546.

[55] Edman K. et al., Deformation of helix C in the low temperature L-intermediate of bacteriorhodopsin, *J.Biol.Chem.* 279 (2004) 2147-2158.

[56] Takeda K. et al., Crystal structure of the M intermediate of bacteriorhodopsin: allosteric structural changes mediated by sliding movement of a transmembrane helix, *J.Mol.Biol.* 341 (2004) 1023-1037.

[57] Yamamoto M. et al., Crystal structures of different substates of bacteriorhodopsin's M intermediate at various pH levels, *J.Mol.Biol.* 393 (2009) 559-573.

[58] Schobert B. et al., Crystallographic structure of the K intermediate of bacteriorhodopsin: Conservation of free energy after photoisomerization of the retinal, *J.Mol.Biol.* 321 (2002) 715-726.

[59] Lanyi J. et Schobert B., Crystallographic structure of the retinal and the protein after deprotonation of the Schiff base: the switch in the bacteriorhodopsin photocycle, *J.Mol.Biol.* 321 (2002) 727-737.

[60] Lanyi J.K. et Schobert B., Mechanism of proton transport in bacteriorhodopsin from crystallographic structures of the K, L, M-1, M-2, and M-2 ' intermediates of the photocycle, *J.Mol.Biol.* 328 (2003) 439-450.

[61] Schobert B. et al., Crystallographic structures of the M and N intermediates of bacteriorhodopsin: assembly of a hydrogen-bonded chain of water molecules between Asp-96 and the retinal Schiff base, *J.Mol.Biol.* 330 (2003) 553-570.

[62] Lanyi J.K. et Schobert B., Structural changes in the L photointermediate of bacteriorhodopsin, *J.Mol.Biol.* 365 (2007) 1379-1392.

[63] Facciotti M.T. et al., Structure of an early intermediate in the M-state phase of the bacteriorhodopsin photocycle, *Biophys.J.* 81 (2001) 3442-3455.

[64] Lanyi J.K., What is the real crystallographic structure of the L photointermediate of bacteriorhodopsin?, *Biochim.Biophys.Acta* 1658 (2004) 14-22.

[65] Seddon A.M. et al., Membrane proteins, lipids and detergents: not just a soap opera, *Biochim.Biophys.Acta* 1666 (2004) 105-117.

[66] Gordeliy V.I. et al., Molecular basis of transmembrane signalling by sensory rhodopsin II-transducer complex, *Nature* 419 (2002) 484-487.

[67] Misquitta Y. et al., Rational design of lipid for membrane protein crystallization, *J.Struct.Biol.* 148 (2004) 169-175.

[68] Misquitta L.V. et al., Membrane protein crystallization in lipidic mesophases with tailored bilayers, *Structure.* 12 (2004) 2113-2124.

[69] Borshchevskiy V. et al., Isoprenoid-chained lipid [beta]-XylOC16+4--A novel molecule for in meso membrane protein crystallization, *Journal of Crystal.Growth* 312 (2010) 3326-3330.

[70] Yamashita J. et al., New lipid family that forms inverted cubic phases in equilibrium with excess water: molecular structure-aqueous phase structure relationship for lipids with 5,9,13,17-tetramethyloctadecyl and 5,9,13,17-tetramethyloctadecanoyl chains, *J.Phys.Chem.* B112 (2008) 12286-12296.

[71] Hato M. et al., Aqueous phase behavior of lipids with isoprenoid type hydrophobic chains, *J.Phys.Chem.* B113 (2009) 10196-10209.

[72] Pebay-Peyroula E. et al., X-ray structure of bacteriorhodopsin at 2.5 angstroms from microcrystals grown in lipidic cubic phases, *Science* 277 (1997) 1676-1681.

[73] Michel H., Crystallization of Membrane Proteins, (1991)

[74] McPherson A., Crystallization of biological macromolecules, (1999)

[75] Caffrey M., Crystallizing membrane proteins for structure determination: use of lipidic mesophases, *Annu.Rev.Biophys.*38 (2009) 29-51.

[76] Qutub Y. et al., Crystallization of transmembrane proteins in cubo: mechanisms of crystal growth and defect formation, *J.Mol.Biol.*343 (2004) 1243-1254.

[77] Luecke H. et al., Structure of bacteriorhodopsin at 1.55 angstrom resolution, *J.Mol.Biol.*291 (1999) 899-911.

[78] Markov I.V., Crystal growth for the begginers: Fundamentals of Nucleation, Crystal Growth and Epitaxy, second (2003)

[79] Grabe M. et al., Protein interactions and membrane geometry, *Biophys.J.*84 (2003) 854-868.

12

Growth of Organic Nonlinear Optical Crystals from Solution

A. Antony Joseph and C. Ramachandra Raja
Department of Physics, Government Arts College (Autonomous),
Kumbakonam, Tamil Nadu,
India

1. Introduction

Investigations on the growth of good quality single crystals play an important role in the development of modern scientific world with advanced technology. Behind the development in every new solid state device and the explosion in solid state device, there stands a single crystal. Crystal growth is an important field of materials science, which involves controlled phase transformation. In the past few decades, there has been a growing interest in crystal growth process, particularly in view of the increasing demand for materials for technological applications (Brice, 1986). Researchers worldwide have always been in the search of new materials and their single crystal growth.

Solids exist in two forms namely single crystals, polycrystalline and amorphous materials depending upon the arrangement of constituent molecules, atoms or ions. An ideal crystal is one, in which the surroundings of any atom would be exactly the same as the surroundings of every similar atom. Real crystals are finite and contain defects. However, single crystals are solids in the most uniform condition that can be attained and this is the basis for most of the uses of these crystals. The uniformity of single crystals can allow the transmission without the scattering of electro magnetic waves. The methods of growing crystals are very wide and mainly dictated by the characteristics of the material and its size (Buckley, 1951).

2. Nucleation

Comprehensive study on the growth of crystals should start from an understanding of nucleation process. Nucleation is the process of generating within a metastable phase, initial fragments of a new and more stable phase. In a supersaturated or super-cooled system when a few atoms or molecules join together, a change in energy takes place in the process of formation of the cluster. The cluster of such atoms or molecules is termed as "embryo". An embryo may grow or disintegrate and disappear completely. If the embryo grows to a particular size, critical size known as 'critical-nucleus' then greater is the possibility for the nucleus to grow into a crystal. Thus nucleation is an important event in crystal growth.

2.1 Kinds of nucleation

Nucleation is broadly classified into two types. These two types are frequently reserved to as primary and secondary nucleation. The former occurs either spontaneously or induced artificially. The spontaneous formation of crystalline nuclei within the interior of parent phase is called homogeneous nucleation. On the other hand if the nuclei form heterogeneously around ions, impurity molecules or on dust particles or on surface of the container or at structural singularities such as dislocation or imperfection, it is called heterogeneous nucleation. If the nuclei are generated in the vicinity of crystals present in supersaturated system then this phenomenon is often referred to as "secondary" nucleation. Nucleation can often be induced by external influence like agitation, mechanical shock, friction, spark, extreme pressure, electric and magnetic fields, UV -rays, X-rays, gamma rays and so on.

2.2 Classical theory of nucleation

The formation of the crystal nuclei is a difficult and a complex process, because the constituent atoms or molecules have to be oriented into a fixed lattice. In practice, a number of atoms or molecules may come together as a result of statistical incidence to form an ordinary cluster of molecules known as embryo.

2.3 Kinetic theory of nucleation

The main aim of the nucleation theory is to calculate the rate of nucleation. Rate of nucleation is nothing but the number of critical nuclei formed per unit time per unit volume. In kinetic theory, nucleation is treated as the chain reaction of monomolecular addition to the cluster and ultimately reaching macroscopic dimensions is represented as follows:

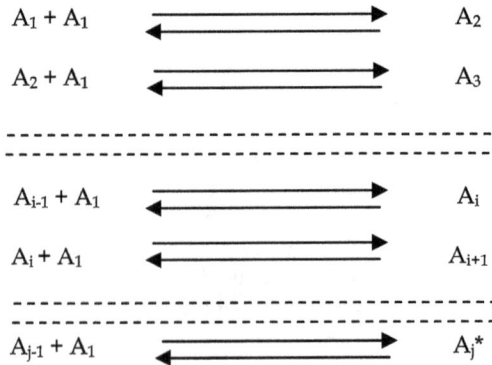

$$A_1 + A_1 \rightleftarrows A_2$$

$$A_2 + A_1 \rightleftarrows A_3$$

$$- -$$

$$A_{i-1} + A_1 \rightleftarrows A_i$$

$$A_i + A_1 \rightleftarrows A_{i+1}$$

$$- -$$

$$A_{j-1} + A_1 \rightleftarrows A_j^*$$

Two monomers collide with one another to form a dimer. A monomer joins with a dimer to form a trimer. This reaction builds a cluster having i-molecules known as i-mer. As the time increases the size distribution in the embryos changes and larger ones have increase in size. As the size attains a critical size A_j^*, the further growth into macroscopic size is guaranteed, as there is a possibility for the reverse reaction i.e., the decay of a cluster into monomers.

3. Stability of nucleus

An isolated droplet of a fluid is most stable when its surface free energy and therefore its area is a minimum. According to Gibbs (Gibbs & Longmans, 1982), the total free energy of a crystal in equilibrium would be minimum for a given volume if its surrounding is at constant temperature and pressure.

If the volume free energy per unit volume is considered to be constant then,

$$\sum a_i \, \sigma_i = \text{minimum} \tag{1}$$

Where, a_i is the area of ith face and σ_i is the corresponding surface energy per unit area.

4. Energy formation of spherical nucleus

The total free energy change associated with the process of homogenous nucleation shall be considered as follows. Let ΔG be the overall excess free energy of an embryo between the two phases mentioned above. ΔG can be represented as a combination of volume and surface energies since an embryo possesses both these energies.

$$\Delta G = \Delta G_s + \Delta G_v \tag{2}$$

Where, ΔG_s is the surface excess free energy and $\Delta G v$ is the volume excess free energy. Assuming the second phase to be spherical.

$$\Delta G = 4 \pi r^2 \sigma + 4/3 \, \pi \, r^3 \Delta G_v \tag{3}$$

Where ΔG_v is the free energy change per unit volume is a negative quantity and 'σ' is the free energy change per unit area. The quantities ΔG, ΔG_s and ΔG_v are represented in Fig 1. The surface excess free energy increases with r^2 and the volume excess free energy decreases with r^3 so the total free energy change increases with increase in size of the nucleus and attains a maximum and then decreases for further increase in the size of nucleus. The size corresponding to the nucleus in which the free energy change is maximum is known as the critical size and can be obtained mathematically by maximizing the equation (3).

$$\text{i.e.,} \qquad \frac{d\Delta G}{dr} = 0$$

Or

$$\frac{d\Delta G}{dr} = 8\pi r \sigma + 4\pi r^2 \, \Delta Gv = 0$$

When $r = r^*$ (radius of critical nucleus), simplifying we have

$$r^* = -2\sigma \; / \; \Delta G_V \tag{4}$$

The free energy change associated with the formation of critical nucleus can be estimated by substituting equation (4) in equation (3)

$$\Delta G^* = 16 \, \pi \, \sigma^3 \; / \; 3 \, \Delta G_V{}^2 \tag{5}$$

$$G^* = 4/3 \, \pi \, r^{*2} \, \sigma$$

$$\Delta G^* = 1/3 \, S.\sigma \tag{6}$$

Where 'S' is the surface area of the critical nucleus. Though the present phase is at constant temperature and pressure, there will be variation in the energies of the molecules. The molecules having higher energies temporarily favour the formation of the nucleus. The rate of nucleation can be given by Arhenius reaction which is a velocity equation since the nucleation process is basically a thermally activated process. The nucleation rate J is given by

$$J = A \exp \left\{ \frac{-\Delta G^*}{KT} \right\} \qquad (7)$$

Where, 'A' is the pre-exponential constant, 'K' is the Boltzman constant and 'T' is the absolute temperature.

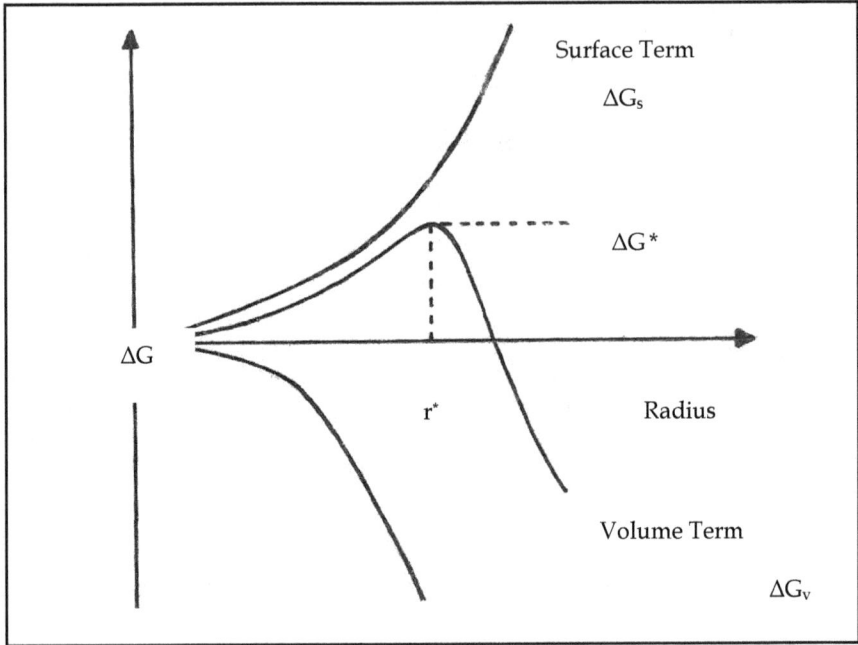

Fig. 1. Free energy diagram.

5. Classification of crystal growth

The growth of the single crystal developed over the years to satisfy the needs of modern technology. The free energy of the growing crystal must be lower than initial stage of the system. It is the common condition for all crystal growth process. The crystal growth method is classified into three types namely growth from melt, from vapour, from the solution. The selection method of crystal growth depending upon the physical properties of material

- Growth from solid   ------>   solid–solid phase transformation
- Growth from liquid ------>   liquid–solid phase transformation
- Growth from vapour ------>   vapour–solid phase transformation

We can consider the conversion of the polycrystalline piece of a material into a single crystal by causing the grain boundaries to sweep through and pushed out of the crystal in the solid–solid growth of crystals. The crystal growth from liquid falls into four categories namely,

i.   Gel growth,
ii. Flux growth,
iii. Hydrothermal growth and
iv. Low temperature solution growth.

Low temperature solution growth is the most widely practiced next to growth from melt. Crystal growth from solution always occurs under condition in which the solvent and crystallizing substance interact. The expression "solution" is most commonly used to describe the liquid which is the result of dissolving a quantity of given substance in a pure liquid known as solvent. Usually water is used as the solvent rarely other liquid is also used as solvent.

6. Solvent selection

The solution is a homogeneous mixture of a solute in a solvent. Solute is the component present in a smaller quantity. For a given solute, there may be different solvents. Apart from high purity starting materials, solution growth requires a good solvent. The solvent must be chosen taking into account the following factors:
1.   A good solubility for the given solution.
2.   A positive temperature co-efficient of solubility.
3.   A small vapour pressure.
4.   Non-corrosiveness.
5.   Non-flammability.
6.   Less viscosity.
7.   Low price in the pure state.

7. Solution preparation and crystal growth

After selecting the desirable solvent with high purity solute to be crystallized, the next important part is preparation of the saturated solution. To prepare a saturated solution, it is necessary to have an accurate solubility-temperature data of the material. The saturated solution at a given temperature is placed in the constant temperature bath. Whatman filter papers are used for solution filtration. The filtered solution is taken in a growth vessel and the vessel is sealed by polythene paper in which 10-15 holes were made for slow evaporation. This solution was transferred to crystal growth vessels and crystallization is allowed to take place by slow evaporation at room temperature or at a higher temperature in a constant temperature bath. As a result of slow evaporation of solvent, the excess of solute which got deposited in the crystal growth vessel results in the formation of seed crystals.

8. Low temperature solution growth methods

Solution growth is the most widely used method for the growth of crystals, when the starting materials are unstable at high temperatures. In general, this method involves seeded growth from a saturated solution. The driving force i.e., the supersaturation is achieved either by temperature lowering or by solvent evaporation. This method is widely used to grow bulk crystals, which have high solubility and have variation in solubility with temperature (James & Kell, 1975).
Low temperature solution growth (LTSG) can be subdivided into the following categories:
i.   Slow cooling method
ii.   Slow evaporation method
iii. Temperature gradient method

8.1 Slow cooling method
In this process, supersaturated solution is prepared by keeping quantity of the solution same as that of the initial stage and temperature of the solution is reduced in small step. By doing so, solution which is just saturated at initial temperature will become supersaturated solution. Once supersaturation is achieved, growth of single crystal is possible. The main disadvantage of slow cooling method is the need to use a range of temperature. Wide range of temperature may not be desirable because the properties of the grown crystal may vary with temperature. Even though this method has technical difficulty of requiring a programmable temperature control, it is widely used with great success.

8.2 Slow evaporation method
In this process the temperature of the solution is not changed, but the solution is allowed to evaporate slowly. Since the solvent evaporates, concentration of solute increased and therefore supersaturation is achieved. For example 40 g of solute in 100 ml solvent is considered as saturated solution at 50°C. Now the solution is allowed to evaporate at the same temperature. The 100 ml of the solution is reduced to some lower level say 70 ml. Then 40 g in 70 ml at 50°C is supersaturated solution. The evaporation technique has an advantage that the crystals grow at a fixed temperature. But inadequacies of the temperature control system still have a major effect on the growth rate. This method can effectively be used for materials having very low temperature coefficient of solubility.

8.3 Temperature gradient method
This method involves the transport of the materials from hot region containing the source materials to be grown to a cooler region, where the solution is supersaturated and the crystal grows. The advantages of this method are that [a] the crystal is grown at fixed temperature, [b] this method is insensitive to changes in temperature provided both the source and the growing crystal under go the same change. [c] economy of the solvent. On the other hand, small changes in temperature difference between the source and the crystal zones have a large effect on the growth rate.

In general, crystal growth from solution is mainly influenced by super saturation. Super saturation may be achieved by any methods (described above) which are based on the principle that solution which is saturated at a particular temperature will behave as unsaturated at high temperature. The disadvantages are the slow growth rate and in many cases inclusion of the solvent in to the growing crystal. Materials having moderate to high solubility in temperature range, ambient to 100°C at atmospheric pressure can be grown by LTSG method. This method is well suitable for those materials which suffer from decomposition in the melt and which undergo structural transformation while cooling from the melting point. The other advantages of LTSG method are the low working temperature, easy operation and feasible growth condition.

9. Nonlinear optical crystals

Non Linear optics deals with the interaction of intense electromagnetic fields in suitable medium producing magnified fields different from the input field in frequency, phase or amplitude Nonlinear optics is now established as an alternative field to electronics for the future photonic technologies. The fast-growing development in optical fiber communication systems has stimulated the search for new highly nonlinear materials capable of fast and efficient processing of optical signals. Organic nonlinear optical (NLO) materials have been intensely investigated due to their potentially high nonlinearities and rapid response in electro-optic effect compared to inorganic NLO materials. In recent years, there has been considerable interest in the study of organic NLO crystals with good nonlinear properties because of their wide applications in the area of laser technology, optical communication, optical information processing and optical data storage technology (Chenthamarai et al., 2000). Among the organic crystals for NLO applications, amino acids display specific features of interest such as (i) molecular chirality, which secures acentric crystallographic structures, (ii) absence of strongly conjugated bonds, leading to wide transparency ranges in the visible and UV spectral regions, (iii) Zwitterionic nature of the molecule, which favours crystal hardness. Further they can be used as a basis for synthesizing organic compounds and derivatives (Eimerl et al., 1990). In our laboratory, we have grown NLO crystals such as L-Alaninium Succinate (LAS), L-Valinium Succinate (LVS), L-Alaninium Fumarate (LAF), L-Valinium Fumarate (LVF) and reported in the journal of repute (Ramachandra Raja, 2009a, 2009b, 2009c, 2010).

In this chapter, we have discussed the growth of organic nonlinear optical crystal. L-Alaninium Succinate (LAS) and L- Valinium Succinate (LVS) which have been grown by slow evaporation solution growth technique in detail. The characteristic studies such as single crystal and powder X-ray Diffraction (XRD) analysis, UV–Vis–NIR spectrum, FT-IR, nuclear magnetic resonance studies, TGA/DTA studies and SHG are also discussed.

10. Growth and characterization of L- Alaninium Succinate (LAS)

10.1 Crystal growth

LAS have been grown from aqueous solution by slow evaporation. The starting material was synthesized from commercially available L-Alanine (AR grade) and Succinic acid (AR grade), taken in the equimolar ratio 1:1. In deionized water, L-Alanine and Succinic acids were allowed to react by the following reaction to produce LAS.

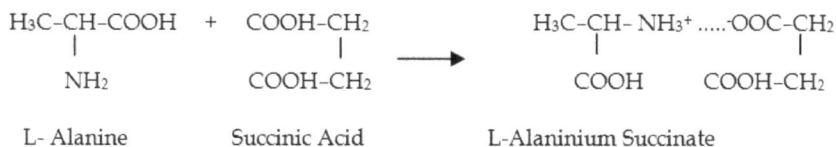

$$H_3C-CH-COOH \quad + \quad COOH-CH_2 \quad \longrightarrow \quad H_3C-CH-NH_3^+ \ldots .^-OOC-CH_2$$
$$\qquad\quad | \qquad\qquad\qquad | \qquad\qquad\qquad\qquad\qquad | \qquad\qquad\quad |$$
$$\qquad\quad NH_2 \qquad\qquad\quad COOH-CH_2 \qquad\qquad\qquad\quad COOH \qquad COOH-CH_2$$

L- Alanine   Succinic Acid   L-Alaninium Succinate

Calculated amount of the reactants were thoroughly dissolved in deionized water and stirred well for about 3 hours using a magnetic stirrer to obtain a homogenous mixture. Then the solution was allowed to evaporate slowly until the solvent was completely dried. The purity of the synthesized salt was further increased by successive recrystallization process. The synthesized powder of LAS was dissolved thoroughly in double distilled water to form a saturated solution. The solution was then filtered twice to remove any insoluble impurities. Growth was carried out by low-temperature solution growth technique by slow evaporation in a constant temperature bath controlled to an accuracy of ±0.01°C. Crystals begin to grow inside the solution and were removed from the solution after 3 weeks, washed and dried in air.

10.2 Characterization studies
10.2.1 Single crystal XRD analysis

In order to estimate the crystal data, the single crystal XRD analysis of grown LAS crystal have been carried out using ENRAF NONIUS CAD-4 X-ray diffractometer equipped with MoKα (λ = 0.71069 Å) radiation. The X-ray diffraction study on grown crystal reveals that the grown crystal belongs to orthorhombic system with the following unit cell parameters: a = 5.77 Å, b = 6.02 Å, c = 12.32 Å and $\alpha = \beta = \gamma = 90\,^\circ$, the cell volume = 428 Å^3. These lattice parameters are tabulated in the Table 1.

10.2.2 Powder XRD analysis

The structural property of the single crystals of LAS has been studied by X-ray powder diffraction technique. Powder X-ray diffraction studies of LAS crystal is carried out, using Rich Seifert diffractometer with CuK$_\alpha$ (λ =1.54060 Å) radiation. The sample is scanned for 2θ values from 10° to 90° at a rate of 2°/min. Figure 2 shows the Powder XRD pattern of the pure LAS crystal. The diffraction pattern of LAS crystal has been indexed by Reitveld index software package. The lattice parameter values of LAS crystal has been calculated by Reitveld unit cell software package and are matched with single crystal XRD data. The comparison of lattice parameters between single crystal and powder XRD is shown in Table 1.

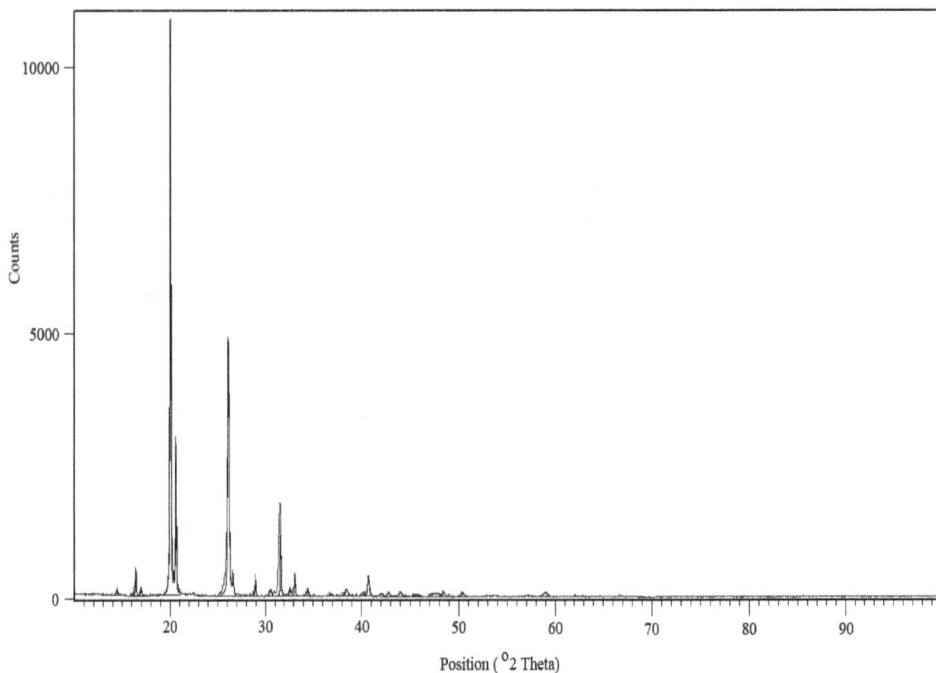

Fig. 2. Powder XRD pattern of LAS crystal.

XRD	a Å	b Å	c Å	α deg	β deg	γ deg	Volume Å³
Single crystal	5.77	6.02	12.32	90	90	90	428
Powder	5.74	5.98	12.53	90	90	90	430

Table 1. The cell parameters of LAS crystal.

Position °2 θ	d- spacing Å	(h k l)
14.4335	6.13689	(0 0 2)
16.4368	5.39317	(0 1 1)
16.9673	5.22574	(1 0 1)
20.0604	4.42642	(0 1 2)
20.6379	4.30384	(1 0 2)
26.1583	3.40675	(0 1 3)
26.6351	3.34684	(1 0 3)
28.9666	3.08254	(0 0 4)
30.5183	2.92926	(1 1 3)
31.5955	2.83181	(2 0 1)
32.5805	2.74841	(0 1 4)
33.0863	2.70753	(0 2 2)
34.4077	2.60652	(2 1 0)
38.4281	2.34257	(2 0 3)
40.6932	2.21726	(1 2 3)
42.0137	2.15057	(0 2 4)
42.789	2.11339	(2 0 4)
43.9448	2.06045	(2 2 1)
45.5981	1.98951	(2 1 4)
47.5815	1.91111	(0 3 2)
48.4142	1.88017	(2 0 5)
50.3422	1.81259	(1 2 5)
58.9467	1.56559	(3 2 2)

Table 2. Powder XRD data of LAS crystal.

It is observed that LAS belongs to orthorhombic system and cell parameters values are in good agreement with the single crystal XRD data. The h, k, l values, d-spacing and 2θ values are tabulated in Table 2.

10.2.3 UV-Vis-NIR analysis

The UV–Vis–NIR transmittance spectrum of grown LAS crystal has been recorded with a Lambda 35 double-beam spectrophotometer in the range 190–1100 nm to find the suitability of LAS crystal for optical applications. The recorded spectrum is shown in Fig. 3. The crystal shows a good transmittance in the visible region which enables it to be a good material for optoelectronic applications. As observed in the spectrum, there is no significant absorption in the entire range tested. A good optical transmittance from ultraviolet to infrared region is very useful for nonlinear optical applications. Most of the nonlinear optical effects are studied using Nd:YAG laser operating at a fundamental wavelength of 1064 nm. Absorption, if any, near the fundamental or the second harmonic at 532 nm, will lead to a loss of conversion efficiency of second harmonic generation (SHG). From the UV–Vis–NIR spectrum, it is clear that the transparency of the grown crystals extends up to UV region.

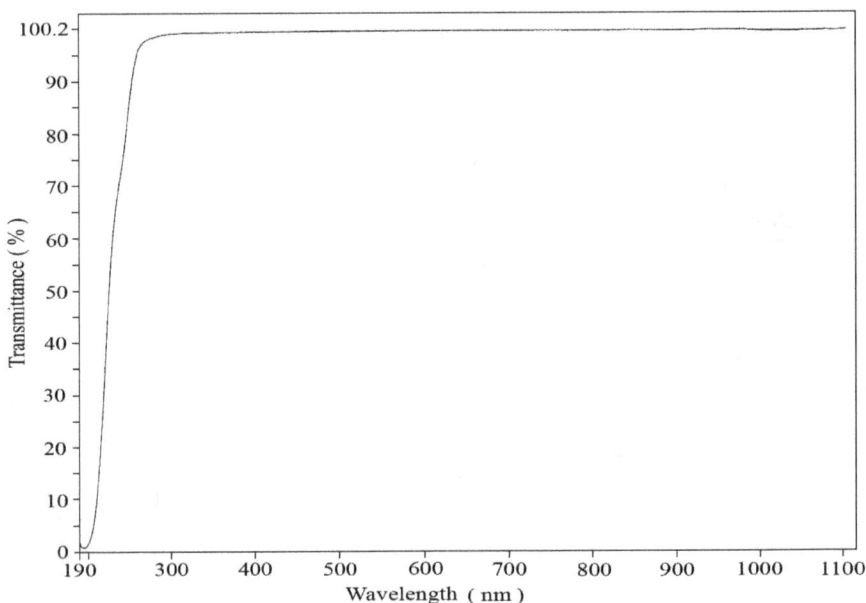

Fig. 3. Transmission spectrum of LAS crystal.

The lower cut-off wavelength is as low at 190 nm. The wide range of transparency suggests that the crystal is a good candidate for nonlinear optical applications (Aravindan et al., 2007). This transmittance window (190–1100 nm) is sufficient for the generation of second harmonic light (λ = 532 nm) from the Nd:YAG laser (λ=1064nm) (Natarajan et al, 2008). The lower cut-off near 190 nm combined with the very good transparency, makes the usefulness of this material for optoelectronic and nonlinear optical applications.

10.2.4 FT-IR analysis

The infrared spectrum of LAS has been carried out to analyse the chemical bonding and molecular structure of the compound. The FT-IR spectrum of the crystal has recorded in the frequency region from 400 cm⁻¹ to 4000 cm⁻¹ with Perkin–Elmer FT-IR spectrometer model SPECTRUMRX1 using KBr pellets containing LAS powder obtained from the grown single crystals. The observed FT–IR spectrum of LAS is as shown in Fig. 4.

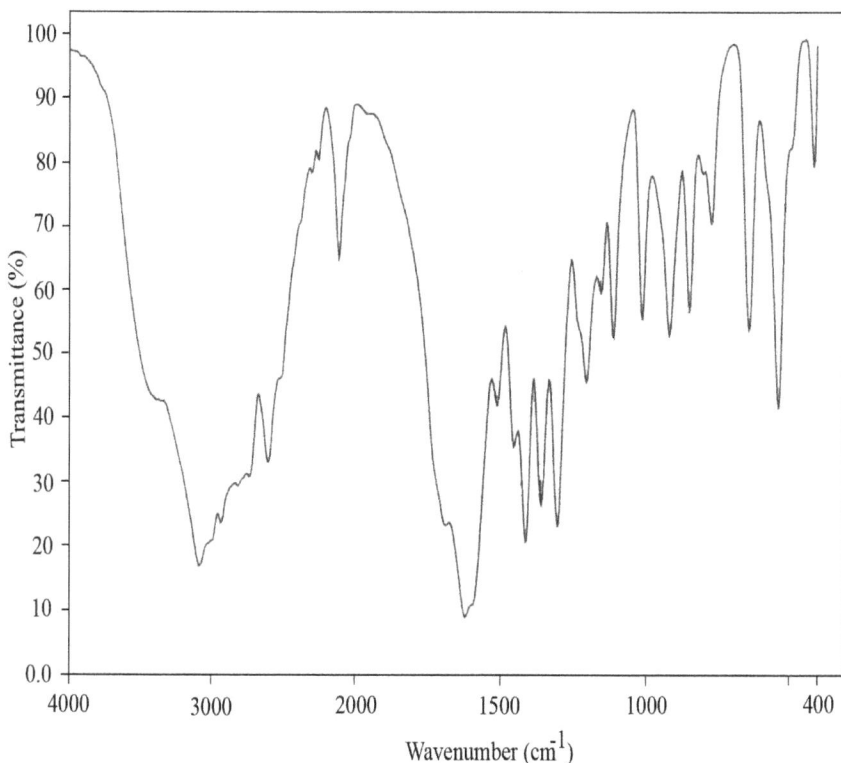

Fig. 4. FT-IR spectrum of LAS crystal.

The characteristics vibration of LAS has been compared with L-alaninium alanine nitrate (Aravindan et al., 2007) and L-alanine cadmium chloride (Dhanuskodi et al., 2007) as shown in Table 3. The asymmetric stretching vibration of NH_3^+ is observed at 3086 cm⁻¹ of LAS is confirming the presence of NH_3^+ in compound. The NH_3^+ absorption range of amino acids (3130–3100 cm⁻¹) is shifted to lower wave number, due to formation of amino salts, and in LAS, it is observed at 3086cm⁻¹.

Amino group absorption bands are noted at 2604 cm⁻¹ (symmetric stretching), 1620cm⁻¹ (bending), and 1111 cm⁻¹ (rocking). These bands are due to NH_3^+ ions. During the formation of amino salts, the NH_2 group in amino acids is converted in to NH_3^+ ion. The strong absorption at 1413 cm⁻¹ indicates the symmetric stretching vibration frequency of carbonyl group. The bending and rocking vibrations of COO⁻ are observed at 772 cm⁻¹ and 539 cm⁻¹, respectively. CH_2 wagging (1304cm⁻¹) and CH_3 stretching (1204 cm⁻¹) vibrations are also observed (Ramachandran & Natarajan, 2007).

Wavenumber (cm^{-1})	Assignment
3086	NH$_3^+$ asymmetric stretching
2604	NH$_3^+$ symmetric stretching
1620	NH$_3^+$ bending
1453	CH$_3$ bending
1413	COO$^-$ symmetric stretching
1360	CH$_3$ symmetric bending
1304	CH$_2$ wagging
1204	CH$_3$ symmetric stretching
1111	NH$_3^+$ rocking
1012	CH$_3$ rocking
917	CCN symmetric stretching
848	C-CH$_3$ bending
772	COO$^-$ bending
539	COO$^-$ rocking
412	COO$^-$ rocking

Table 3. Assignments of FT-IR bands observed for LAS crystal.

10.2.5 NMR studies

The ^{1}H- and ^{13}C-NMR spectra of LAS have been recorded using D$_2$O as solvent on a Bruker 300MHz (Ultrashield) TM instrument at 23°C (300 MHz for ^{1}H-NMR and 75 MHz for ^{13}C-NMR) to confirm the molecular structure. The spectra are shown in Figures 5 and 6 respectively and the chemical shifts are tabulated with the assignments in Table 4.

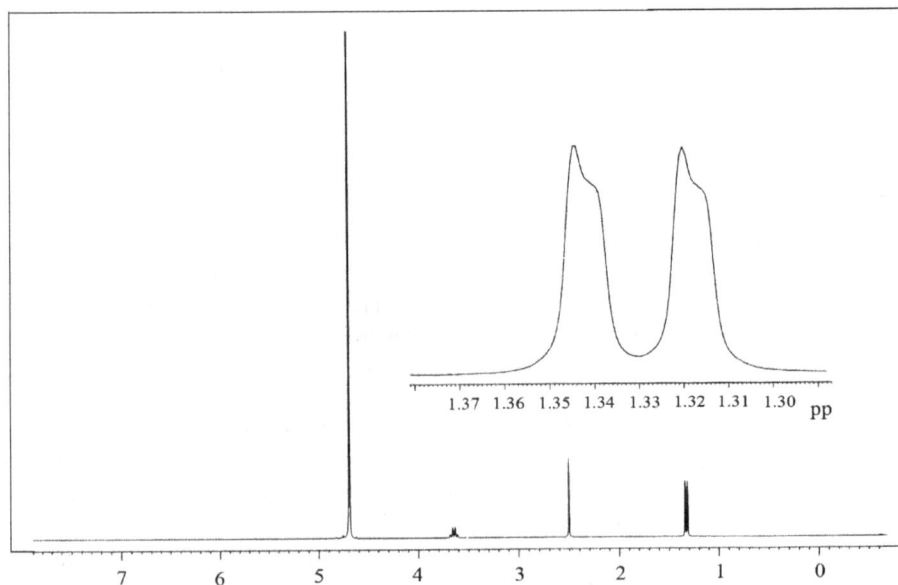

Fig. 5. ^{1}H-NMR spectrum of LAS crystal.

The resonance peaks at δ = 1.33 ppm of ^{1}H-NMR spectrum is due to the CH$_3$ group and peaks observed at δ = 3.65 ppm is due to the CH group of L-Alanine. The methyl proton signal at δ = 1.33 ppm is split into a proton doublet due to the coupling of the neighboring proton (CH) and the signal at δ = 3.65 ppm is split into a proton quartet due to the coupling of three neighboring protons (CH$_3$). The resonance peak observed as a singlet at δ = 2.57 ppm exhibits the presence of methylene (CH$_2$) proton of succinic acid.

The signal at δ = 4.69 ppm is due to the solvent (D$_2$O). The signals due to NH and COOH do not show up because of fast deuterium exchange reactions takes place in these two groups, with D$_2$O being used as solvent (Bruice, 2002). Because of the presence of the methylene (CH$_2$) groups of LAS, electron contributions towards the rest of the compound get enhanced, so that the protons are more protected in LAS. Such property is not noticed in L-alaninium oxalate (LAO), due to the absence of methylene groups so that the proton groups in LAS absorbs signals at the values lesser than the value of LAO (Dhanuskodi & Vasantha, 2004). The ^{13}C NMR spectrum of LAS contains five signals. The resonance peaks observed at δ = 16.00 ppm and at δ = 50.33 ppm are due to the carbon environments of CH$_3$ and CH groups of L-alanine respectively. The signal at δ = 29.06 ppm is due to the presence of two methylene (CH$_2$) groups of succinic acid. The resonance signal observed at δ = 175.52 ppm is due to the free carboxylic acid from L -alanine. In solution, the two carboxyl groups of succinic acid are equivalent due to the fast exchange of H$^+$ between them and give rise to a single signal at δ = 177.41 ppm.

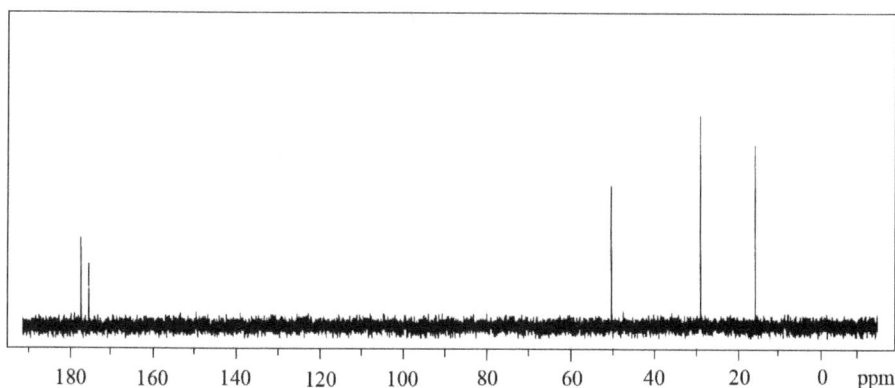

Fig. 6. ^{13}C-NMR spectrum of LAS crystal.

Spectra	Chemical Shift (ppm)	Group identification
^{1}H NMR	1.33	- CH$_3$-
	2.57	- CH$_2$ -
	3.65	- CH -
	4.69	- D$_2$O
^{13}C-NMR	16.00	- CH$_3$
	29.06	- CH$_2$ -
	50.33	- CH -
	175.52	COOH of L-Alanine
	177.41	COOH of Succinic Acid

Table 4. The chemical shifts in ^{1}H-NMR and ^{13}C-NMR spectra of LAS.

10.2.6 Thermal analysis

The Thermo Gravimetric Analysis (TGA), Differential Thermal Analysis (DTA) spectra of grown LAS crystal have been obtained using the instrument NETSZCH SDT Q 600 V8.3 Build 101. The TGA and DTA have been carried out in nitrogen atmosphere at a heating rate of $20^{\circ}C/min$ from $0^{\circ}C$ to $1000^{\circ}C$. The TGA curve is presented in Fig. 7.

The initial mass of the materials to analysis was 2.5720 mg and the final mass left out after the experiment was only 1.729 % of initial mass. The TGA trace shows that the material exhibit very small weight loss of about 1.17 % in the temperature up to $155^{\circ}C$ due to loss of water. TGA curve shows that there is the weight loss (85%) between $178^{\circ}C$ and $274^{\circ}C$ indicating that the decomposition of LAS crystals. From the Fig. 7, the appearance of endothermic in the DTA at $178^{\circ}C$ corresponds to TGA results. From the TGA, DTA analyses, it is clearly understood that the LAS is thermally stable upto $178^{\circ}C$.

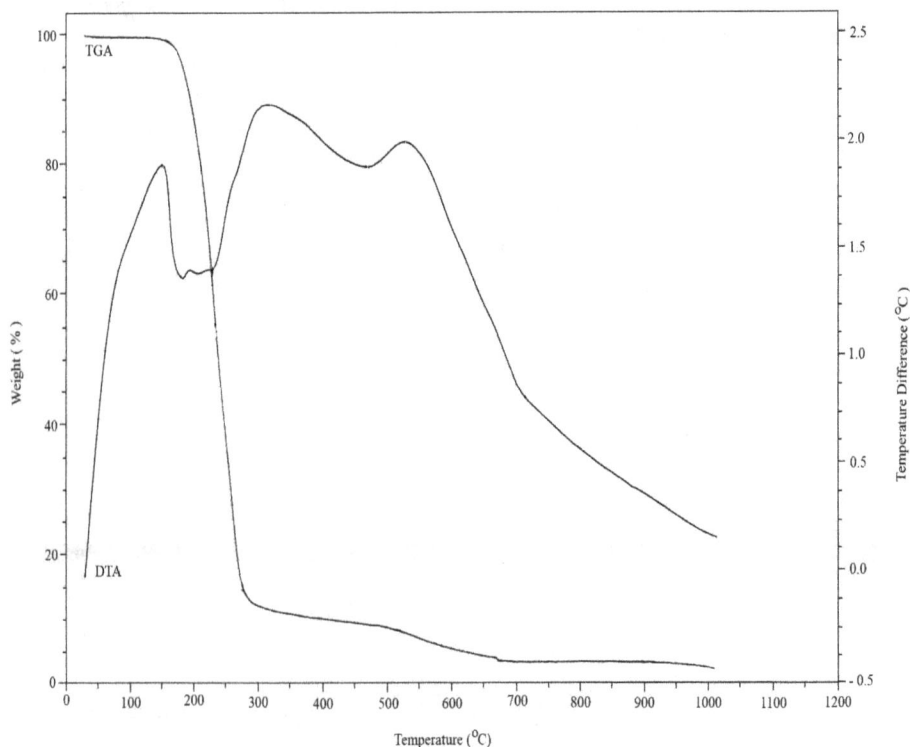

Fig. 7. TGA / DTA curve of LAS crystal.

10.2.7 Second harmonic generation analysis

A preliminary study of the powder SHG measurement of LAS has been performed using Kurtz powder technique (Kurtz & Perry, 1968) with 1064nm laser radiations. An Nd:YAG laser producing pulses with a width of 8 ns and a repetition rate of 10 Hz was used. The crystal was crushed into powder and densely packed in a capillary tube. It is observed that the crystal converts the 1064 nm radiation into green (532 nm) while passing the Nd:YAG laser output into the sample which confirms the SHG. The observed intensity of output light is obtained as 12 mV and for the same incident radiation, the output of KDP is observed as 52 mV. It was found that the efficiency of SHG is 23% when compared with that of the standard KDP (Ramachandra Raja , 2009c).

11. Growth and characterization of L- Valinium Succinate (LVS)

11.1 Experimental procedure
The nonlinear optical crystal L-Valinium Succinate (LVS) were grown by slow evaporation solution growth method. The LVS was synthesized from analar grade L-Valine and Succinic acid which were taken in equimolar ratio 1:1 using following reaction:

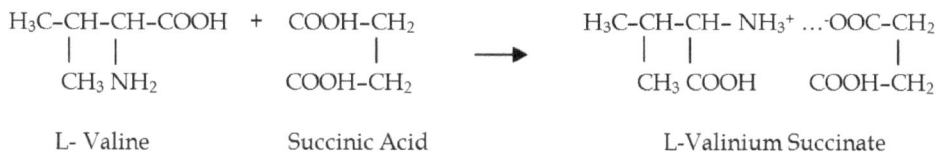

$$H_3C-CH-CH-COOH \quad + \quad COOH-CH_2 \longrightarrow H_3C-CH-CH- NH_3^+ ...^-OOC-CH_2$$

$$\begin{array}{ccc} | \quad | & | & | \quad | & | \\ CH_3 \ NH_2 & COOH-CH_2 & CH_3 \ COOH & COOH-CH_2 \end{array}$$

L- Valine   Succinic Acid   L-Valinium Succinate

The calculated amounts of reactants were thoroughly dissolved in double distilled water and stirred continuously using magnetic stirrer. The saturated solution may contain impurities such as solid and dust particles and therefore it was filtered using filter paper. Then the filtered solution was covered by polythene paper in which 10 to 15 holes were made for slow evaporation. This solution was transferred to crystal growth vessels and crystallization was allowed to take place by slow evaporation at a temperature range of 35°C in a constant temperature bath of accuracy ±0.01°C. As a result of slow evaporation of water, the excess of solute has grown as LVS crystals in the period of two weeks.

11.2 Characterization studies
11.2.1 Single crystal XRD analysis
The X-Ray diffraction pattern of LVS crystals have been studied by ENRAF NONIUS CAD4 single crystal X-Ray diffractometer with MoKα radiation (λ=0.71069 Å). The single crystal X-ray diffraction study of crystals is used to identify the cell parameters. It is observed that the LVS crystal belongs to orthorhombic system with following cell parameters: a = 9.85 Å, b = 5.35 Å, c = 12.26 Å and α = β = γ = 90 °, the cell volume = 646 Å³. From the lattice parameters it is clear that for grown crystal a ≠ b ≠ c and α = β = γ = 90 ° and they are compared with powder XRD data and tabulated in Table 5.

XRD	a Å	b Å	c Å	α deg	β deg	γ deg	Volume Å³
Single crystal	9.85	5.35	12.26	90	90	90	646
Powder	9.99	5.36	12.19	90	90	90	652

Table 5. The cell parameters of LVS crystal.

11.2.2 Powder XRD analysis
The powder X-ray diffraction (XRD) pattern of LVS crystals has been obtained using Rich Seifert X-ray diffractometer. The crushed powder sample was subjected to intense X-rays of wavelength 1.54060 Å (CuK$_\alpha$) at a scan speed of 1°/minute. The powder X-ray pattern of LVS is shown in Fig. 8. The observed powder XRD pattern has been indexed by Rietveld Index software package. The lattice parameters have been calculated by Rietveld Unit Cell

software package and they are shown in Table 5. It is observed that LVS belongs to orthorhombic system and cell parameters values are agreed with the single crystal XRD data. The h, k, l values, d-spacing and 2θ values are tabulated in Table 6.

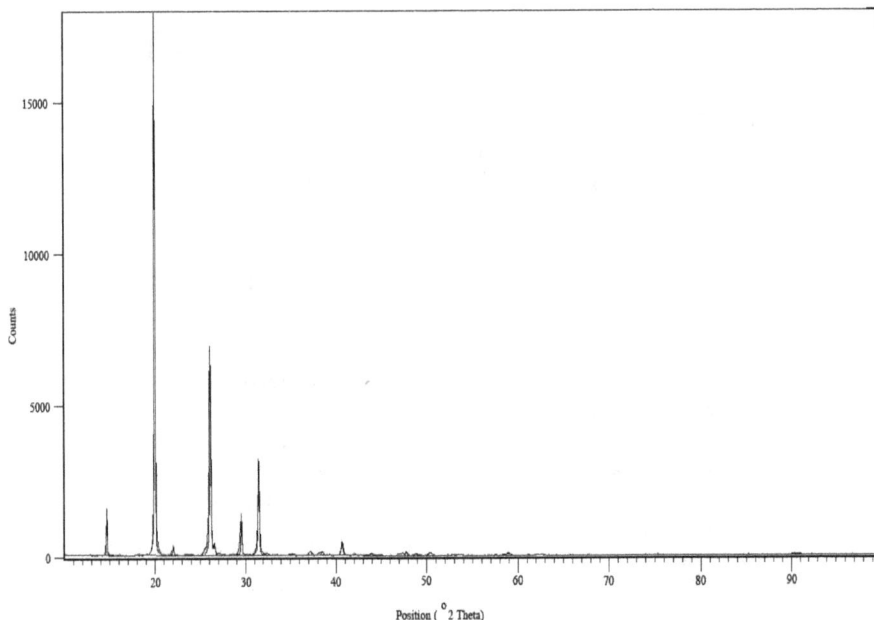

Fig. 8. Powder XRD pattern of LVS crystal.

Position°2 θ	d- spacing Å	(h k l)
14.6613	6.04207	(0 0 2)
20.0051	4.43853	(0 2 1)
22.0352	4.03399	(1 0 2)
23.6856	3.75650	(0 1 3)
26.1127	3.41259	(1 2 1)
29.5373	3.02427	(0 0 4)
31.4841	2.84157	(0 3 2)
35.0908	2.55733	(0 3 3)
37.1320	2.42131	(2 0 2)
38.3887	2.34489	(2 2 0)
40.6179	2.22120	(1 0 5)
43.8867	2.06304	(1 3 4)
47.7577	1.90447	(1 0 6)
48.8085	1.86590	(2 2 4)
50.3581	1.81206	(1 4 4)
58.9110	1.56775	(3 3 0)
90.4585	1.08503	(2 2 10)

Table 6. Powder XRD data of LVS crystal.

11.2.3 UV-Vis-NIR analysis

To find the optical transmission range of LVS crystals, the UV-Vis-NIR spectrum has been recorded using Lambda 35 double beam spectrophotometer in the range between 190 nm and 1100 nm and it is shown in Fig. 9. When the transmittance is monitored from longer to shorter wavelengths, LVS is transparent from 190 nm to 1100 nm. Optical absorption with lower cut-off wavelength near 190 nm makes the crystal suitable for UV tunable laser and SHG device applications.

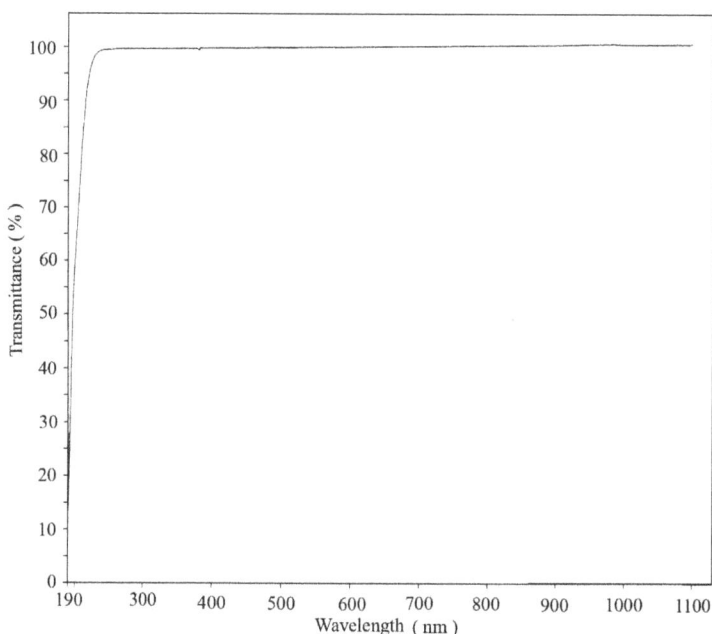

Fig. 9. UV-Vis-NIR spectrum of LVS crystal.

11.2.4 FT-IR analysis

The Fourier transform infrared spectrum of LVS have been recorded in between the region 400 – 4000 cm^{-1} using Perkin Elmer Fourier transforms infrared spectrometer (model SPECTRUM RX1) with the help of KBr pellets as shown in Fig. 10. The presence of functional groups was identified and they are stacked in Table 7. The presence of NH_3^+ group in LVS has confirmed by peaks at 3429 cm^{-1} and 3156 cm^{-1}. It is due to protonation of NH_2 group by the COOH group of succinic acids (Nakamo, 1978; Sajan et al., 2004). The symmetric and asymmetric bending of NH_3^+ was obtained at 1587 and 1508 cm^{-1} respectively. The strong absorption at 1393 cm^{-1} indicates that the symmetric bending of CH_2. The CH_2 wagging vibration produces a sharp peak at 1327 cm^{-1}. The C-CH bending vibration produced its characteristic peak at 1270 cm^{-1}. The rocking vibration at 1177 cm^{-1} establishes the presence NH_3^+ group. The peak at around 1137 cm^{-1} is assigned to NH_3^+ wagging. The stretching vibration of C-O-C, C-C-N and C-C are positioned at 2108 cm^{-1}, 1063 cm^{-1} and 1029 cm^{-1} respectively. Meanwhile, for the peaks at 945 cm^{-1} is due to CH_2

rocking. The bending vibration of COO- is observed at 662 cm^{-1}. The bending and rocking vibration of COO- are observed at 714 cm^{-1} and 430 cm^{-1} respectively.

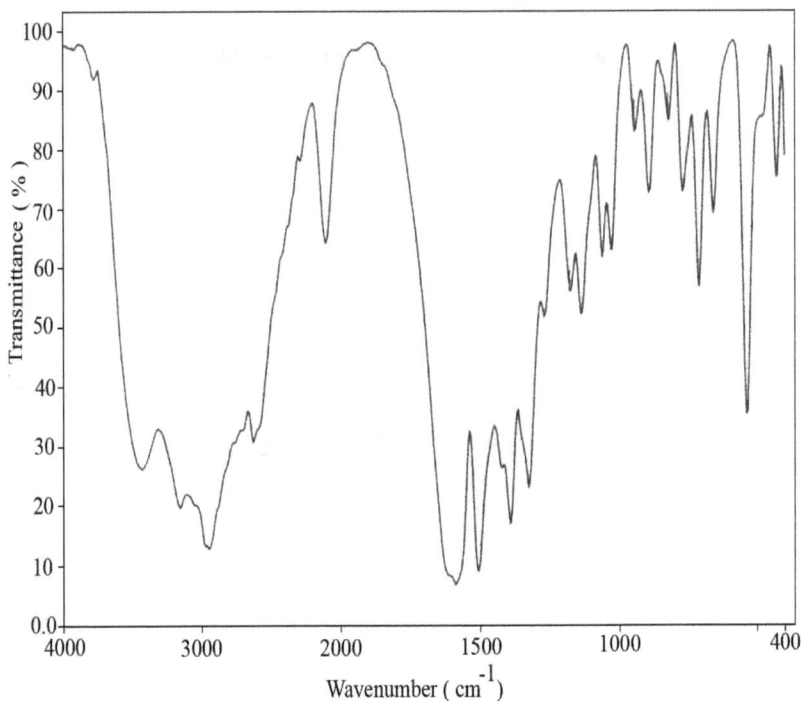

Fig. 10. FT-IR spectrum of LVS crystal.

11.2.5 NMR studies

The ^{1}H- and ^{13}C-NMR spectra of LVS have been recorded for the crystals dissolved in water (D$_2$O) using BRUKER 300 MHz (Ultrashield)™ instrument at 23°C (300 MHz for ^{1}H-NMR and 75 MHz for ^{13}C-NMR) for the confirmation of molecular structure. The ^{1}H – NMR spectrum of LVS is shown in the Fig. 11. The deuterium exchange proton NMR spectrum of LVS crystal is found to contain resonance signals integrated for a total of 12 protons. From the spectrum, it is observed that the two methyl proton signal is split into two doublets due to the coupling of neighbouring (-CH) proton which is confirmed from the signal at δ = 0.84 ppm, δ = 0.89 ppm respectively. The –CH group signal is split into a multiplet due to the hyperfine splitting of neighbouring three (-CH$_3$) protons is confirmed from the signal of LVS crystal centered at = δ 2.13 ppm. The doublet signal observed at δ 3.48 ppm is attributed to a (-CH) proton next to carboxylic acid. There is one peak found at δ = 2.51 ppm due to the -CH$_2$- group of succinic acid. The signal at δ = 4.69 ppm is due to the solvent D$_2$O. The signals due to N-H and COOH do not show up because of fast deuterium exchanges which took place in those two groups, where the D$_2$O was used as the solvent (Bruice 2002). The chemical shift values of LVS with assignments are tabulated in Table 8.

Wavenumber (cm^{-1})	Assignment
3429	NH$_3^+$ symmetric stretching
3156	NH$_3^+$ asymmetric stretching
2946	CH$_2$ asymmetric stretching
2626	NH$_3^+$ symmetric stretching
2108	C-O-C stretching
1587	NH$_3^+$ symmetric bending
1508	NH$_3^+$ asymmetric bending
1393	CH$_2$ symmetric bending
1327	CH$_2$ wagging
1270	C-CH bending
1177	NH$_3^+$ rocking
1137	NH$_3^+$ wagging
1063	C-C-N stretching
1029	C-C stretching
945	CH$_2$ rocking
893	C-C-N stretching
823	COO$^-$ rocking
773	NH wagging
714	COO$^-$ bending
662	COO$^-$ bending
541	C-C=O wagging
430	COO$^-$ rocking

Table 7. Assignments of FT-IR bands observed for LVS crystal.

The ^{13}C-NMR spectrum is shown in Fig. 12. The characteristic absorption peaks of ^{13}C-NMR spectrum of LVS are explained as follows. The signals at δ = 17.82 ppm and δ = 16.55 ppm are attributed to the two methyl group of LVS. An intense signal is observed at δ = 28.93 ppm is due to presence of two methylene groups of succinic acid. The signal of (CH) at δ = 29.00 ppm is integrated for one carbon due to presence of carboxylic acid isopropyl carbon. The peaks at δ = 60.12 ppm is due to tertiary carbon connected to amino group. The peak at δ = 173.97 ppm and δ = 177.28 ppm are due to deuterium exchange of carbon in carbonyl group. A peak with higher intensity at δ = 177.28 ppm can be safely attributed to carbonyl carbons of two COOH groups of succinic acid present in the same chemical environment.

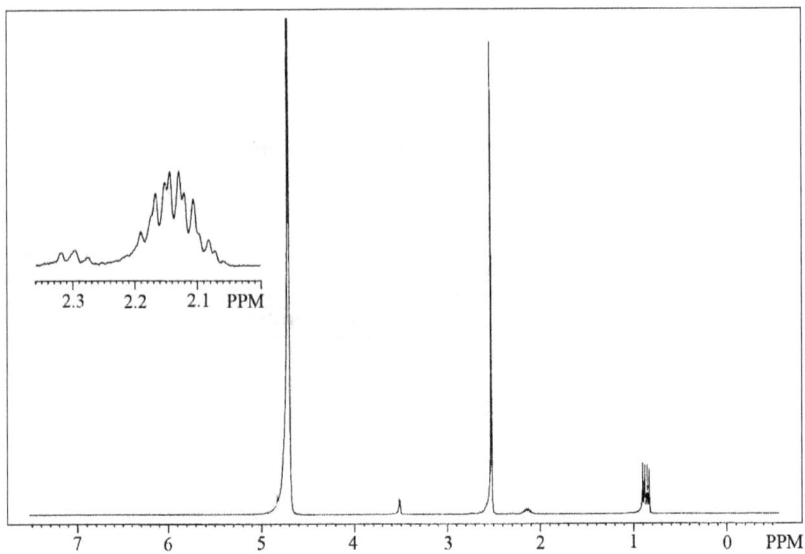

Fig. 11. ¹H-NMR spectrum of LVS crystal.

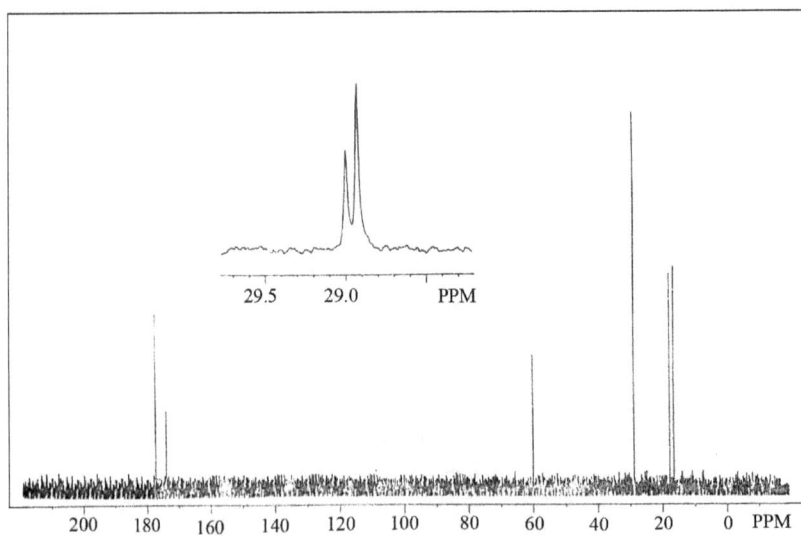

Fig. 12. ¹³C-NMR spectrum of LVS crystal.

Spectra	Chemical Shift (ppm)	Group identification
¹H NMR	0.84 & 0.89	- (CH₃) -
	2.13	- CH -
	2.51	- CH₂ -
	3.48	- CH -
	4.69	D₂O
¹³C-NMR	16.55 & 17.82	- (CH₃) -
	29.00	- CH –
	28.93	- CH₂ -
	60.12	- CH -
	173.97	COOH of L-Valine
	177.28	COOH of Succinic Acid

Table 8. The chemical shifts in ¹H-NMR and ¹³C-NMR spectra of LVS.

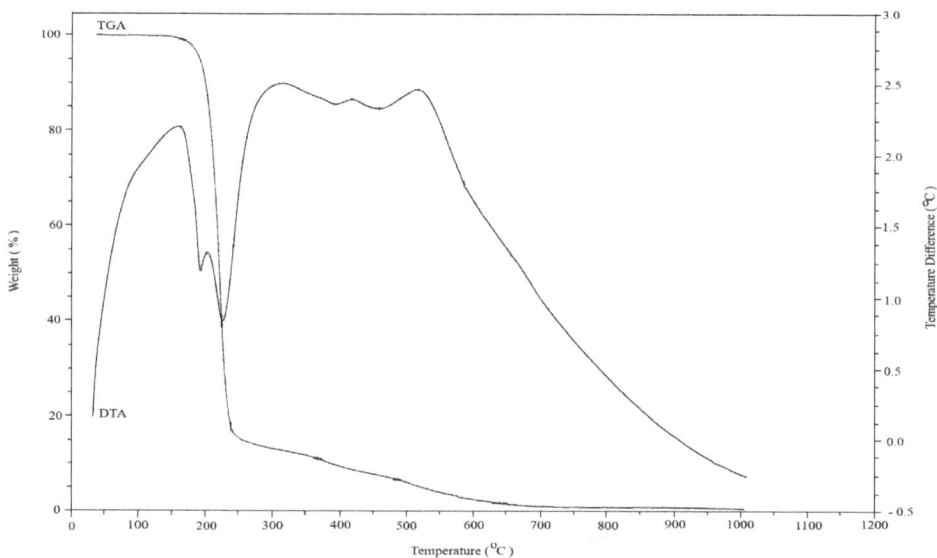

Fig. 13. TGA / DTA curve of LVS crystal.

11.2.6 Thermal analysis

The Thermo Gravimetric Analysis (TGA), Differential Thermal Analysis (DTA) spectra of grown LVS crystal have been obtained using the instrument NETSZCH SDT Q 600 V8.3 Build 101. The TGA and DTA have been carried out in nitrogen atmosphere at a heating rate of 20°C/min from 0°C to 1000°C. The TGA curve is presented in Fig. 13. The initial mass of the materials to analysis was 3.0160 mg and the final mass left out after the experiment was only 0.8631 % of initial mass.

The TGA trace shows that the material exhibit very small weight loss of about 1.04 % in the temperature up to 160°C due to loss of water. TGA curve shows that there is a weight loss of about 92 % between 160°C to 500°C indicating that the decomposition of LVS crystals. From the Fig. 13, the appearance of sharp endothermic in the DTA at 222°C corresponds to TGA results. From the TGA, DTA analyses, it is clearly understood that the LVS is thermally stable upto 222°C.

11.2.7 Second harmonic generation analysis

The nonlinear optical susceptibility of grown LVS crystals have been measured through second harmonic generation using standard Kurtz and Perry method (Kurtz & Perry, 1968). The output of laser beam having the bright green emission of wavelength 532 nm confirms the second harmonic generation output. The observed intensity of output light is 31 mV. For the same incident radiation, the output of KDP was observed as 55 mV. The second harmonic efficiency of LVS is 0.56 times that of KDP.

12. Conclusions

Thus the chapter fully discussed about solution crystal growth methods and nucleation. Then the growth and characterization of Single crystals of L- Alaninium Succinate (LAS) and L-Valinium Succinate (LVS) have been grown by slow evaporation method from saturated solution are also discussed. From X-Ray diffraction, it is observed that LAS and LVS crystal belongs to orthorhombic system. The UV-Vis-NIR spectral studies confirm that the grown crystals have wider transparency range in the visible and UV spectral regions and both LAS and LVS crystals have lower cut-off at 190 nm. The good transparency shows that LAS and LVS crystal can be used for nonlinear optical applications. The modes of vibration of the molecules and the presence of functional groups have been identified using FT-IR technique. The chemical structure of the grown crystals is established by ^{1}H and ^{13}C NMR techniques. From Thermal analysis, the melting point of LAS and LVS are identified is 178 °C and 222°C respectively. The SHG output proves that LAS and LVS crystals can be used as nonlinear optical materials.

13. Acknowledgment

The authors thank Dr. P.K. Das, Indian Institute of Science, Bangalore for the measurement of powder SHG efficiency. The authors are thankful to St. Joseph's College, Trichy, India, and SASTRA University, Thanjavur, India and Central Electro Chemical Research Institute (CECRI), Karaikudi, India for spectral facilities. The authors also express their gratitude to, Indian Institute of Technology, Chennai, India for XRD facilities

14. References

Aravindan, A., Srinivasan, P., Vijayan, N., Gopalakrishnan, R., & Ramasamy P. (2007). Investigations on the growth, optical behaviour and factor group of an NLO crystal: L-alanine alaninium nitrate. *Cryst. Res. Technol.*, Vol. 42, No.11, (November 2007), pp. 1097–1103, ISSN 0232-1300

Brice, J.C. (January 1987). *Crystal Growth Processes*, John Wiley and Sons, ISBN 0-470-20268-8, New York.

Bruice P.Y. (2002). *Organic Chemistry*, Pearson Education Pvt. Ltd, Singapore.

Buckley H.E. (4 May 1951). *Crystal Growth*, Wiley, New York.

Chenthamarai, S., Jayaraman, D., Ushasree, P.M., Meera, K., Subramanian, C., & Ramasamy P. (2000). Experimental determination of induction period and interfacial energies of pure and nitro doped 4-hydroxyacetophenone single crystals. *Mater. Chem. Phys.*, Vol. 64, No. 3, (May 2000), pp. 179–183, ISSN 0254-0584

Dhanuskodi, S., & Vasantha, K. (2004). Structural, thermal and optical characterizations of a NLO material: L-alaninium oxalate. *Cryst. Res. Technol.*, Vol. 39, No. 3, (March 2004), pp. 259–265, ISSN 0232-1300

Dhanuskodi, S., Vasantha, K., & Angeli Mary, P.A. (2007). Structural and thermal characterization of a semiorganic NLO material: l-alanine cadmium chloride. *Spectrochim. Acta A* , Vol. 66, No. 3, (March 2007) pp. 637-642, ISSN 1386-1425

Eimerl, D., Velsko, S., Davis, L., & Wang, F. (1990). Progress in nonlinear optical materials for high power lasers. *Prog. Cryst. Growth Charact.*, Vol. 20, No. 1, pp. 59-113, ISSN 0960-8974

Kurtz, S.K., & Perry, T.T. (1968). A Powder Technique for the Evaluation of Nonlinear Optical Materials. *J. Appl. Phys.*, Vol. 39, pp.3798 – 3813, ISSN 0021-8979

Milton , B., Boaz, Leyo Rajesh, A., Xavier Jesu Raja, S., & Jeromedas S. (2004). Growth and characterization of a new nonlinear optical semiorganic lithium paranitrophenolate trihydrate (NO_2–C_6H_4–OLi·3H_2O) single crystal. *J. Cryst. Growth*, Vol. 262, pp. 531-535, ISSN 0022-0248

Natarajan, S., Shanmugam, G., and Martin Britto Dhas, S.A. (2008), Growth and characterization of a new semi organic NLO material: L-tyrosine hydrochloride', *Cryst. Res. Technol.*, Vol. 43, pp. 561–564.

Ramachandra Raja, C. & Antony Joseph, A. (2009) Crystal growth and characterization of new non linear optical single crystals of L- alaninium fumarate. *Materials Letters*, Vol. 63, No. 28 (November 2009) 2507- 2509, ISSN 0167-577X

Ramachandra Raja, C. & Antony Joseph, A. (2009). Crystal growth and comparative studies of XRD, spectral studies on new NLO crystals: L- valine and L- valininium succinate. *Spectrochim. Acta. A*, Vol. 74, No. 3 (October 2009), pp. 825-828, ISSN1386-1425

Ramachandra Raja, C., Gokila, G. & Antony Joseph, A. Growth and spectroscopic characterization of a new organic nonlinear optical crystal: L-alaninium succinate. *Spectrochim. Acta. A* Vol. 72, No. 4, (May 2009) pp. 753-756, ISSN1386-1425

Ramachandra Raja, C, & A. Antony Joseph (2010). Synthesis, spectral and thermal studies of new nonlinear optical crystal: L-valinium fumarate. *Materials Letters* Vol. 64, No.2. (January 2010), pp. 108-110, ISSN 0167-577X

Part 3

Theory of Crystal Growth

13

Colloidal Crystals

E. C. H. Ng[1], Y. K. Koh[2] and C. C. Wong[1]
[1]Singapore-MIT Alliance, Nanyang Technological University,
[2]DSO National Laboratories,
Singapore

1. Introduction

A colloidal system consists of insoluble particles well-dispersed in a continuous solvent phase, with dimensions (generally less than 1μm in at least one important dimension) that are relatively larger than the molecules of the solvent. When the particles in this system are arranged in periodic arrays, analogous to a standard atomic crystal with repeating subunits of atoms or molecules, they form colloidal crystals (Pieranski, 1983). Gem opal (silica particles in close packed arrangement), iridescent butterfly wings made of periodic and spongelike pepper-pot structure (Biró et al, 2003) and sea mouse with hexagonal close packed structure of holes (McPhedran et al, 2003) are typical examples of colloidal crystals found in nature.

In 1935, discovery of tobacco and tomato virus by Stanley (Kay, 1986) provided excellent examples of naturally occurring monodisperse colloidal crystals. By centrifuging the dilute suspension of virus particles, crystals formed at the bottom of centrifuge tube can be examined by X-ray or light diffraction. The ease to obtain close-packed arrangement of colloidal particles has fascinated many, especially researchers working in chemical sensors, photonic band gap (PBG[1]) crystals, nanopatterning and sensors. 3D Colloidal crystals with periodicity ranging from 100nm to 1μm diffract visible wavelength according to Bragg's law, serving as waveguide and reflective surfaces for many useful devices in communication (Avrutsky et al, 2000; Liu, 2005) and solar harvesting (Mihi et al, 2011). Monolayers of colloidal arrays can serve as lithography mask (Lee et al, 2010) and physical mask for nanoimprinting. Periodicity can then be manipulated with the right particle size.

In this chapter, we discuss briefly the concepts of a colloidal system and types of interaction that give rise to crystallization, growth techniques and characterization tools available to help readers who are interested and new to colloidal systems. However, we limit the scope to simple colloidal particles which are spherical and of identical size, despite the possibility to self-assemble colloidal structures of complicated dimensions (Chen et al, 2006; Lu et al, 2001). Of the many growth methods, we place particular emphasis on capillary growth and its dependence on interparticle interactions, the substrate, and the manipulation of the solvent meniscus.

---

[1]Photonic band gap crystals are periodic structures which are able to block light propagation through the crystal in one or more directions. (Yablonovitch, 1987)

2. General concepts of a colloidal system

2.1 Colloidal system as a model for condensed matter

Colloidal systems share several similar cooperative phenomena with condensed matter (atomic crystal): ordering, phase transitions and stability of the resulting phases. As the ratio of colloidal particle size to atomic dimension is huge ($\sim 10^3$), parameters like the time scales of diffusion and lattice distances are also scaled up appropriately to milliseconds and micrometers respectively. This allows one to probe into real time processes that are otherwise inaccessible in atomic systems (Arora & Tata, 1996). As a result, a suspension of colloidal particles has provided fascinating models to investigate the physics of nucleation and growth (Gasser et al, 2001; Habdas & Weeks, 2002), phase transitions (Bartlett et al, 1992; Gast & Russel, 1998; Sirota et al, 1989), and fundamental problems of crystallization kinetics (Auer & Frenkel, 2001; Yethiraj et al, 2004).

A stable suspension of monodisperse colloidal particles normally appears milky white and will only become iridescent when interparticle distance shrinks to submicron range and satisfy Bragg's diffraction. Extensive study of equilibrium phase behavior of monodisperse colloidal system has shown that face-centered cubic (fcc) ordering is the equilibrium phase (Hales, 1997), above a threshold volume fraction (Pusey & Megan, 1986). Interestingly, body centered cubic (bcc) system was also discovered at ionic strength lower than 2.7×10^{-6} M KCl and volume fraction less than 0.008 (Monovoukas & Yiannis, 1989). The ability to change interparticle forces using electrolytes soon becomes a major advantage of colloidal system in modeling condensed matter.

Understanding all the interaction forces involved for particles in close proximity is nontrivial, as their magnitude or strength will decide their stability in suspension (no aggregation), ease of crystallization and final packing arrangement. Since many practical applications in interface science and colloidal science revolve around the problems of controlling forces between colloidal particles and between surfaces of different curvatures, many have devoted considerable effort to model surface forces and engineer their interactions in either short range or long range. It is impossible to give a detail perspective for all proposed models regarding colloids, only a few important concepts are introduced here, chosen based on the authors' preference for convenience. A more comprehensive review of all the interactions involved can be found in "Ordering and Phase transitions in Charged Colloids" (Arora & Tata, 1996).

2.2 Hard sphere model

One can first treat colloidal particles as hard and electrically neutral particles. Besides van der Waals forces, they only interact by steric repulsion when they are brought into physical contact. For monodisperse spherical particles, the close-packing limit is 0.74, which is equivalent to atomic packing factor of fcc atomic crystal. In another words, fcc packing arrangement would be the most stable form of colloidal crystal at equilibrium.

In practice, high concentration of colloidal suspension tends to flocculate before maximum packing limit is reached. Brownian motion allows particles to gain thermal energy easily and collide with each other to form clusters. If many aggregates form upon collision (especially at high temperature and high concentration), there is a high chance that amorphous colloidal aggregates will form, preventing further packing. In order to obtain FCC crystal structure, slow sedimentation of large particles in less concentrated suspension (volume fraction, v.f. << 0.5) by gravitational forces was explored (Crandall & Williams,

1977; Mayoral et al, 1997; Míguez et al, 1997). A phase transition from fluid to crystal can be observed as the local volume fraction increases beyond 0.49 (Pussey & Megan, 1986). This process is rather slow and may take up to a few days to weeks, depending on the initial concentration and desired volume of crystal grown.

In 1968, Hoover and Francis confirmed the existence of first-order melting transition for colloidal hard spheres and reported the densities of coexisting phases via Monte Carlo simulation (Hoover & Ree, 1968). It was found that melting and crystallization occur at reduced pressure of p = 8.27. The corresponding volume fractions, Φ is thus in the range of 0.50 to 0.55.

2.3 Charged spheres

Almost all colloidal particles are not electrically neutral. They may contain a large number of acid, base or other functional groups which are susceptible to dissociation in a polar solvent. Their functional groups are normally determined by the synthesis path or catalysts used. For instance, polystyrene beads synthesized using KPS (Potassium persulfate) catalyst will give rise to negatively charged particles. Hence, hard sphere model using exclusively steric repulsion is only a good approximation, but not an accurate approach to address real colloidal system.

Fig. 1 depicts the charge distribution surrounding negatively charged particles in water suspension. Given a neutral system of suspension, each colloidal particle of radius R carries negative charges of Ze. The water solvent contains an equivalent amount of counterions as well as possible stray ions such as salt (Na^+ and Cl^-). In a pure colloidal suspension with no salt present, the negative charges on particle surfaces will form thick double layer (~10^1 nm) and exert repulsion on each other, preventing formation of particle clusters. Part of the counterions will be attracted to near surface of negatively charged particles, while the remaining can move around in bulk suspension. If a small amount of salt (e.g. NaCl) is added into the colloidal system, positive ions (Na^+) will be attracted to particle surface, screening the repulsions between negatively charged particles. As a result, interparticle distance decreases and the attraction between particles could be enhanced. Super high salt concentration will collapse the double layer, and if the particles happen to be in physical contact by chance (Brownian motion), they will form irreversible aggregates immediately, giving no chance of forming ordered structures.

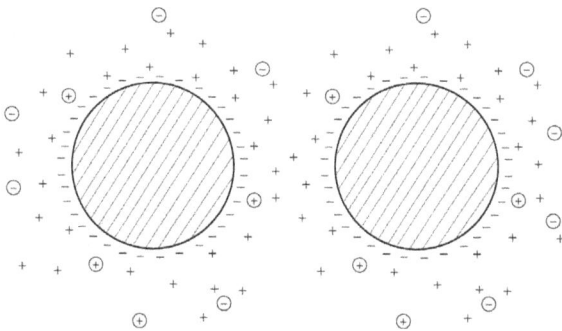

Fig. 1. Colloidal particles of radius R carrying negative charge -, with counterions + present in vicinity. + and - signs which are enclosed in circles represent stray ions.

2.4 DLVO model

To further assist reader, a model based on DLVO theory is introduced. It is named after Derjaquin, Landau, Verwey and Overbeek. The theory describes the interaction forces between charged surfaces in a liquid medium. It combines both the van der Waals forces and electrostatic interaction of charged surface, computed in mean field approximation in the limit of low surface potentials.

Here, it is assumed that:

a. Each colloidal particle is surrounded by its own double layer and behaves as if it was electrically neutral unless it approaches another particle closely enough

b. Only small and very mobile ions are first allowed to move. The mean field must be consistent with the boundary conditions on the surface of colloidal particles in suspension and the distribution obeys Boltzmann statistics (Brownian motion).

c. The potential energy of an elementary charge on a surface is much smaller than the thermal energy, kBT.

d. The particles are allowed to change position and find their own equilibrium states.

For DLVO potential,

$$V_{DLVO} = \frac{64\pi R c_0 \Gamma_0^2 kT}{k^2} \exp(-\kappa r) - \frac{AR}{12r} \tag{1}$$

R is particle radius, c_0 is ion concentration, k is Boltzmann constant, T is temperature, κ is inverse double layer thickness, r is separation between particles, and A is Hamaker constant. Surface potential of particles can be described by

$$\Gamma_0 = \tanh(\frac{ze\Phi_0}{4kT}) \tag{2}$$

where z is valence of counter ions, e is the electronic charge, and Φ is the surface potential.

It is obvious that the first part of equation (1) is related to electric double layer interaction (repulsion by similar charge of particles) where as the second part can be referred to attractive van der Waals interaction. As discussed earlier, electric double layer provides an energy barrier against irreversible agglomeration. This energy barrier, as indicated in Fig. 2 (right) as V_{max}, normally has a magnitude ranging from 0 ~ 100 kT, depending on the suspension parameters. Unlike the atomic model, two minima exist between two approaching colloidal particles, which are irreversible (primary) and reversible (secondary) respectively. In atomic crystals, atoms are held strongly in the single equilibrium position (as indicated in Fig. 2 (left)). High energy barrier or chemical bonds must be broken to disrupt the ordering. A brief comparison of interaction potential between atoms and colloidal particles in close proximity is best illustrated in Fig. 2 for clarity.

Surprisingly, the existence of secondary minimum allows scientists to "anneal" or remove defects in ordered arrays of wet colloidal crystal at this metastable state, before reaching the final irreversible dry state of colloidal crystal. The related works of particulate mobility in wet colloidal crystal will be further discussed in section 3.2.

For crystallization to occur, van der Waals forces must exceed double layer repulsion. This could only happen when particles are so close enough, overcoming the energy barrier V_{max}. This is possible as double-layer interaction energy is finite or increases slowly when r approaches zero, while V_{vdw} decreases exponentially when r → 0. Unlike electrostatic interaction, van der Waals interaction is highly insensitive to the change of pH and electrolyte concentration. Small addition of salt may change the magnitude of V_{max} significantly and remove the secondary minimum. Thus, no ordering in long range is ever possible if this situation happens.

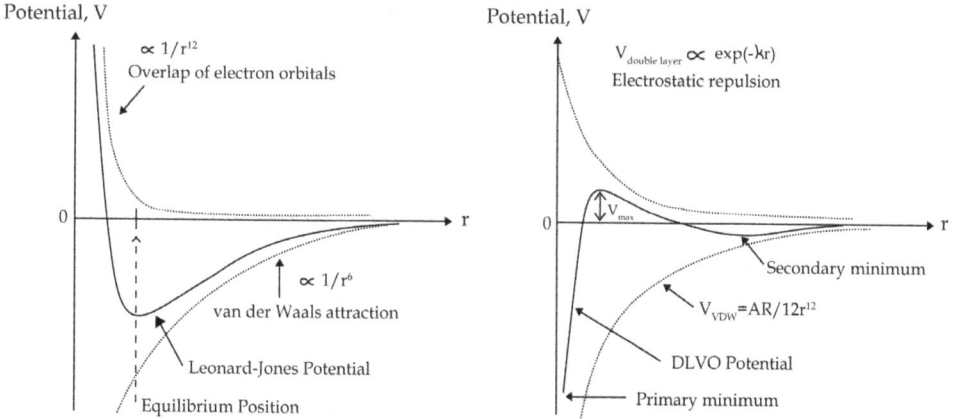

Fig. 2. A comparison of interaction potentials between atoms (Left) and colloidal particles (Right) in close proximity, described by Leonard-Jones potential and DLVO potential respectively.

3. Ordering and kinetics of colloidal particles

The ordering and kinetics of colloidal particles could be quite different from atomic crystals. The medium of colloidal dispersion gives rise to the dynamic motion of particles (Brownian) and exerts capillary forces on the particles near the interfaces (Kralchevsky & Nagayama, 1994). Fluid instability at interface (e.g. dewetting) further complicates the kinetics of crystallization process (Davis, 1980; Troian, 1989). All these contribute to several technical issues in achieving perfect crystal growth.

Hence, it is important to address the relevant critical forces involved in particle ordering. First, the proposed mechanism will be presented here with supporting research works from literature review and authors' publications. However, the discussion will be mostly focused on assembly process induced by capillary forces. Other assembly methods like e-field induced assembly (Holgado et al, 1999; Prieve et al, 2010; Yethiraj and Blaaderen, 2003; Zhang & Liu, 2004) and spin-coating assembly (Jiang & McFarland, 2004) can be referred to the recommendations in the reference list.

As mentioned in section 2, self-assembly via gravitation force is rather slow, and limits its application. Besides, the particles will only settle if their size and density is sufficiently high. This sedimentation tendency is best characterized by the Peclet number (see equation 3). Hence, this practical issue led to the studies of filtration and centrifugation (Park et al, 1999; Velev & Lenhoff, 2000), which however do not necessarily offer good control over packing quality and thickness of crystal grown. The formation of polycrystalline domains with different lattice orientation in colloidal crystals is commonly found, possibly due to nucleation at different locations of the specimen surface and the subsequent growth of crystal domains in different directions (Pusey et al, 1989).

Here we describe colloidal self-assembly under the effect of evaporation and capillary force, a phenomenon observed in the "coffee-ring" experiment [Deegan et al, 1997]. When a drop of coffee is spilled onto a solid substrate, a dense ring of stain will be formed upon drying, leaving the center of the initial droplet empty. The dense deposition of coffee solids near the outer ring signifies an intense movement of solid particles from the droplet interior to the drying perimeter. This driving force is attributed to evaporation, causing outward capillary flow (both solvent and solids) to the pinned contact line of the drying droplet, in order to replenish solvent loss from the edge (Dushkin et al, 1993). Deegan and his group then confirmed a power-law growth of ring mass with time in coffee-ring deposition, a mechanism which is independent of substrate type, carrier fluid and deposited solids. In short, the deposition process could be very fast, depending only significantly on the evaporation rate.

Dating back to 1992, the mechanism and stages of evaporation-induced assembly was first investigated in detail using an experimental cell containing a thin well of monodisperse micrometer-size latex particles, which allowed in-situ microscopic observation (Denkov et al, 1992). Similar to coffee-ring experiment, the particles were brought convectively to the edge of evaporation front, and the ordering of particles was found to be initiated when the thickness of water layer approached the underlying diameter of particles. Examples of particles moving to the drying edge of water film are shown in Fig. 3 and Fig. 4. Self-assembly starts when water thickness is about one particle diameter (860 nm) and it is normally monolayer near the perimeter of drying droplet.

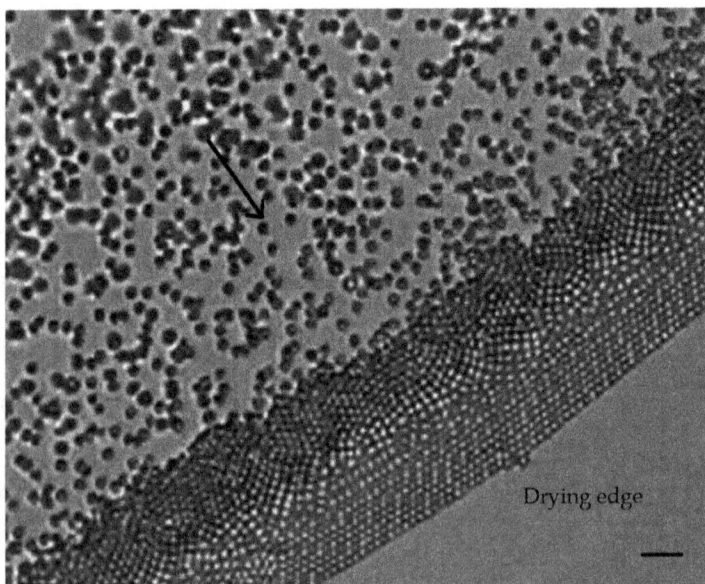

Fig. 3. Evaporation driven self-assembly in sessile droplet. Polystyrene particles (860 nm) move convectively to the drying edge, forming monolayer crystal at the outer perimeter. Transitional phases (e.g. buckling, square) are observed few particle diameters away from the perimeter (copyright reserved). Arrow indicates the direction of particle movement.

In evaporation-induced self-assembly, much attention is paid to the drying stage when the thickness of water layer is close to particle diameter. As evaporation procceds, thinning of water layer causes deformation of menisci (shown in Fig. 4) between particles at drying front. Further evaporation from the concave menisci increases the curvature and local capillary pressure, driving more water influx from thicker water layer to thinner region. Partially immersed particles also experience capillary attraction, leading to further packing of particles to form close-packed crystal. It is also worth noting that large and partially submerged particles can be immobilized by the thinning water layer as the vertical component of surface tension force pressing the particles against the horizontal substrate is huge. This phenomenon also serves as a condition for contact line pinning, which we will discuss later in section 4.2. Besides, formation of multilayer colloidal crystal is possible if wetting angle is large or thickness of water layer is large (Jiang et al, 1999).

Due to the almost fixed contact line formed by colloidal droplet on horizontal substrate, a variation of colloidal crystal thickness is commonly observed, from one single layer near the edge to multilayer further away from the contact line (Fig. 3). In spite of evaporation, slight reduction of water layer thickness over time does not help in the uniformity. Hence, an approach to let contact line move along deposition direction was introduced. Fig. 5 illustrates an experimental setup of vertical deposition, where a substrate is submerged vertically or at an inclined angle inside a colloidal suspension of dilute concentration (< 0.1 volume fraction). The temperature of the suspension can be controlled by a water bath surrounding the suspension, driving the convective flow of particles to contact line for controlled deposition. As long periods of evaporation may change suspension concentration over time, the temperature is normally adjusted to be very low, close to room temperature. Then the moving contact line is driven accordingly by withdrawing the substrate or pumping out the colloidal suspension at a controlled speed (Zhou & Zhao, 2004). The speed of the moving contact line will then determine the thickness of colloidal crystal obtained.

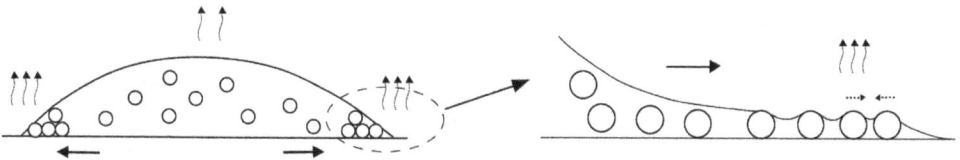

Fig. 4. Sessile droplet of colloidal suspension is shown on the left and a magnified version of thinning water layer of colloidal particles is depicted on the right. The scaling is for illustration purpose only. Curvy arrows indicate evaporation or water loss, which is faster at the edge. Straight arrows indicate convective flow of particles to the edge of sessile droplet. And dashed arrows on the right figure indicate capillary attraction between two partially submerged particles.

Other parameters such as incline angle of submerging substrate, surfactant addition, evaporation rate, ambient temperature, solvent volatility, particle concentration and substrate hydrophilicity have been explored to control the thickness of deposition (Denkov et al, 1992; Dimitrov & Nagayama, 1996; Im et al, 2003; Jiang et al, 1999; Kralchevsky & Denkov, 2001; Mclachlan, 2004). All these parameters must be optimized together with the speed of moving contact line, in order to obtain large area of uniform colloidal crystal.

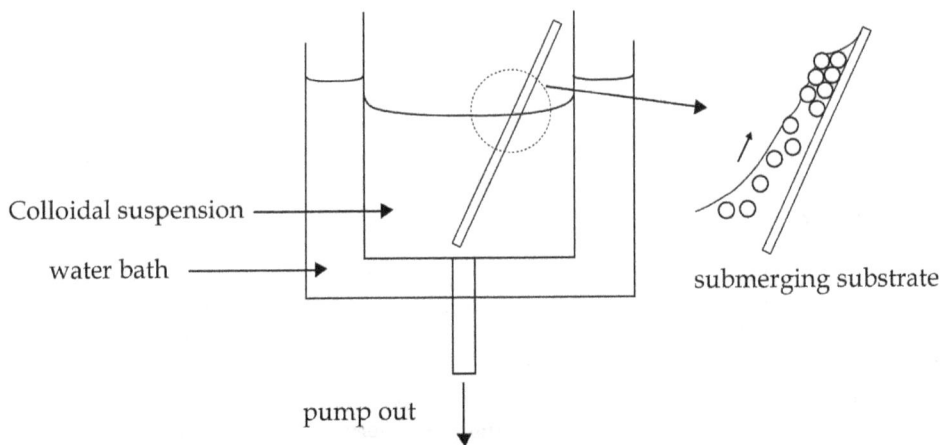

Fig. 5. Figure depicts an experimental setup for flow-controlled vertical deposition, an assembly method to control movement of contact line along crystal growth direction (Zhou & Zhao, 2004). Inset on the right shows capillary rise region near the submerged substrate.

3.1 Kinetic stages of colloidal crystallization

Since capillary forces can be significant in influencing the final assembly during drying process, systems with continuous environmental changes must be taken into account if one aims to design a process to obtain long-range ordered structures. There was also a desire to correlate the microscopic interaction and macroscopic boundary forces to provide a more complete model of colloidal self-assembly. The macroscopic forces are primarily capillary forces experienced by partially immersed particles under the thinning of water layer. The microscopic forces are contributed by electrostatic, steric interactions and van der Waals forces between closely arranged particles. In the case of larger particles, they will experience stronger effect of macroscopic boundaries before they have a chance to interact by microscopic forces. One can assume dominant function of capillary forces in this type of system. On the contrary, particles which are relatively smaller than the volume in which they are confined will be driven by microscopic interaction forces before experiencing capillary forces.

In order to understand the transition dynamics from well-dispersed colloids in suspension to dry colloidal crystalline film by evaporation-induced self-assembly, in situ transmission spectroscopy was introduced to monitor the assembly stages (Koh & Wong, 2006; Koh et al, 2008). Treating the assembly process as an emergent photonic crystal, several sequential stages of colloidal ordering with different lattice parameters in wet suspension could be interpreted based on Bragg's law (equation 4). Fig. 6 depicts an experimental setup by Koh et al, to investigate the self-assembly stages at the meniscus near the contact line formed on vertical substrate in vertical deposition.

Due to the good wettability of the hydrophilic glass substrate, thin meniscus formed near the substrate was higher than the bulk meniscus, providing a clear line of sight for transmitted beam. As evaporation proceeds, consistent arrival flux of particles at the thin meniscus could either stay or sediment back into bulk suspension. The tendency to sediment can be described by Peclet number, Pe:

$$Pe = m_B g R / kT \qquad (3)$$

where m_B is buoyancy mass of a particle with radius R and g is the gravitational acceleration. This number is also a good indicator to determine the stability of a colloid suspension. For Pe >> 1, the particles will tend to sediment and form agglomerates. In vertical deposition, commonly used particles are small, having Pe of unity or smaller. For instance, polystyrene spheres of 0.1 μm and 10 μm in water solvent at a temperature of 23°C give Peclet number of 5 x 10^{-5} and 5030 respectively.

A sequence of transmission spectra were taken from time t_1 to t_5, as shown in Fig. 7. The diffraction features shown in Fig. 7 can be correlated to photonic band gap structures (Ho et al, 1990; Koh et al, 2008). Under slow evaporation, it can be assumed that ordering in the suspension will lead to the equilibrium FCC structure (Monovoukas & Yiannis, 1989). Besides, (111) plane of colloidal crystal obtained is confirmed to be parallel to the substrate, providing an important reference for photonic band gap calculation

The spectra was at first featureless as light was hardly transmitted due to random scattering by the disordered structure. As local volume fraction at thin meniscus increases with evaporation, first feature A was observed at t_1 (300 min). This indicates the onset of order in the colloidal suspension: the first kinetic stage of ordering. Here, the rising local concentration shrinks the average distance between particles. Since particles are fully submerged in solvent, it is believed that the interaction forces between the particles are the only driving force for ordering, overwhelming the randomizing Brownian forces. It is also found that feature A corresponds to a larger lattice parameter (368nm) of fcc colloidal crystal, compared to 276 nm of the expected equilibrium colloidal crystal formed by hard-sphere packing of the polystyrene particles with diameter of 195 nm. This transition structure is stabilized by the interparticle forces, in which the existence of DLVO potential barriers prevents the particles from coming into direct contact. Here, we treat the interaction potential between two particles in a solvent using DLVO theory (see section 2.4) where the total potential is taken as the sum of the repulsive and attractive forces.

With continued growth of the colloidal crystal, a second feature (B) appears at a shorter wavelength (from t_2 onwards in Fig. 7). This corresponds to wet FCC colloidal crystal with the particles in direct physical contact (zero separation) with water in the interstices. Water-retaining capillary pores are normally formed in the interstices of deposited colloidal crystal as the meniscus recedes below the self-ordered crystal. The subsequent loss of water upon drying of wet colloidal crystal immediately gives rise to a blue shift, feature C at t_4 (770 min) and t_5 (800 min), which can be correlated to the change of dielectric contrast as water in the interstices is replaced by air. In the transitions from t_1 to t_3 and t_3 to t_5, coexistences of the two features A and B are observed at t_2 and t_4 simultaneously. These are likely due to the simultaneous existence of the corresponding transition structures where not all region of wet colloidal crystal shrink and dry at the same time. The growth of intensity for feature B from t_2 to t_3 indicates the area increase of double-layer collapse across the studied area of colloidal crystal (area of the incident light beam). Also, the intensity increase from t_4 to t_5 explains the continual evaporation of water from the wet colloidal crystal to form dry crystal, revealing feature C.

The essence of this work is the demonstration of three distinctive stages in colloidal self-assembly, as shown in Fig. 8. This model is further supported with sequential changes of lattice parameters derived from the transmission spectra in Fig. 7. It was then confirmed that feature A had a lattice parameter of 368 nm, which was larger than the equilibrium colloidal crystal with lattice parameter of 276 nm. The larger interparticle distance in wet suspension during self-assembly indicates the existence of DLVO potential barrier, which prevents particles from coming into direct contact. This finding highlights the important role of interaction forces for small particles, despite the earlier understanding that colloidal self-assembly at liquid menisci is driven solely by capillary forces. This could only be true for the sizes of the particles that are small relative to the volume that is confined by the macroscopic boundaries. It is believed that the capillary forces are only brought into action from the second stage onwards, where the forces collapse the electric double layer, thus bringing

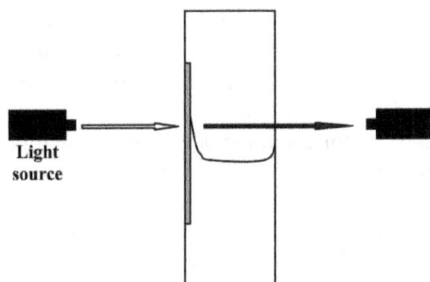

Fig. 6. Schematic shows an in situ transmission spectroscopy of colloidal self-assembly. A glass substrate is located in a plastic cuvette of colloidal suspension. This whole apparatus is kept in a temperature-controlled chamber, which controls evaporation from the suspension. Reprinted with permission from *Langmuir*, Vol. 24, No. 10 (May 2008), pp. 5245-5248. Copyright 2008 American Chemical Society.

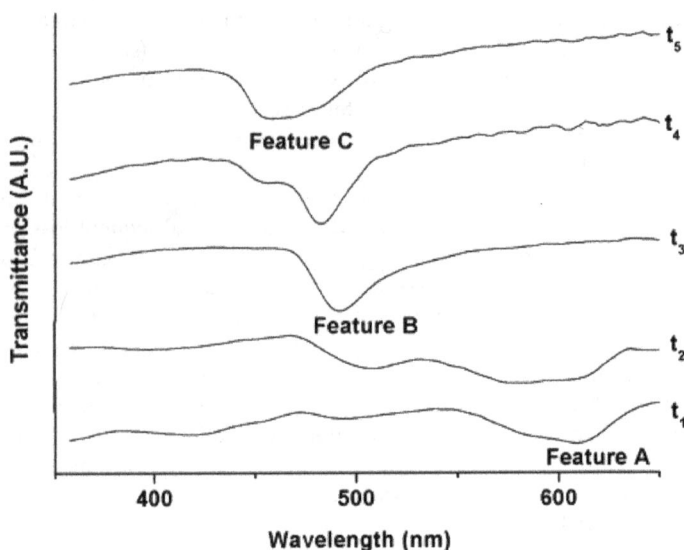

Fig. 7. Transmission spectra show emerging features A (608 nm), B (491 nm), and C (462 nm) at different time interval. Two features are observed simultaneously at times t_2 and t_4. The spectra are offset vertically for clarity. Spectra are taken at t_1 (300 min), t_2 (320 min), t_3 (400 min), t_4 (770 min) and t_5 (800 min). Reprinted with permission from *Langmuir*, Vol. 24, No. 10 (May 2008), pp. 5245-5248. Copyright 2008 American Chemical Society.

particles into direct contact, with solvent trapped in the interstices (Fig. 8, middle). The assembly process then ends with the final replacement of water with air, bringing a change in periodic dielectric contrast as water evaporates from wet colloidal crystal (Fig. 8, Top). Once this stage is achieved, the colloidal crystal obtained will be stable and strong enough to resist structural changes against liquid infiltration process. This robustness enables infiltration with other functional materials to obtain inverse opal structures or opals with different material properties via double templating (Yan et al, 2009).

Understanding the dynamic transition of colloidal crystal may provide some insights towards improving long-range quality of colloidal crystal. First, the electric double layer around each particle must be as thin as possible, and yet prevent premature aggregation. This is because large double layer thickness will give rise to large shrinkage stress during the final collapse of double layer (A to B), resulting in macroscopic cracks (Jiang et al, 1999). In the next section, the manipulation of DLVO potential to study particulate mobility of self-assembly process will be discussed.

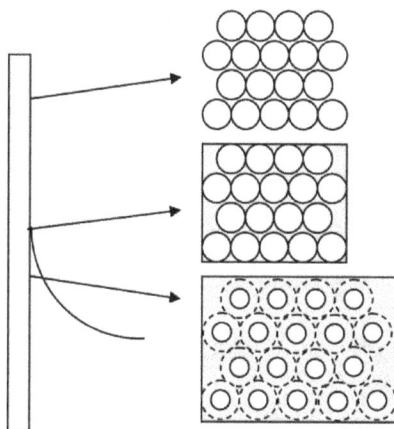

Fig. 8. Three distinct stages in colloidal self-assembly process. First stage (bottom) shows a transition structure with a large lattice parameter, corresponding to feature A in Fig. 7. As the meniscus moves, this structure collapses to a smaller lattice parameter (middle), with water retained in its interstices, giving rise to feature B in the transmission spectra. Finally, a dried colloidal crystal (top) corresponding to feature C is obtained. Reprinted with permission from *Langmuir*, Vol. 24, No. 10 (May 2008), pp. 5245-5248. Copyright 2008 American Chemical Society.

3.2 Particulate mobility in vertical deposition

The absence of repulsion forces between charged particles and oppositely charged surfaces often lead to disordered clustering (Yan et al, 2008). At high particle concentration in a confined volume (e.g. thin meniscus), repulsion force is the key to prevent irreversible clustering and the prerequisite for sufficient particulate mobility to obtain higher packing quality, if sufficient time of ordering is given. Since understanding the collective behavior of the particles in an environment of high mobility is indispensable, the mobility and electrostatic interactions of negatively charged substrates and positive colloids were studied, with optimized parameters of ionic strength, volume fraction, and solvent evaporation temperature in vertical deposition (Tan et al, 2010).

3.2.1 Stick-slip behavior of colloidal deposition

When the positive polystyrene colloids are self-assembled on a negatively charged substrate, a uniform array of alternating linear patterns is usually obtained with limited widths (Ray et al 2005; Tan et al, 2010). As shown in Fig. 9, these 1D particulate bands are deposited at relatively regular intervals across the entire substrate. A magnified view of SEM photos reveals close-packed ordering within each band. However, the ordering quality is poor with abundant point and planar defects.

This alternating band is no different to the case of negatively charged particles being self-assembled on glass surface of same charge (Teh et al, 2004). The only difference is the presence of tiny clusters scattered within the so called "empty band" region. As discussed by Teh, the alternating bands can be attributed to the stick-slip motion of meniscus growth front during deposition, while the presence of scattered clusters in the "empty bands" could be caused by electrostatic attractions between colloids and substrate. Since the surface charge density of silica glass was determined to be $-0.32 mC/m^2$ (Behrens & Grier, 2001), strong electrostatic attraction will immobilize positive particles if they happen to come close to the glass surface. This could explain the disordered random ordering of positive colloids in the "empty-band" region.

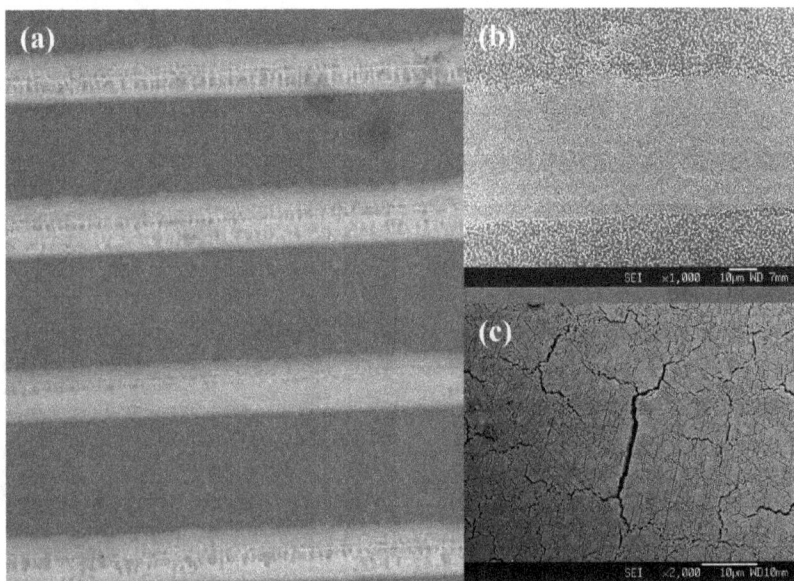

Fig. 9. (a) Microscopic picture shows interplay of stick-slip deposition and attractive deposition of positive colloids on negatively charged glass surface. (b) and (c) are magnified SEM photos showing alternating bands of ordered region and scattered clusters of random ordering. Particle concentration = 0.5 vol%, temperature = 35°C. Reprinted with permission from *Langmuir*, Vol. 26, No. 10 (May 2010), pp. 7093-7100. Copyright 2010 American Chemical Society.

At the pinned contact line of meniscus, meniscus will first recede and deform with the consistent withdrawal of solvent in flow-controlled vertical deposition. Then the continual flux of particles to the thinning region will increase the local volume fraction and decrease interparticle distance, causing the ordering to nucleate in the confined meniscus. If the meniscus thickness is larger than one particle diameter and recedes much slower than particle deposition, multilayer band with hcp orientation will be obtained. When the solution recedes further with contact angle reaching the minimum receding angle, the meniscus contact line will slip rapidly to a lower pinning level with a new contact angle. This process is then repeated with dynamic change of angle and alternating bands of deposition. More details can be found in Teh's published work (Teh et al, 2004).

3.2.2 Effect of volume fraction and ionic strength

Multilayer colloidal crystal shown in Fig. 9 was obtained with 0.5 vol% of colloidal suspension at 35oC. When a more dilute concentration of 0.1 vol% was used (not shown), multilayered bands can be replaced by monolayers with locally ordered configurations. This can be explained by the lower particle flux to the drying edge during deposition, at the same withdrawal rate of suspension. At the same time, longer time is required for colloids to reach threshold concentration in the confined thin meniscus, implying that particles will have sufficient time to self-assemble in an orderly manner. Larger interparticle distance and lower frequency of collision due to Brownian motion are believed to slow down irreversible aggregation and random electrostatic adsorption of particles onto charged substrate. Hence, lower volume fraction of colloidal suspension is normally used to obtain long-range ordered crystal as slow increment of local volume fraction at thin meniscus is expected to impart greater inplane colloidal mobility during assembly process. Another comparison of concentration effect (0.05 and 0.01 vol%) on monolayer crystalline quality is given in Fig. 10. Dilution leads to better ordering quality, in agreement with other work (Zhou & Zhao, 2004).

Besides volume fraction, inplane mobility is also affected by electrostatic interaction between ordering particles in thin meniscus layer (Maskaly, 2006; Tan et al, 2008). For positive colloids, addition of salt (ionic strength increases) will reduce Debye screening length of the electric double layer. The result of this is twofold. First, positive colloids of similar charge will approach each other closer, and can be configured into stable in-plane ordered array with minimal cracks upon drying. Second, shorter Debye length will give positive colloids extra time to form ordered array, before being adsorbed onto negatively charged surface by electrostatic attraction. For example, the addition of 10 µM KCl was reported to give highest density of hcp domains, further supported by the distinctive 6-fold coordinated diffraction spots of a hexagonal lattice (Fig. 10 b). Detailed evidence can be referred to the relevant publication (Tan et al, 2010).

Besides, Tan also postulated that the assembly of charged colloids may achieve an intricate balance between particle-particle repulsion and particle-substrate attraction, when a colloidal suspension of low volume fraction and low ionic strength is used. This is a condition where the particles are sufficiently far apart to reorient themselves into geometrically and thermodynamically favored close-packed arrangement.

Fig. 11 shows the phase behavior of positive polystyrene colloids assembled on a negative silica glass substrate at 25°C, at various ionic strength and volume fraction. It indicates that aggregates are likely to occur at high ionic strength across the whole studied range of volume fractions (0 to 0.1 vol%). For low ionic strength (< 10µM), long-range hcp ordering could be obtained with the use of low volume fraction (< 0.6 vf%).

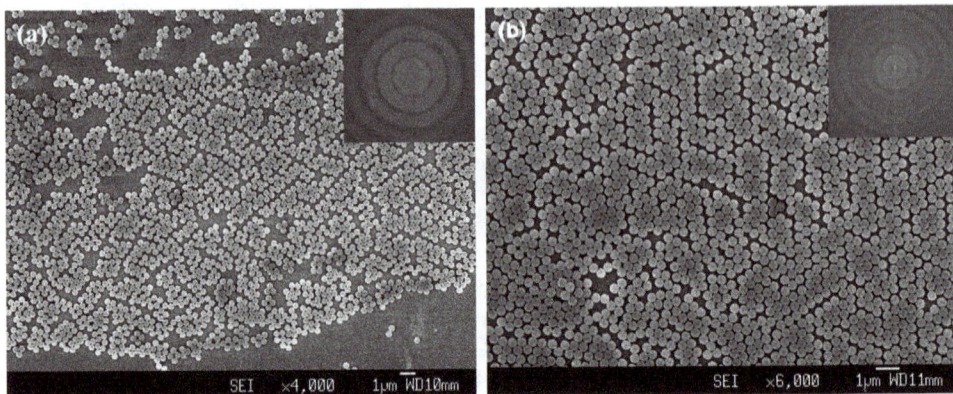

Fig. 10. Hexagonal close-packed domains and corresponding FFT inset obtained from vertical depositon at 25°C and 10 μM KCl. (a) Particle concentration = 0.05 vol % (b) Particle concentration = 0.01 vol%. Reprinted with permission from *Langmuir*, Vol. 26, No. 10 (May 2010), pp. 7093-7100. Copyright 2010 American Chemical Society.

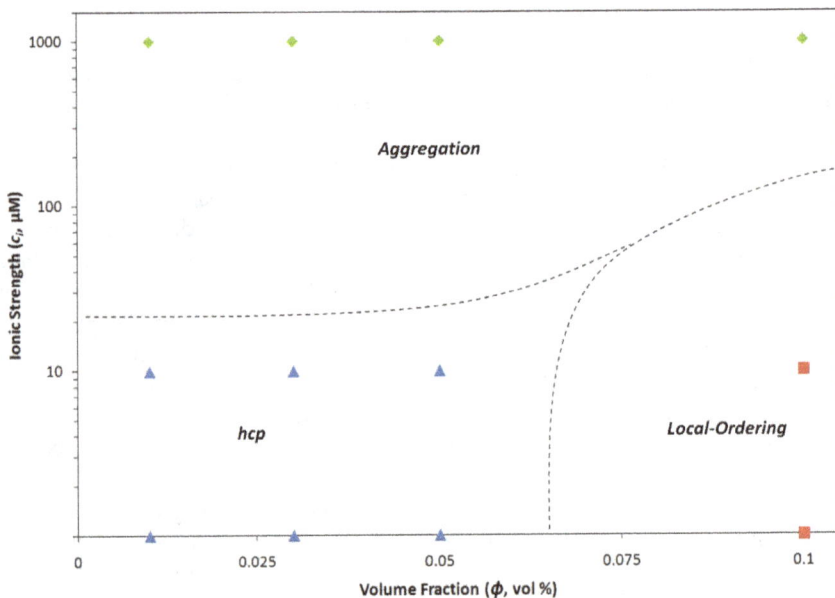

Fig. 11. Phase behavior of self-assembled colloidal structure, by deposition of positively charged polystyrene colloids on a negative borosilicate glass substrate at 25°C. Approximate boundaries between the ordered and glassy phases at different conditions are indicated by dashed lines. The axis of ionic strength is plotted in logarithm scale. Reprinted with permission from *Langmuir*, Vol. 26, No. 10 (May 2010), pp. 7093-7100. Copyright 2010 American Chemical Society.

3.2.3 Effect of temperature

There are two major effects brought by temperature increase (Dimitrov & Nagayama, 1996; McLachlan et al, 2004). First, high evaporation rate drives higher flux of particles to the drying edge, leading to faster crystal growth rate. Larger thickness of colloidal deposition is possible if solution withdrawal and meniscus deformation rate (thickness reduction) is much slower than particle flux. Second, kinetically active particles will bump into each other with high frequency. If they are of similar charge with substrate, higher ordering could be obtained since mobility is sufficiently large for further packing. However, substrate of opposite charge will likely immobilize colliding particles, giving no chance of good ordering (irreversible clustering).

3.2.4 Mobility in binary colloid

Earlier discussion has been devoted to the fine control of PS colloidal mobility in single-component crystallization. However, close-packed lattices have limited available symmetries (FCC or HCP) and associated properties which are too restrictive for diverse potential applications, especially in photonics. Using a mixed suspension of two particle types, colloidal crystals with lower symmetries can be made possible to provide novel properties in photonics, sensing and filtering (Bartlett et al, 1992; Kitaev & Ozin, 2003; Sharma et al, 2009; Tan et al, 2008).

Layer-by-layer growth is commonly used to self-assemble these structures, with right conditions of surfactants, temperature and ionic strength. Unlike the uniform negatively charged glass substrate used in self-assembly of positive colloids (see section 3.2.2), an underlying negatively charged L-colloidal template can be used to provide a periodic potential landscape to assist ordering of the next layer of colloids (named S-colloids) with opposite charge. These steps could be repeated to get additional layers of LS_n structures.

In LS_2 structures (see inset in Fig. 12a), each interstice in the first layer of hexagonally close packed (hcp) particles (L) is filled by one small particle (S). LS_6 structure is also possible where each interstice of first layer is filled by three particles (S) instead (Sharma et al, 2009). Unfortunately, both lower and higher densities of binary structures are usually observed together in LBL experiments. This led to an extended work to study the particulate mobility of colloids over an ordered potential landscape of an assembled hcp monolayer of opposite charge (Tan et al, 2010).

In previous work, Tan reported that a low ionic strength of 10 µM KCl could vastly improve the ordering of attractive binary colloidal structures in layer-by-layer (LbL) growth, presumably because the S-colloids possess sufficient mobility to self-assemble into a highly symmetrical LS_2 2D-superlattice. Besides ionic strength, crystalline quality also depends strongly on evaporation temperature and the most uniform LS_2 was obtained at low temperature (25°C). Fig. 13 shows a comparison of temperature effects (25°C vs 35°C) over LS_2 assembly using 10µM of KCl electrolyte. Similar to the case of single-component crystallization on glass surface of opposite charge, lower evaporation rate and slow crystal growth rate provide additional time for S-particles to reorient and stabilize into a thermodynamically favorable in-plane LS_2-superlattice (in suspension), before settling onto the oppositely charged template (L-layer). On the other hand, a slight increase of temperature to 35°C disrupts the inplane ordering structure, resulting in disordered binary arrays. This can be explained by greater Brownian motion of particles and faster loss of water due to evaporation, in which water is the key to mobility during ordering.

Thus a consensus can be drawn that an intricate balance of the particulate mobility by all three parameters, (1) evaporation temperature, (2) volume fraction, and (3) ionic strength, is critical for the quality of self-assembled colloidal crystals. Other than these, the delivery speed of the particles to crystal growth front should equal the crystal growth rate. These can be optimized by evaporation control and meniscus receding rate. Regardless of the type of substrates and the charge of particles used, particulate mobility must be assured to guarantee sufficient time of reordering to achieve final irreversible crystallization. Next we will discuss various templating efforts to obtain perfect single crystal and crystals with complex symmetry, which are impossible to be achieved by conventional self-assembly on bare substrates.

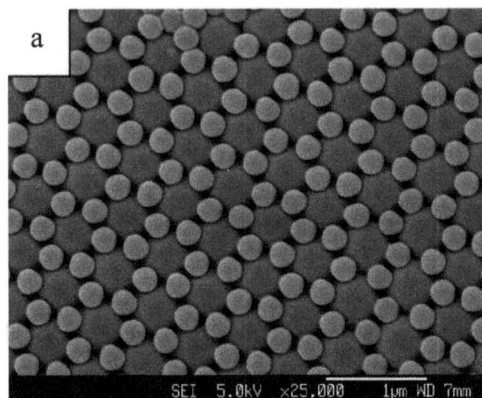

Fig. 12. Layer-by-layer assembly of positive colloids (250 nm) onto a hcp monolayer of negatively charged particles (550 nm) in flow-controlled vertical deposition, revealing LS_2 structure. Reprinted with permission from *Langmuir*, Vol. 26, No. 10 (May 2010), pp. 7093-7100. Copyright 2010 American Chemical Society.

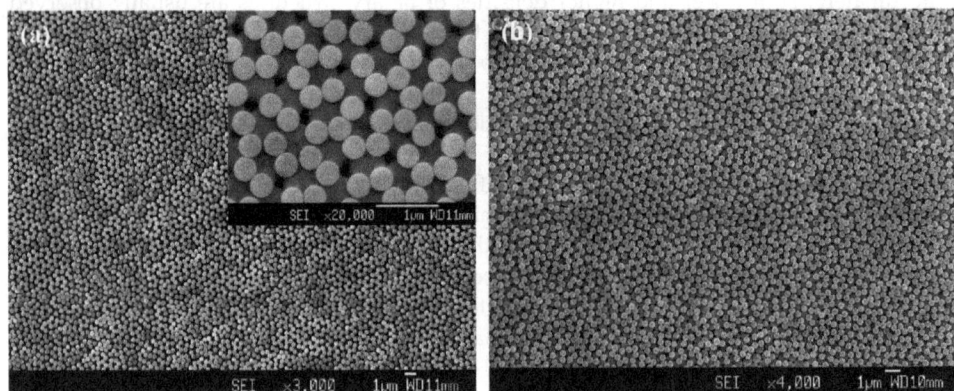

Fig. 13. Layer-by-layer assembly of positive colloids (371 nm) onto a hcp monolayer of negatively charged particles (604 nm) in flow-controlled vertical deposition, at temperature of (a) 25°C and (b) 35°C. Reprinted with permission from *Langmuir*, Vol. 26, No. 10 (May 2010), pp. 7093-7100. Copyright 2010 American Chemical Society.

4. Towards perfect crystallization

4.1 Template-assisted self-assembly

Despite many efforts to grow large-area perfect crystal (> 1000μm³) that are useful for optical devices, most do not offer sound practicality for large scale integration. Sedimentation is extremely slow, limited by absence of control over number of layers and uniformity over topology. The method based on vertical deposition requires strict control of surface charge density of particles or substrate, particle and electrolyte concentration, temperature, and humidity.

Geometrical confinement has been long studied to affect the phase behavior of colloidal particle ordering (Schmidt & Löwen, 1997; Ramiro-Manzano et al, 2007). By using thin parallel plates or a wedge cell of few particle diameters in gap, crystal transitions (e.g. buckling) can be observed with changes of cell thickness. Similar confinement approach was also extended by Park et al to obtain much better ordering and orientation compared to the colloidal crystals grown from bare substrate (Park et al, 1997). Using pressure, they injected a suspension of colloidal particles into a well-confined rectangular cubic cell, with solvents being drained out through the channels (< particle diameter) built lithographically along the side walls of the cell. This left behind accumulating particles at the bottom of the cell and rapid crystallization of particles over 1 cm² with well-controlled number of layers (1 ~ 50 layers) could be easily attained. With sonication (Sasaki & Hane, 1996), the packing quality of close-packed lattice could be further improved under flow. This cell can then be dried off in an oven and dismantled later. The perfect colloidal crystal confined by these physical walls could then be integrated into device-making. Crytals grown this way have 3D domain size of 12 μm x 0.5 cm x 2 cm, which is almost equivalent to the cell size of 12μm x 2 cm x 2cm.

Coupling with the laminar flow of colloidal particles, flow-driven organization of particles in microchannels (rectangular grooves) was studied (Kumacheva et al, 2003). Using template with periodic rectangular grooves, the influence of the width and size of such rectangular grooves on colloidal self-assembly was investigated. Depending on the commensurability of the particle into the grooves, various structures like close-packed hexagonal, rhombic and disordered structures were reported. If a large mismatch (> 15%) exists between the ideal and experimental ratios, defects would be introduced in the crystal structure. In a subsequent study, a much narrower groove was explored, in which only parts of a colloidal sphere could (snugly) fit into. Precise ordering of colloidal lines was thus obtained (Allard et al, 2004).

Fig. 14. (a) Anisotropically etched V-shaped channels into a Si (100) wafer obtained using lithography. (b) SEM photo depicts six-layer (100) single crystal made of silica particles, sitting in a V-grooved channel. (c) Vacancy defects observed in the top layer of micro-spheres. (Yang & Ozin, 2000) – Reproduced by permission of The Royal Society of Chemistry.

In order to enable realization of colloidal crystals in devices, they must be fabricated into planarized microphotonic crystal chips. Ozin and his group pioneered the fabrication of planarized microphotonic structures by the deposition of micro-spheres into V-shaped grooves (apex angle = 70.6°) (Ozin & Yang, 2000 & 2001). The nucleation was believed to first occur along the apex of the V-grooves, followed by the subsequent layers of growth. The depth of such grooves will determine the number of layers of colloidal crystal formed and the close-packing tendency ensures projection of (100) crystal orientation, terminating at the crystal-air interface (see Fig. 14).

Emerging micro- and nanofabrication technologies in template structuring allows one to shrink the pattern size to tens of nanometers (e.g. e-beam lithography). The ability to create nano- or microfeatures consistently enables precise deposition of each particle and control over its packing symmetry, packing efficiency and packing quality of the resulting crystal. Using topologically patterned templates, Van Blaaderen et al showed that slow sedimentation of colloidal particles into the "holes" of topologically patterned templates enables formation of fcc colloidal crystals with crystal orientation of (100) planes parallel to the patterned surface (Blaaderen et al, 1997). This is different from the usual (111) plane orientation obtained, with respect to the surface. They succeeded to achieve large oriented crystals at which the defect structures were tailored by surface graphoepitaxy[2] approach. As the template used has a known orientation, photonic crystals grown could be sliced such that the exposed (001) and (110) facets of the fcc crystal structure could be integrated into specific applications. It was also shown that intentional mismatch of hole pitch and particle diameter can give rise to defect structures, such as randomly stacked (111) planes above the first few layers from the surface of template. This is similar to the case of growing epitaxial layer of CdTe (111) on GaAs (001) substrate, with a mismatch of about 14.6% (Bourret et al, 1993). Other than this, other cubic packing system like body-centered cubic (bcc) and simple cubic (sc) colloidal crystals have also been reported using similar approach (Hoogenboom et al, 2004).

It is obvious that templating offers a remedy to the shortcomings of spontaneous colloidal self-assembly, especially in manufacturing crystals tailored for realistic photonic applications. Other than the potential to obtain defect-free colloidal crystal of fcc structure with different plane orientation, it is also possible to assemble monodisperse colloids into complex structures or subunits (Romano & Sciortino, 2011; Vinothan et al, 2003), and then lead them to complex crystal symmetries of lower packing density.

According to photonic band structures calculated for various crystal symmetries and dielectrics (Ho et al, 1990; Pradhan et al, 1997; Busch & John, 1998; Moroz & Sommers, 1999; Vlasov et al, 2000), it was confirmed that an fcc colloidal structure has a PBG only in the second Brillouin zone (second-order Bragg diffraction), not in the first Brillouin zone (Blanco et al, 2000). Besides, a sufficiently high refractive index contrast (> 2.8) between the building blocks of the fcc crystal (colloidal particles and the interparticle space) is required to obtain a full omnidirectional band gap. Furthermore, photonic properties of commonly found fcc crystals are very sensitive to structural disorder (Vlasov et al, 2000). In this regard, nonspherical particles also offer immediate advantages in applications that require lattices with lower symmetries and higher complexities.

---

[2] Growth of crystal by substrate topology as opposed to atomic lattice in which a material is crystallized onto an existing crystal of another material, resulting in effective continuation of the crystal structure of the substrate.

Templates used for "nucleation" of complex colloidal clusters or crystal layers are usually engineered by modification of the surfaces via lithography (Chen et al, 2000; Choudhury, 1997; Rijn et al, 1998) and chemical patterning (Bertrand et al, 2000; Delamarche et al, 1998; Ulman, 1996; Xia & Whitesides, 1998). The control of surface chemistry like charge and functional reactivity can be obtained by coating an adhesive monolayer which is specific to the substrate, called self-assembled monolayers (SAM) (Ulman, 1996). For example, one can use thiol functional groups for gold surface, and silanes for silica substrate. The functional groups of this self-assembled molecular layer will then adhere to the corresponding surface with the other desired ends (e.g. charge for particle interactions) projected outwards. Besides, one can also use SAMs to coat different charges (positive, negative or neutral) on each crystal layer grown via layer-by-layer method to obtain binary colloidal crystal in vertical deposition.

Other than direct photolithography or e-beam lithography to create paterns on these SAM layers, SAM patterns can also be transferred to a flat substrate by soft lithography (Xia & Whitesides, 1998) and nanoimprint lithography (Hu & Jonas, 2010; Torres, 2003). Subsequent ordering of colloidal particles on such chemically defined patterns can be achieved in vertical deposition via electrostatic interaction and capillary forces. As discussion in section 3.2, charged microspheres can be self-assembled on the oppositely charged areas of the patterns when the substrate is slowly taken out from the colloidal suspension (Fustin et al, 2004). Depending on the size ratio of colloidal particle and patterned area, complex colloidal aggregates can be grown (Lee et al, 2002).

Despite progressive advancements in lithography systems, high facility cost and maintenance impede practical use of templating in colloidal self assembly. Hence, it is worth exploring simple templating like V-groove, and low-cost templating system like soft lithography and nanoimprinting, to enable innovative ways for template-assisted self-assembly. Next, we will discuss our approach to utilize meniscus pinning to control positional nucleation and inplane-oriented growth of large area monolayer colloidal crystal from one straight surface relief.

4.2 Controlling inplane orientation of large area monolayer colloidal crystal

Among various top-down and bottom-up methods discussed earlier, capillary forces induced convective self-assembly is attractive, requiring only a simple and economical setup. However, common nonidealities like thickness nonuniformity, restricted domain size and empty bands or voids are frequently reported. When a substrate is submerged into a liquid, a wavy contact line is commonly observed at air-liquid-solid interface, due to Rayleigh instability (Davis, 1980). In vertical deposition, it is believed that the trapping of colloidal particles along this wavy contact line will first lead to accumulation of particles, and then multidirectional initiation of colloidal crystal growth (Fig. 15c). Since the domain growth directions tend to be different along the wavy contact line; this eventually limits the final domain size of colloidal crystals obtained. Other than this, dynamic change of receding contact angle of colloidal suspension during liquid or substrate withdrawal will produce colloidal stripes or alternating colloidal and empty bands via the stick-slip mechanism (Adachi et al, 2005; Teh, 2004; Thomson et al, 2008).

Here, we demonstrate the usage of meniscus pinning by surface relief boundaries to control in-plane orientation of monolayer colloidal crystals without the interruption of grain disorientation. By printing a straight surface relief which has a strong affinity to water molecules (common solvent for colloidal particles), a straight wetting line of colloidal suspension could be pinned along the surface relief patterned (Fig. 15a). The photoresist SU-8 has been shown to work in this context. As most photoresists do not have good affinity to water (hydrophobic), their surface can be treated with UV ozone to improve wetting. A small addition of surfactant (SDS) and trapping of colloidal particles along wetting line do give enhanced pinning by almost 100%.

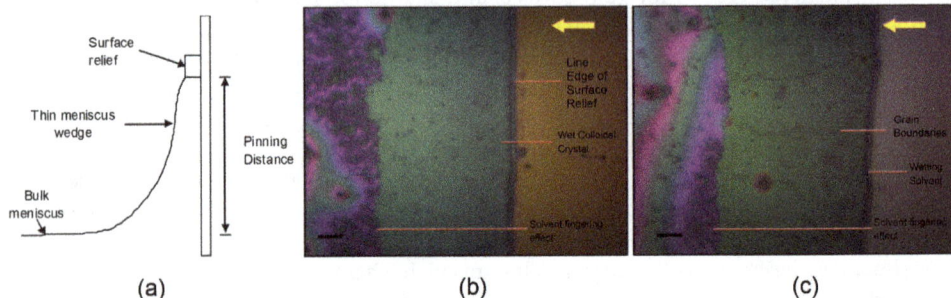

(a)   (b)   (c)

Fig. 15. (a) Schematic shows how water film is pinned and stretched near surface relief when the substrate is being pulled out of colloidal suspension. (b) Optical micrograph shows straight liquid pinning and nucleation along surface relief, and the subsequent inplane oriented growth. (c) Optical micrograph shows that colloidal nucleation on bare substrate always starts with wavy contact line and wavy nucleation line, giving rise to polydomain growth. As indicated, fingering effect of wetting solvent is observed above the wet assembled colloidal crystal. Yellow arrows show liquid receding direction. The scale bar is 8μm. Reprinted with permission from *Langmuir*, Vol. 27, No. 6 (March 2011), pp. 2244-2249. Copyright 2011 American Chemical Society.

The initial establishment of liquid pinning along straight surface relief will allow colloidal particle deposition along the thin meniscus wedge (Fig. 15b). Fast substrate withdrawal or receding bulk meniscus relative to colloidal deposition speed will pull the pinned contact line, either causing depinning or contact-line movement in a fingering pattern (Sharma & Reiter, 1996; Troian et al, 1989), together with the pinned colloidal domains. Hence, depinning of the initial contact line must be avoided.

By optimizing the pinning boundary and withdrawal speed, a well-controlled linear meniscus contact line allows a straight nucleation edge of monolayer crystal growth front, which then acts as a crystal growth seed, permitting the most close-packed direction <11> or <10> (as in 2D hexagonal lattice) to assemble along the surface relief. As a result, this unidirectional growth can give rise to single domain crystals with only twins and vacancies present as residual defects (see Fig. 16). More evidence can be referred to the supporting documents provided at the publication site (Ng et al, 2011).

Conservatively, the domain crystal size obtained can be as large as 1 mm², with residual defects of vacancies, twin boundaries and 'small misoriented domains. Despite these imperfections, the domain orientation of large crystal domain remains similar. More evidence in the form of SEM photos scanned sequentially can be found in the supporting documents published (Ng et al, 2011). To conclude, this novel approach could offer the desired ease of integration for device making, as the inplane-orientation of crystal grown can be easily identified by referring to the engineered surface relief.

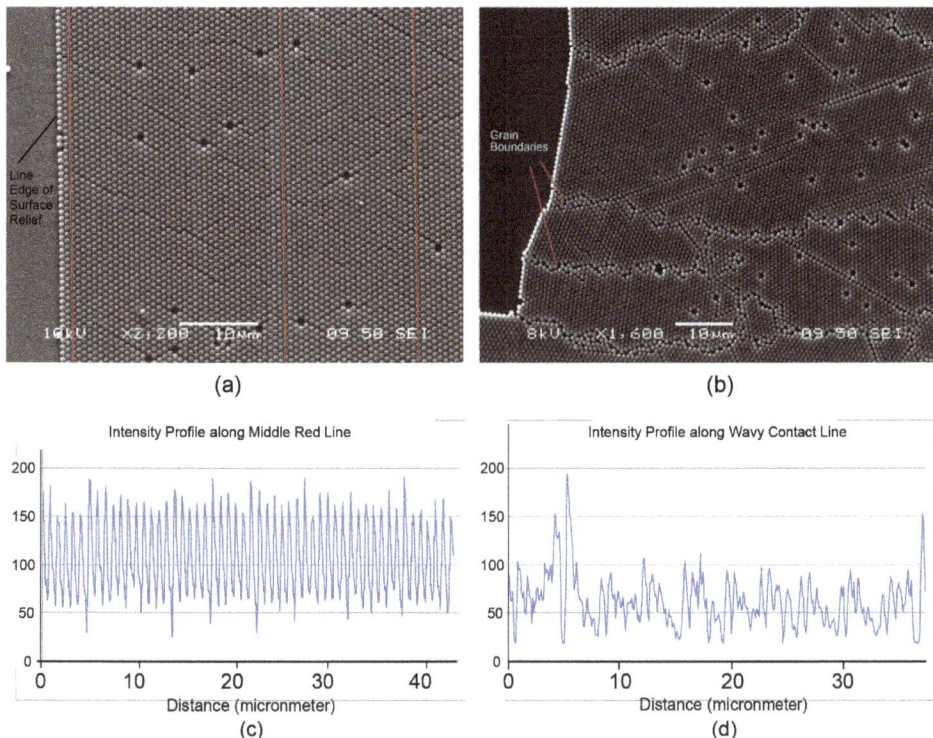

(a)   (b)

(c)   (d)

Fig. 16. SEM photo (a) shows inplane domain-oriented growth of colloidal crystal from the edge of straight surface relief, producing high degree of directionality. Particles are lined up in the close-packed direction <10> or <11>, along the surface relief. Red lines drawn serve as a guide to illustrate the perfect orientation under the straight pinning effect. (b) For comparison, colloidal assembly on bare substrate is shown, explaining the effect of wavy lines which result in domain growth of various directions. The desired growth directions are from left to right. (c) and (d) show line-plot profiles generated using ImageJ[3], along middle red line in part a and across central region in part b, respectively. Reprinted with permission from *Langmuir*, Vol. 27, No. 6 (March 2011), pp. 2244-2249. Copyright 2011 American Chemical Society.

5. Characterization of colloidal crystal

The most commonly used modern instruments in imaging dried colloidal crystals are scanning electron microscope (SEM) and transmission electron microscope (TEM). These types of imaging provide a quick view on the periodic structures grown via colloidal self-assembly; it does not however give quantitative data like crystal parameters in three dimensions. They are only good for 2D and topology scanning with smaller field of view

---

[3] ImageJ is a Java application popular for SEM and TEM image processing and analysis. More on http://rsbweb.nih.gov/ij/

(short range). If the particles are large enough (> 1µm diameter), their motion and ordering can also be observed with optical microscopes (Denkov et al, 1992; Pieranski, 1983; Yan et al, 2008).

Fortunately, the ability of colloidal crystal to diffract light allows one to characterize the crystal structure and quality with ease (Hiltner & Krieger, 1969). For colloidal crystals with lattice spacings in the order of visible-light wavelengths, diffraction method can be used in transmission mode or reflectance mode (Imura et al, 2009; Koh et al, 2006). As discussed earlier, in order for electromagnetic waves to diffract, it must obey the following Bragg's law. Direct reflection peak or transmission peak can then be recorded using UV-vis spectrometer.

$$\lambda = 2n_{eff}d_{111} \tag{4}$$

$$n_{eff} = \sqrt{(\Psi n_{ps}^2 + (1 - \Psi)n_{air}^2} \tag{5}$$

$$d_{111} = (\frac{2}{3})^{\frac{1}{2}}D \tag{6}$$

where n_{eff} is effective refractive index of the colloidal suspension, d_{111} is the interlayer spacing between (111) plane, D is particle diameter, n_{ps} and n_{air} are refractive indices of polystyrene particle and air respectively, and ψ is the volume fraction of particles in suspension.

Unfortunately, strong interaction between light and the crystals could result in multiple scattering. It was observed that the Bragg spacings derived from diffraction measurements could deviate strongly from the real lattice spacings (Los et al, 1996). Besides, the available optical spectrum limits the number of Bragg reflections to be observed. These issues could be remedied by small angle X-ray scattering. First, X-ray interacts weakly with colloidal particles, serving as an excellent tool to probe internal structure of photonic crystal. Second, since there is a dramatic difference between X-ray wavelength (~1Å) and the particle diameter (e.g. 1µm), a tiny diffraction angle (narrow focus range) in the order of 10^{-4} rad will be able to supply sufficient information regarding the crystal (Thijssen et al, 2006). To conclude, radius, size distribution and internal structure of particle, crystal structure, lattice parameter and average orientation of colloidal crystal can be investigated via X-ray scattering (Megens et al, 1997; Vos et al, 1997).

Kossel lines, previously used in X-ray diagrams or electron diffraction experiments (Kikuchi lines) had also been used to examine phase transformation of colloidal crystal, crystal structures and their lattice parameters (Clark et al, 1979; Pieranski et al, 1981; Yoshiyama et al, 1984). Fig. 17 demonstrates a simple setup to obtain Kossel diagram, which is either projected on a spherical screen, V or a flat one, F. The colloidal crystal suspension is held in a glass or quartz cuvette, which is immersed in a spherical vessel V (diameter ~ 10 cm) filled with pure water. This water-filled vessel serves to minimize the refraction at the surface of the crystal. A divergent laser beam (normally He-Ne, λ = 632.8 nm) is then focused through a window and the diffracted beam will be projected on the spherical projection screen V or on a flat screen at distance X.

It should be noted that a 2D map of diffraction spots could be printed on the projected screen in Fig. 17, if a collimated white light is used. This offers a distinct advantage in immediate identification of the wavelength diffracted, based on the colour of the projected Laue spots. Other than white light, collimated laser and X-rays can also be used to obtain diffraction spots, by vary the scanning angle of the incoming beam (Williams & Crandall, 1974; Clark & Hurd, 1979).

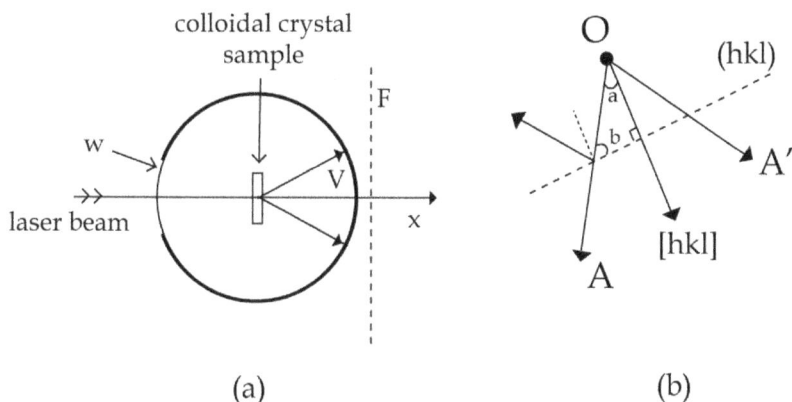

(a)   (b)

Fig. 17. (a) Experimental set-up of Kossel analysis, where a divergent laser beam is shone through thin colloidal crystal sample and the diffracted beam is projected on the curved screen V or flat screen F. (b) Diagram shows formation of Kossel cone in (a), where the Kossel lines are the intersection between the Kossel cone and the projection screen shown in (a). (Pieranski et al, 1981)

6. Conclusion

In spite of recent progress in colloidal self-assembly, large scale ordering of colloidal crystals with controlled packing symmetry, periodicity, crystal orientation and packing quality in a practical and economical application still remains an active scientific and engineering activity. This interest is intensively fuelled by the many fundamental scientists, application engineers and researchers working in tailored colloidal crystals exhibiting novel functions in applications. This chapter highlights not only our own work in colloidal self-assembly, but also the fundamental concepts and key approaches used to grow colloidal crystals. We think that particle-particle interactions in the process of assembly, mobility and space confinement remain the three crucial keys in fabricating performance-sound colloidal crystals.

For example, one could think of giving directional properties to colloidal particles and control the particle interactions in short range to form tetrahedral clusters. The clusters can then be directed into long range assembly via necessary confinement (e.g. meniscus, template, etc) to achieve equilibrium ordering of a diamond lattice. Simulations can also be deployed to assess the feasibility of related colloidal crystal structures for specific tailored synthesis, assembly and performance. However, the common inherent problem remains over the practicalities: speed, cost, area of crystal grown and ease of integration. In conclusion, the potential is huge, and there is still much room for future research.

7. Acknowledgment

We acknowledge all the works contributed by our group members, including those who have left for good cause. They are Mr. Tan Kwan Wee, Dr. Yan Qing Feng, Dr. Teh Lay Kuan and Dr. Yip Chan Hoe. We also thank Singapore-MIT alliance for the funding support throughout the years.

8. References

Adachi, E., Dimitrov, A. S. & Nagayama, K. (1995). Stripe Patterns Formed on a Glass Surface during Droplet Evaporation. *Langmuir*, Vol.11, No.4, (April 1995), pp. 1057-1060

Allard, M., Sargent, E. H., Lewis, P. C. & Kumacheva, E. (2004). Colloidal Crystals Grown on Patterned Surfaces. *Advanced Materials*, Vol.16, No.15, (August 2004), pp. 1360-1364

Arora, A. K. & Tata, B. V. R. (1996). *Ordering and Phase Transitions in Charged Colloids* (1st ed), Wiley-VCH, ISBN 0471186309

Auer, S. & Frenkel, D. (2001). Prediction of Absolute Crystal-nucleation Rate in Hard-sphere Colloids. *Nature*, Vol.409, No.6823, (February 2001), pp. 1020-1023

Avrutsky, I., Kochergin, V. & Zhao, Y. (2000). Optical Demultiplexing in a Planar Waveguide with Colloidal Crystal. *IEEE Photonics Technology Letters*, Vol.12, No.12, (December 2000), pp. 1647-1649

Bartlett, P., Ottewill, R. H. & Pusey, P. N. (1992). Superlattice Formation in Binary Mixtures of Hard-sphere Colloids. *Physical Review Letters*, Vol.68, No.25, (June 1992), pp. 3801-3805

Behrens, S. H. & Grier, D. G. (2001). The Charge of Glass and Silica Surfaces. *The Journal of Chemical Physics*. Vol.115, No.14, (2001), pp. 6716-6721

Bertrand, P., Jonas, A., Laschewsky, A. & Legras, R. (2000). Ultrathin Polymer Coatings by Complexation of Polyelectrolytes at Interfaces: Suitable Materials, Structure and Properties. *Macromolecular Rapid Communications*, Vol.21, No.7, (April 2000), pp. 319-348

Biró, L. P., Bálint, Z., Kertész, K., Vértesy, Z., Márk, G. I., Horváth, Z. E., Balázs, J., Méhn, D., Kiricsi, I., Lousse, V. & Vigneron, J. P. (2003). Role of Photonic-crystal-type Structures in the Thermal Regulation of a Lycaenid Butterfly Sister Species Pair. *Physical Review E*, Vol.67, No.2, (February 2003), pp. 021907-1 – 021907-6

Blanco, A., Chomski, E., Grabtchak, S, Ibisate, M., John, S., Leonard, S., Lopez, C., Meseguer, F., Miguez, H, Mondian, J. Orzin, G. A., Toader, O. & van Driel, H. M. (2000). Large-scale Synthesis of a Silicon Photonic Crystal with a Complete Three-dimensional Bandgap near 1.5 Micrometres. *Nature*, Vol.405, No.6785, (May 2000), pp. 437-440

Bourret, A., Fuoss, P., Feuillet, G. & Tatarenko, S. (1993). Solving an Interface Structure by Electron Microscopy and X-ray Diffraction: The GaAs(001)-CdTe(111) Interface. *Physical Review Letters*, Vol.70, No.3, (January 1993), pp. 311-315

Busch, K. & John, S. (1998). Photonic Band Gap Formation in Certain Self-organizing Systems. *Physical Review E*, Vol.58, No.3, (September 1998), pp. 3896-3908

Chen, K. M., Jiang, X., Kimerling, L. C. & Hammond, P. T. (2000). Selective Self-Organization of Colloids on Patterned Polyelectrolyte Templates. *Langmuir,* Vol.16, No.20, (October 2000), pp. 7825-7834

Chen, M., Kim, J., Liu, J. P., Fan, H. & Sun, S. (2006). Synthesis of FePt Nanocubes and Their Oriented Self-Assembly. *Journal of the American Chemical Society,* Vol.128, No.22, (June 2006), pp. 7132-7133.

Clark, N. A., Hurd, A. J. & Ackerson, B. J. (1979). Single Colloidal Crystals. *Nature,* Vol.281, No.5726, (September 1979), pp. 57-60

Crandall, R. S., & Williams, R. (1977). Gravitational Compression of Crystallized Suspensions of Polystyrene Spheres. *Science,* Vol.198, No.4314, (October 1977), pp. 293-295

Davis, S. H. (1980). Moving Contact Lines and Rivulet Instabilities. Part 1. The Static Rivulet. *Journal of Fluid Mechanics,* Vol.98, No.2, (1980), pp. 225-242

Delamarche, E., Schmid, H., Bietsch, A., Larsen, N. B., Rothuizen, H., Michel, B. & Biebuyck, H. Transport Mechanisms of Alkanethiols during Microcontact Printing on Gold. *The Journal of Physical Chemistry B,* Vol.102, No.18, (April 1998), pp. 3324-3334

Denkov, N., Velev, O., Kralchevski, P., Ivanov, I., Yoshimura, H. & Nagayama, K. (1992). Mechanism of Formation of Two-dimensional Crystals from Latex Particles on Substrates. *Langmuir,* Vol.8, No.12, (December 1992), pp. 3183-3190

Dushkin, C. D., Yoshimura, H. & Nagayama, K. (1993). Nucleation and Growth of Two-dimensional Colloidal Crystals. *Chemical Physics Letters,* Vol.204, No. 5-6, (March 1993), pp. 455-460

Fustin, C., Glasser, G., Spiess, H. W. & Jonas, U. (2004). Parameters Influencing the Templated Growth of Colloidal Crystals on Chemically Patterned Surfaces. *Langmuir,* Vol.20, No.21, (October 2004), pp. 9114-9123

Gasser, U., Weeks, E. R., Schofield, A., Pusey, P. N. & Weitz, D. A. (2001). Real-Space Imaging of Nucleation and Growth in Colloidal Crystallization. *Science,* Vol.292, No.5515, (April 2001), pp. 258 -262

Gast, A. P. & Russel, W. B. (1998). Simple Ordering in Complex Fluids. *Physics Today,* Vol.51, No.12, (1998), pp. 24-30

Habdas, P. & Weeks, E. R. (2002). Video Microscopy of Colloidal Suspensions and Colloidal Crystals. *Current Opinion in Colloid & Interface Science,* Vol.7, No.3-4, (August 2002), pp. 196-203

Hales, T. C. (1997). Sphere packings, I. *Discrete & Computational Geometry,* Vol.17, No.1, (January 1997), pp. 1-51

Hiltner, P. A. & Krieger, I. M. (1969). Diffraction of Light by Ordered Suspensions. *The Journal of Physical Chemistry,* Vol.73, No.7, (July 1969), pp. 2386-2389

Ho, K. M., Chan, C. T. & Soukoulis, C. M. (1990). Existence of a Photonic Gap in Periodic Dielectric Structures. *Physical Review Letters,* Vol.65, No.25, (December 1990), pp. 3152-3155

Holgado, M., García-Santamaría, F., Blanco, A., Ibisate, M., Cintas, A., Míguez, H., Serna, C. J., Molperceres, C., Requena, J., Mifsud, A., Meseguer, F. & López, C. Electrophoretic Deposition To Control Artificial Opal Growth. *Langmuir,* Vol.15, No.14, (July 1999), pp. 4701-4704

Hoogenboom, J. P., Rétif, C., de Bres, E., van der Boes, M., van Langen-Suurling, A. K. & Romijn, J. (2004). Template-Induced Growth of Close-Packed and Non-Close-Packed Colloidal Crystals during Solvent Evaporation. *Nano Letters*, Vol.4, No.2, (February 2004), pp. 205-208

Hoover, W. & Ree, F. (1968). Melting Transition and Communal Entropy for Hard Spheres. *The Journal of Chemical Physics*, Vol.49, No.8, (October 1968), pp. 3609-3617, ISSN 00219606

Hu, Z. & Jonas, A. M. (2010). Control of Crystal Orientation in Soft Nanostructures by Nanoimprint Lithography. *Soft Matter*, Vol.6, No.1, (2010), pp. 21-28

Im, S. H., Kim, M. H. & Park, O. O. (2003). Thickness Control of Colloidal Crystals with a Substrate Dipped at a Tilted Angle into a Colloidal Suspension. *Chemistry of Materials* Vol.15, No.9 (May 2003), pp. 1797-1802

Imura, Y., Nakazawa, H., Matsushita, E., Morita, C., Kondo, T. & Kawai, T. (2009). Characterization of Colloidal Crystal Film of Polystyrene Particles at the Air-Suspension Interface. *Journal of Colloid and Interface Science*, Vol.336, No.2, (August 2009), pp. 607-611

Jiang, P., Bertone, J. F., Hwang, K. S., Colvin, V. L. (1999). Single-Crystal Colloidal Multilayers of Controlled Thickness. *Chemistry of Materials*, Vol.11, No.8, (1999), pp. 2132-2140

Jiang, P. & McFarland, M. J. (2004). Large-Scale Fabrication of Wafer-Size Colloidal Crystals Macroporous Polymers and Nanocomposites by Spin-Coating. *Journal of the American Chemical Society*, Vol.126, No.42, (October 2004), pp. 13778-13786

Kay, L. E. (1986). W. M. Stanley's Crystallization of the Tobacco Mosaic Virus, 1930-1940. *Isis*, Vol.77, No.3, (September 1986), pp. 450-472

Kitaev, V. & Ozin, G. A. (2003). Self-Assembled Surface Patterns of Binary Colloidal Crystals. *Advanced Materials*, Vol.15, No.1, (January 2003), pp.75-78

Koh, Y. K. & Wong C. C. In Situ Monitoring of Structural Changes during Colloidal Self-Assembly. *Langmuir*, Vol.22, No.3, (January 2006), pp. 897-900

Koh, Y. K., Yip, C. H., Chiang, Y. M. & Wong, C. C. (2008). Kinetic Stages of Single-Component Colloidal Crystallization. *Langmuir* Vol. 24, No. 10 (May 2008), pp. 5245-5248

Kralchevsky, P. A. & Nagayama, K. (1994). Capillary Forces between Colloidal Particles. *Langmuir*, Vol.10, No.1, (January 1994), pp. 23-36

Kralchevsky, P. A. & Denkov, N. D. (2001). Capillary Forces and Structuring in Layers of Colloid Particles. *Current Opinion in Colloid & Interface Science*, Vol.6, No.4, (August 2001), pp. 383-401

Kumacheva, E., Garstecki, P., Wu, H. & Whitesides, G. M. (2003). Two-Dimensional Colloid Crystals Obtained by Coupling of Flow and Confinement. *Physical Review Letters*, Vol.91, No.12, (September 2003), pp. 128301-1 –128301-4

Lee, I., Zheng, H., Rubner, M. F. & Hammond, P. T. (2002). Controlled Cluster Size in Patterned Particle Arrays via Directed Adsorption on Confined Surfaces. *Advanced Materials*, Vol.14, No.8, (April 2002), pp. 572-577

Lee, K. H., Chen, Q. L., Yip, C. H., Yan, Q. & Wong, C. C. (2010). Fabrication of Periodic Square Arrays by Angle-resolved Nanosphere Lithography. *Microelectronic Engineering*, Vol.87, No.10, (October 2010), pp. 1941-1944

Liu, J. (2005). *Photonic devices*, Cambridge University Press, ISBN 0521551951

Lu, Y., Yin, Y. & Xia, Y. (2001). Preparation and Characterization of Micrometer-Sized 'Egg Shells'. *Advanced Materials*, Vol.13, No.4, (February 2001), pp. 271-274

Manoharan, V., Elsesser, M & Pine, D. J. (2003). Dense Packing and Symmetry in Small Clusters of Microspheres. *Science*, Vol.301, No.5632, (July 2003), pp. 483-487

Maskaly, G. R., García, R. E., Carter, W. C. & Chiang, Y. M. (2006). Ionic Colloidal Crystals: Ordered, Multicomponent Structures via Controlled Heterocoagulation. *Physical Review E*, Vol.73, No.1, (January 2006), pp. 011402-1 – 011402-8

Mayoral, R., Requena, J., Moya, J., López, C., Cintas, A., Miguez, H., Meseguer, F., Vásguez, L., Holgado, M. & Blanco, A. (1997). 3D Long-range ordering in an SiO_2 Submicrometer-sphere Sintered Superstructure. *Advanced Materials*, Vol.9, No.3 (March 1997), pp. 257-260

Megens, M., van Kats, C. M., Bösecke, P. & Vos, W. L. (1997). Synchrotron Small-Angle X-ray Scattering of Colloids and Photonic Colloidal Crystals. *Journal of Applied Crystallography*, Vol.30, No.5-2, (October 1997), pp. 637-641

McLachlan, M. A., Johnson, N. P., Rue, R. M. & McComb, D. W. (2004). Thin Film Photonic Crystals: Synthesis and Characterisation. *Journal of Materials Chemistry*, Vol.14, No.2, (2004), pp. 144-150

Míguez, H., Meseguer, F., López, C., Mifsud, A., Moya, J. S. & Vázquez, L. (1997). Evidence of FCC Crystallization of SiO2 Nanospheres. *Langmuir*, Vol.13, No.23, (November 1997), pp. 6009-6011

Mihi, A., Zhang, C. & Braun, P. V. (2011). Transfer of Preformed Three-Dimensional Photonic Crystals onto Dye-Sensitized Solar Cells. *Angewandte Chemie International Edition*, Vol.50, No.25, (June 2011), pp. 5712-5715

Moroz, A & Sommers, C. (1999). Photonic Band Gaps of Three-dimensional Face-centred Cubic Lattices. *Journal of Physics: Condensed Matter*, Vol.11, No.4, (February 1999), pp. 997-1008

Norris, D. J. & Vlasov, Y. A. (2001). Chemical Approaches to Three-Dimensional Semiconductor Photonic Crystals. *Advanced Materials*, Vol.13, No. 6, (March 2001), pp. 371-376

Ozin, G. A. & Yang, S. M. (2001). The Race for the Photonic Chip: Colloidal Crystal Assembly in Silicon Wafers. *Advanced Functional Materials*, Vol.11, No.2, (April 2001), pp. 95-104

Park, S. H., Qin, D. & Xia, Y. (1998). Crystallization of Mesoscale Particles over Large Areas. *Advanced Materials*, Vol.10, No.13, (September 1998), pp. 1028-1032

Park, S. H., Gates, B. & Xia, Y. (1999). A Three-Dimensional Photonic Crystal Operating in the Visible Region. *Advanced Materials*, Vol.11, No.6, (April 1999), pp. 462-466

Pieranski, P. (1983). Colloidal Crystals. *Contemporary Physics*, Vol.24, No.1, (January 1983), pp. 25-73, ISSN 00107514

Pradhan, R. D., Bloodgood, J. A. & Watson, G. H. (1997). Photonic Band Structure of bcc Colloidal Crystals. *Physical Review B*, Vol.55, No.15, (April 1997), pp. 9503-9507

Prieve, D. C., Sides, P. J. & Wirth, C. L. (2010). 2-D Assembly of Colloidal Particles on a Planar Electrode. *Current Opinion in Colloid & Interface Science*, Vol.15, No.3, (June 2010), pp. 160-174

Pusey, P. N. & van Megen, W. (1986). Phase Behaviour of Concentrated Suspensions of Nearly Hard Colloidal Spheres. *Nature*, Vol.320, No.6060, (March 1986), pp. 340-342.

Pusey, P. N., van Megen, W., Bartlett, P., Ackerson, B. J., Rarity, J. G. & Underwood, S. M. (1989). Structure of Crystals of Hard Colloidal Spheres. *Physical Review Letters*, Vol.63, No.25, (December 1989), pp. 2753-2756

Rai-Choudhury, P. (1997). *Handbook of Microlithography, Micromachining, and Microfabrication: Microlithography*, SPIE Press, ISBN 9780819423788

Ray, M. A., Kim, H. & Jia, L. Dynamic Self-Assembly of Polymer Colloids to Form Linear Patterns. *Langmuir*, Vol.21, No.11, (May 2005), pp. 4786-4789

Ramiro-Manzano, F., Bonet, E., Rodriguez, I. & Meseguer, F. (2009). Layering Transitions in Colloidal Crystal Thin Films between 1 and 4 Monolayers. *Soft Matter*, Vol.5, No.21, (2009), pp. 4279-4282

Romano, F. & Sciortino, F. (2011). Colloidal Self-assembly: Patchy from the Bottom Up. *Nat Mater*, Vol.10, No.3, (March 2011), pp. 171-173

Sasaki, M. & Hane, K. (1996). Ultrasonically Facilitated Two-dimensional Crystallization of Colloid Particles. *Journal of Applied Physics*, Vol.80, No.9, (1996), pp. 5427-5431

Schmidt, M. & Löwen, H. (1997). Phase Diagram of Hard Spheres Confined between Two Parallel Plates. *Physical Review E*, Vol.55, No.6, (June 1997), pp. 7228-7241

Sharma, A. & Reiter, G. (1996). Instability of Thin Polymer Films on Coated Substrates: Rupture, Dewetting, and Drop Formation. *Journal of Colloid and Interface Science*, Vol.178, No.2, (March 1996), pp. 383-399

Sharma, V., Yan, Q., Wong, C. C., Carter, W. C. & Chiang, Y. M. (2009). Controlled and Rapid Ordering of Oppositely Charged Colloidal Particles. *Journal of Colloid and Interface Science*, Vol.333, No.1, (May 2009), pp. 230-236

Sirota, E. B., Ou-Yang, H. D., Sinha, S. K., Chaikin, P. M., Axe, J. D. & Fujii, Y. (1989). Complete Phase Diagram of a Charged Colloidal System: A Synchro-tron X-ray Scattering Study. *Physical Review Letters*, Vol.62, No.13, (March 1989), pp. 1524-1527

Tan, K. W., Li, G., Koh, Y.K., Yan, Q. & Wong, C. C. (2008). Layer-by-Layer Growth of Attractive Binary Colloidal Particles. *Langmuir*, Vol.24, No.17, (July 2008), pp. 9273-9278

Tan, K. W., Koh, Y. K., Chiang, Y.M. & Wong, C. C. (2010). Particulate Mobility in Vertical Deposition of Attractive Monolayer Colloidal Crystals. *Langmuir*, Vol.26, No.10 (May 2010), pp. 7093-7100.

Teh, L. K., Tan, N. K., Wong, C. C. & Li, S. Growth imperfections in three-dimensional colloidal self-assembly. *Applied Physics A: Materials Science & Processing*, Vol. 81, No. 7, (November 2005), pp. 1399-1404

Thijssen, J. H. J., Petukhov, A. V., 't Hart, T. C., Imhof, A., van der Werf, C. H. M., Schropp, R. E. I. & van Blaaderen, A. (2006). Characterization of Photonic Colloidal Single Crystals by Microradian X-ray Diffraction. *Advanced Materials*, Vol.18, No.13, (July 2006), pp. 1662-1666

Thomson, N. R., Bower, C. L. & McComb, D. W. (2008). Identification of Mechanisms Competing with Self-assembly during Directed Colloidal Deposition. *Journal of Materials Chemistry*. Vol.18, No.21, (2008), pp. 2500-2505

Tien, J., Terfort, A. & Whitesides, G. M. (1997). Microfabrication through Electrostatic Self-Assembly. *Langmuir*, Vol.13, No.20, (October 1997), pp. 5349-5355

Torres, C. M. S. (2003) *Alternative lithography: unleashing the potentials of nanotechnology*, Springer, ISBN 9780306478581

Troian, S. M., Herbolzheimer, E., Safran, S. A. & Joanny, J. F. (1989). Fingering Instabilities of Driven Spreading Films. *Europhysics Letters (EPL)*, Vol.10, No.1, (September 1989), pp. 25-30

Ulman, A. (1996). Formation and Structure of Self-Assembled Monolayers. *Chemical Reviews*, Vol.96, No.4, (January 1996), pp. 1533-1554

van Blaaderen, A., Ruel, R. & Wiltzius, P. (1997). Template-directed Colloidal Crystallization. *Nature*, Vol.385, No.6614, (January 1997), pp. 321-324

van Rijn, C. J. M., Veldhuis, G. J. & Kuiper, S. (1998). Nanosieves with Microsystem Technology for Microfiltration Applications. *Nanotechnology*, Vol.9, No.4, (December 1998), pp. 343-345

Velev, O. D. & Lenhoff, A. M. (2000). Colloidal Crystals as Templates for Porous Materials. *Current Opinion in Colloid & Interface Science*, Vol.5, No.1-2, (March 2000), pp. 56-63

Vlasov, Y. A., Astratov, V. N., Baryshev, A. V., Kaplyanskii, A. A., Karimov, O. Z. & Limonov, A. F. (2000). Manifestation of Intrinsic Defects in Optical Properties of Self-organized Opal Photonic Crystals. *Physical Review E*, Vol.61, No.5, (May 2000), pp. 5784-5793

Vos, W. L., Sprik, R., van Blaaderen, A., Imhof, A., Lagendijk, A. & Wegdam, G. H. (1996). Strong Effects of Photonic Band Structures on the Diffraction of Colloidal Crystals. *Physical Review B*, Vol.53, No.24, (June 1996), pp. 16231-16235

Vos, W. L., Megens, M., van Kats, C. M. & Bösecke, P. (1997). X-ray Diffraction of Photonic Colloidal Single Crystals. *Langmuir*, Vol.13, No.23, (November 1997), pp. 6004-6008

Williams, R. & Crandall, R. S. (1974). The Structure of Crystallized Suspensions of Polystyrene Spheres. *Physics Letters A*, Vol.48, No.3, (June 1974), pp. 225-226

Xia, Y & Whitesides, G. M. (1998). Soft Lithography. *Angewandte Chemie International Edition*, Vol.37, No.5, (March 1998), pp. 550-575

Yablonovitch, E. (1987). Inhibited Spontaneous Emission in Solid-State Physics and Electronics. *Physical Review Letters*, Vol.58, No.20, (May 1987), pp. 2059-2062

Yan, Q., Gao, L., Sharma, V., Chiang, Y.M. & Wong, C. C. (2008). Particle and Substrate Charge Effects on Colloidal Self-Assembly in a Sessile Drop. *Langmuir*, Vol. 24, No.20, (October 2008), pp. 11518-11522

Yan, Q., Pavan, N., Chiang, Y. M. & Wong, C. C. (2009). Three-dimensional Metallic Opals Fabricated by Double Templating. *Thin Solid Films*, Vol.517, No.17, (July 2009), pp. 5166-5171

Yang, S. M. & Ozin, G. A. (2000). Opal Chips: Vectorial Growth of Colloidal Crystal Patterns inside Silicon Wafers. *Chemical Communications*, No.24, (2000), pp. 2507-2508

Yethiraj, A. & van Blaaderen, A. (2003). A Colloidal Model System with an Interaction Tunable from Hard Sphere to Soft and Dipolar. *Nature*, Vol.421, No.6922, (January 2003), pp. 513-517

Yethiraj, A., Wouterse, A., Groh, B. & van Blaaderen, A. (2004), Nature of an Electric-Field-Induced Colloidal Martensitic Transition. *Physical Review Letters*, Vol.92, No.5, (February 2004), pp. 058301-1 – 058301-4

Yoshiyama, T., Sogami, I. & Ise, N. (1984). Kossel Line Analysis on Colloidal Crystals in Semidilute Aqueous Solutions. *Physical Review Letters,* Vol.53, No.22, (November 1984), pp. 2153-2158

Zhang, K. & Liu, X. Y. (2004). In Situ Observation of Colloidal Monolayer Nucleation Driven by an Alternating Electric Field. *Nature,* Vol.429, No.6993, (June 2004), pp. 739-743

Zhou, Z. & Zhao, X. S. (2004). Flow-Controlled Vertical Deposition Method for the Fabrication of Photonic Crystals. *Langmuir* Vol.20, No.4, (February 2004), pp. 1524-1526

14

Simulation of CaCO₃ Crystal Growth in Multiphase Reaction

Pawel Gierycz
Faculty of Chemical and Process Engineering, Warsaw University of Technology,
Institute of Physical Chemistry, Polish Academy of Sciences,
Warsaw,
Poland

1. Introduction

Calcium carbonate formation and aggregation processes have been studied from many years and there are widely described in the literature (i.e. Kitano et al., 1962, Montes-Hemandez et al., 2007, Reddy & Nancollas, 1976). However, the mechanism of the process, which depends on the way of reaction conducting (i.e. Dindore et al., 2005, Feng et al., 2007, Jung et al., 2005, Schlomach et al., 2006) is till now not fully understood and investigated due to increasing application of CaCO₃ in commercial production of new materials, pharmaceuticals and many others.

Calcium carbonate occurs in nature in three polymorphic modifications: rhombohedral calcite, orthorhombic aragonite usually with needle-like morphology and hexagonal vaterite with spherical morphology. The most needed from the practical point of view is the most stable thermodynamically calcite. One of its important applications area is connected with fabrication of functional solids where the fully controlled precipitation process must be applied. The big interest in this field is due to the fact that application of produced materials is determined by many strictly defined parameters.

In recent years many researchers deal with application of organic additives (Bandyopadhyaya, 2001) as a template to produce inorganic materials and conduction of reaction in a macro- or microemulsions or in sol-gel matrixes. Such methods give an opportunity to control of precipitation process or to modify product properties but unsolved problem remains purity of the obtained powder. There are also many ways of CaCO₃ precipitation conducting without any additives (i.e. Cafiero et al., 2002, Sohnel & Mullin, 1982, Rigopoulos & Jones, 2003a). Although a lot of investigations described in the literature (i.e. Chakraborty & Bhatia, 1995, Chen et al., 2000, Cheng et al., 2004) there are still many questions about the full mechanism of crystals nucleation and growth of freshly precipitated particles.

Generally, crystallization is a particle formation process by which molecules in solution or vapor are transformed into a solid phase of regular lattice structure, which is reflected on the external faces. Crystallization may be further described as a self-assembly molecular building process. So, crystallographic and molecular factors are thus very important in affecting the shape, purity and structure of crystals (Colfen & Antonietti, 2005, Collier &

Hounslow, 1999, Mullin, 2001). There are two established mechanisms of crystals growth described in literature (Jones et al., 2005, Judat & Kind, 2004, Spanos & Koutsoukos, 1998). The Ostwald ripening involves the larger crystals formation from smaller crystals which have higher solubility than larger ones (the smaller crystals act as fuel for the growth of bigger crystals). Another important growth mechanism revealed in recent years is nonclassical crystallization mechanism by aggregation, i.e. coalescence of initially stabilized nanocrystals which grow together and form one bigger particle (Judat & Kind, 2004, Myerson, 1999).

There are some papers dealing with calcium carbonate formation through oriented aggregation of nanocrystals (Collier & Hounslow, 1999, Myerson, 1999, Wang et al., 2006) or through self assembled aggregation of nanometric crystallites followed by a fast recrystalization process (Judat & Kind, 2004). This way of particles formation control to fabricate ordered structures is inspired by processes observed in biological systems and is one of top topics of modern colloid and materials chemistry (Judat & Kind, 2004, Myerson, 1999, Wang et al., 2006).

Each way of reaction conduction needs its own modeling. In the literature there are many different models and simulations (i.e. Bandyopadhyaya et al., 2001, Hostomsky & Jones, 1991, Malkaj et al., 2004,) done for different particular reactions.

Each model describing the crystal formation (i.e. Quigley & Roger, 2008, Tobias & Klein, 1996, Wachi & Jones, 1991) has to take into account both particulate crystal characteristics and fluid-particle transport processes. The crystals formation and further solid–liquid separation of particulate crystals from solution involves suspension and sedimentation. During these processes solid matter may change phase from liquid to solid or vice versa. New particles may be generated and existing ones can be lost. Thus, both the liquid and solid phases are subject to the physical laws of change: conservation of mass and flow. The crystals may be also separated from fluids by flow through reactor. So, any model well-describing the particular crystallization process has to take into account the conditions of reaction leading in the reactor.

2. Modelling and simulation

The behaviour of real crystallization processes is determined by the interaction of multiple process phenomena, which all have to be modelled to fully describe the process. Over the past decades simulation has become a standard tool for solution of these model equations. Different tools have been developed to solve many typical chemical engineering problems particularly for standard fluid phase processes. Also for more complex processes, as population balance models for crystallization processes (Randolph & Larson, 1988) and computational fluid dynamics (Ferziger & Perić, 1996) problems based on the Navier–Stokes equations, commercial simulation tools are available.

However, still not every kind of crystallization process models can be solved with the available tools. Moreover, every particular crystallization process needs a specific treatment taking into account process parameters, hydrodynamic conditions, crystallizer construction, etc. A separate problem, which also has to be considered is connected with the accuracy of the simulation. The accurate calculations are time consuming and the accuracy is strongly connected with the way in which the simulation is performed. So, for each accurate simulation of a particular crystallization process it is necessary to elaborate both the appropriate physico-chemical description of the process and the proper way of simulation performance.

2.1 Conservation equations - Computational fluid dynamics

Conservation relates to accounting for flows of heat, mass or momentum (mainly fluid flow) through control volumes within vessels and pipes. This leads to the formation of conservation equations which enables to predict results of operation performed in defined equipment.

In continuum mechanics the general equation for all conservation laws can be expressed in the following form (Spiegelman, 2004):

$$\frac{\partial \Phi}{\partial t} + \nabla (\mathbf{F} + \Phi \mathbf{V}) - H = 0 \tag{1}$$

where: Φ - any quantity (in units of stuff per unit volume), $\mathbf{F}$ - the flux of Φ in the absence of fluid transport, $\mathbf{V}$ – velocity of transport, H - a source or sink of Φ.

To derive an equation for *conservation of mass* it is necessary to substitute $\Phi = \rho$ (density - the amount of mass per unit volume), $\mathbf{F} = 0$ (mass flux can be only change due to transport) and $H = 0$ (mass cannot be created or destroyed). After that we get the following equation, called, *the continuity equation*:

$$\frac{\partial \rho}{\partial t} + \nabla (\rho \mathbf{V}) = 0 \tag{2}$$

In the case of *conservation of energy (heat)* in a single phase material (Spiegelman, 2004), the amount of heat per unit volume is $\Phi = \rho\, c_P\, T$ where c_P is the specific heat at constant pressure (energy per unit mass per degree Kelvin) and T is the temperature. The heat flux consists of two components due to conduction (the heat flux is $\mathbf{F} = -k \nabla T$ where k is the thermal conductivity, "-" because heat flows from hot to cold) and transport ($\rho\, c_P\, T\, \mathbf{V}$). Heat can be also created in a investigated volume due to viscous dissipation, radioactive decay, shear heating etc. (H – all the terms creating heat). Thus the conservation of heat equation assumes the form:

$$\frac{\partial \rho c_P T}{\partial t} + \nabla (\mathbf{F} + \rho c_P T \mathbf{V}) = \nabla \cdot k \nabla T + H \tag{3}$$

For constant c_P and k, after introducing thermal diffusivity $\kappa = k / (\rho\, c_P)$, the equation can be rewritten in the form:

$$\frac{\partial T}{\partial t} + \mathbf{V} \cdot \nabla T = \kappa \nabla^2 T + H / (\rho c_P) \tag{4}$$

Conservation of momentum (or force balance) can be derived (Spiegelman, 2004) assuming that the amount of momentum per unit volume is $\Phi = \rho \mathbf{V}$ and the forces which can change the momentum are connected with the stress that acts on the surface ($\mathbf{F} = -\sigma$) and gravity ($H = \rho g$, where g – terrestrial acceleration). So, the equation has the following form:

$$\frac{\partial \rho \mathbf{V}}{\partial t} + \nabla (\rho \mathbf{V} \mathbf{V}) = \nabla \sigma + \rho g \tag{5}$$

which can rewritten in the following way:

$$\frac{\partial \mathbf{V}}{\partial t} + (\mathbf{V} \cdot \nabla) \mathbf{V} = \frac{1}{\rho} \nabla \sigma + g \tag{6}$$

For an isotropic incompressible fluids the above equation can be rewritten into Navier-Stokes equation:

$$\frac{\partial \mathbf{V}}{\partial t} + (\mathbf{V} \cdot \nabla)\mathbf{V} = -\frac{1}{\rho}\nabla P + \nu \cdot \nabla^2 \mathbf{V} + g \tag{7}$$

where: ν is the dynamic viscosity, P – fluid pressure.

In the case of crystallization a further conservation equation is required to account for particle numbers. This is *the population balance*. It is another transport equation which, based on particle formation (nucleation, growth, agglomeration, breakage, etc.), allows for prediction of particle size distribution, $n(L, t)$ in the defined crystallizers. Generally the population balance equation (PBE) can be written in the following form (Jones et al., 2005)

$$\frac{\partial n}{\partial t} + \nabla(\mathbf{v}_i n) + \nabla(\mathbf{v}_e n) = B_0 + B - D \tag{8}$$

where: B_0 is nucleation rate, B and D are the "Birth" and "Death" functions for agglomeration and breakage of crystals, where $\mathbf{v}_i$ is the "internal" velocity and $\mathbf{v}_e$ is the "external" velocity.

The "internal" velocity describes the change of particle characteristic, e.g. its size, volume or composition, and the "external" velocity, the fluid velocity, in the crystallizer. The "internal" velocity, for well-mixed systems, is approximated by the crystal growth rate (G):

$$\nabla(\mathbf{v}_i n) = \nabla(Gn) \tag{9}$$

and usually assumes the following form:

$$\nabla(\mathbf{v}_i n) = \nabla(Gn) = \frac{\partial(Gn)}{\partial L} \tag{10}$$

where: L is a crystal size.

For non well-mixed systems, the velocity derivatives, in addition to crystal growth, have to be included to the equation.

The population balance is a partial integro-differential equation that can be normally solved by numerical methods, except for some simplified cases. Different numerical discretization schemes for solution of the population balance (Kumar & Ramkrishna, 1996, Nicmanis & Hounslow, 1998, Ramkrishna, 2000) and compute correction factors in order to preserve total mass are widely described in the literature (Hostomsky & Jones, 1991, Rigopoulos & Jones, 2003a, Wojcik & Jones, 1998, Wuklow et al., 2001).

Computation Fluid Dynamics, (CFD) is the numerical analysis and solution of system involving all transport processes via computer simulation (Jones et al., 2005,). It is strongly dependent on the development of computer related technologies and on the advancement of our understanding and solving of ordinary and partial differential equations. Direct numerical solving of complex flows in real conditions requires a huge amount of computational power and is very much dependent on the physical models applied. That is why, an ideal model applied for such calculations should introduce minimum amount of complexity into the model equations, while capturing the essence of the relevant physics.

One of the most important flow phenomena is turbulence. If it is present in a certain flow it appears to be the dominant over all other flow phenomena. That is why successful modelling of turbulence greatly increases the quality of numerical simulations. Although, all analytical and semi-analytical solutions of simple flow cases were solved at the end of 1940s, there are still many open questions on modelling of turbulence and properties of turbulence it-self. Till now, no universal turbulence model exists yet.

In the case of crystallization, CFD involves the numerical solution of conservation continuity, momentum and energy equations coupled with constitutive laws of rate (kinetic) processes together with the population balance accounting the solid particles formed and destroyed during crystallization. So, the CFD model solution comprises both the flow properties and a particle size distribution what leads to the formation of conservation equations which enables to predict results of operation performed in defined equipment. Attempts to generate a theoretical model-based description of the interaction of fluid dynamics and crystallization face the multi-scale nature of this interaction.

Usually, the population balance is represented by a partial differential equation of particle size and time and the mass balance, in most cases, is expressed as ordinary differential equations. On the other hand, the growth and nucleation kinetics of particles are often based on empirical correlations.

The main problem connected with a numerical simulation is a problem of discretization of the all coordinates (Euclidean space, particle size, time). Discretization significantly affects the accuracy, the computational costs and even convergence properties of numerical algorithms. Therefore, the selection of the proper discretization grids has to be carefully considered in the context of the characteristic scales of the modelled phenomena.

Usually, for the fluid flow calculation, the Euclidean space is divided into a number of CFD grid cells with elementary volumes. The size of these cells is above the Kolmogorov turbulence scale (order of magnitude 10^{-4} m) but small enough to well resolve the convective flows and energy transport within the unit (Ferziger & Perić, 1996). Such discretization is sufficient to resolve the most of the phenomena occurring in mass crystallization but needs to be improved in the case of reactive crystallization processes where micromixing phenomena play the significant role. The time coordinate of the CFD problem is also discretized using small time steps (seconds) to resolve fast fluctuations.

The particle size coordinate (the population balance equation) has to be also discretized. There are many methods available to perform this discretization (Ramkrishna, 2000, Hounslow, 1990, Hounslow et al., 1988, Hill & Ng, 1995) but in all cases, the most important in the proper evaluation of the size of CFD cell with the appropriate number of particles. If the CFD cells are too small or have too low number of particle the statistical requirements of the population balance is not fulfilled. This may result in an incorrect solution. The next problem connected with the discretization is necessity of solution of several dozens of the equations in each CFD grid cell what would certainly result in prohibitive computational cost and possibly introduce convergence problems. Therefore, some means of model reduction must be employed to allow a numerical simulation. Moreover, all these methods have been developed with a focus on the way in which the systems are mixed.

Generally, CFD models can be implemented to "well-mixed" and "non well-mixed" systems. Assumption of well-mixing is commonly used for the modelling of crystallization processes, what simplifies the simulation and reduces its time. Such approach can be accepted in the case of theoretical calculations and small, laboratory scale, cristallizers. However, even in a stirred tank with impellor (Rielly & Marquis, 2001) we deal with very inhomogeneous fluid mechanical environment. The turbulence quantities and the relevant mean-flow may vary by orders of magnitude throughout the vessel, especially around the impellor. Therefore it is clear, that the 'well-mixed' assumption will lead to significant errors on the rates of growth, nucleation and agglomeration, and consequently, on the crystal size distribution. In these cases, information of the solid concentration distribution, as well as local velocities, shear rates and energy dissipation rates would be needed for the proper design of the process. Crystallization systems frequently show also high levels of supersaturation around the points where it is generated (cooling surfaces, evaporation interfaces, etc.) causing as well as suspension significant local density variations (Sha & Palosaari, 2000). Consequently, crystallization rates locally vary throughout the crystallizer even in case when no reactive crystallization occurs. Therefore, assumption of uniform conditions throughout the reactor volume can not be accepted.

Moreover, many crystallization processes are directly affected by the local fluid dynamic state. One of the most important factors is the shear rate which strongly influences both the frequency and the efficiency of particle collisions (agglomeration (Hounslow et al., 2001)) as well as particle-impeller collisions (Gahn & Mersmann, 1999) which are depended on the relative velocity of the particle and local streamlines around the impeller blade.

The mixing problem increases with increasing of the scale of operation. Typically, fluid dynamics phenomena act on 'micro-scale' (CDF grid), a much smaller scale compared to crystallization phenomena which are usually considered on the 'macroscale' (unit). To solve the problem one can compute the population balance in each CFD grid, accounting for the full locality of the crystallization kinetics or use the scale or spatial resolution for the population balance. The first approach is not recommended because it can violate the statistical assumptions used for the formulation of the population balance equation and needs a long, tome consuming calculations (computational costs). The second method enables for selection of some compartments, representing a certain region in the crystallizer, which can be treated as homogenous (well-mixed) and well described by CFD. Such approach can be a compromise between one single, well mixed unit and the over-detailed system. The use of this model requires the exchange of information between the two scales of calculations: "inner" inside the compartment and "outer" between the compartments. Taking into account these assumptions some compartmental mixing models (Wei & Garside, 1997) for modelling precipitation processes based on the engulfment theory (Baldyga & Bourne, 1984a, 1984b, 1984c) has been elaborated. Also, several mesomixing and micromixing models have been proposed to describe the influence of mixing on chemical reactions on the meso- and molecular scale (Villermaux & Falk, 1994, Baldyga et al., 1995).

2.2 Batch reactor
Batch (or semi-batch) reactor is one of the most popular reactors widely used in chemical and pharmaceutical (especially batch crystallizers) industry. Batch crystallization processes, commonly investigated, are still not well understood because the process is strongly influenced by fluid mixing, particle aggregation and particle breakage. For a batch crystallizer with nucleation, aggregation, breakage and growth occurring (Wan & Ring, 2006). the population balance equation is given by (Randolph & Larson, 1988) as:

$$\frac{\partial n(v,t)}{\partial t} + \frac{\partial [G(v)n(v,t)]}{\partial v} = b(v) - d(v) \tag{11}$$

where: $n(v)$ is the number-based population of particles in the crystallizer being a function of the particle volume v, $G(v)$ is the volume dependent growth rate, $b(v)$ is the volume dependent birth rate and $d(v)$ is the volume dependent death rate. For the initial condition, $n(v,t{=}0) = n_o(v)$.

For aggregation, the birth $b_a(v)$ and death $d_a(v)$ rate terms can be given by (Hulburt & Katz, 1964):

$$b_a(v) - d_a(v) = \int_0^{1/v} \beta(v-u,u)n(v-u)n(u)\,du - n(v)\int_0^{\infty} \beta(v,u)n(u)\,du \qquad (12)$$

where $\beta(v,u)$ is the aggregation rate constant (a measure of the frequency of collision of particles of size v with those of size u).

In the case of breakage, the birth $b_b(v)$ and death $d_b(v)$ rate terms can be given by (Prasher, 1987):

$$b_b(v) - d_b(v) = \int_v^{\infty} S(w)\rho(v,w)n(w)\,dw - S(v)n(v) \qquad (13)$$

where $S(v)$ is the breakage rate that is a function of particle size v, $\rho(v,w)$ is the daughter distribution function defined as the probability that a fragment of a particle of size w will appear at size v. The population balance equations can be solved by the use of the standard method of moments (SMOM) and the quadrature method of moments (QMOM) (Wan & Ring, 2006). Using these methods the population balance can be simplified into a series of a few discrete moment equations (some of them as number of particles ($_vm_0$), volume of particles ($_vm_1$), etc. have physical significance) defined, for k-th volume-dependent moment, in the following way:

$$_vm_k = \int_v^{\infty} v^k n(v)\,dv \qquad (14)$$

Such calculated $_vm_0$ and $_vm_1$ represent the total number and total volume of particles in the system

Because in the CFD code, the particle density function is described as a function of particle size x, instead of particle volume v and the population balance is written in terms of $n(x)$ instead of $n(v)$ the population balance, eq. [11], can be rewritten as:

$$\frac{\partial n(x,t)}{\partial t} + \frac{\partial [G(x)n(x,t)]}{\partial x} = b(x) - d(x). \qquad (15)$$

and the k-th length-dependent moment as:

$$_Lm_k = \int_v^{\infty} x^k n(x)\,dx \qquad (16)$$

The both SMOM and QMOM models has been tested (Wan & Ring, 2006) using numerical cases with nucleation, growth, aggregation and breakage and the obtained results have been compared with the analytical measurements. For all cases the OMOM model gave the very good (the accuracy < 1 %) description of the particle size distribution in the batch reactor.

The particle size distribution in a batch crystallizer can be also simulated in different way. As an example can be given a process of obtaining of calcium carbonate (Kangwook et al., 2002) when we deal with the following overall precipitation reaction:

$$Ca(OH)_2 + Na_2CO_3 = CaCO_3 + 2NaOH \tag{17}$$

where the feeds are a solution of sodium carbonate and a solution of calcium hydroxide at certain, defined concentrations, and the main product is calcium carbonate. The main variable which is to be estimated is particle size distribution of precipitated $CaCO_3$.

The precipitation occurs, when the calcium ions and carbonate ions are present at supersaturated concentration levels. Supersaturation implies that the ionized species are present in the solution where the solubility of the species is exceeded. If we assume that the ionization reactions are fast compared to the precipitation i.e. the ionization reactions reach equilibrium instantaneously and that the perfect mixing in the reactor is obtained we can write the mass balance of the precipitation reactor as follows:

$$\frac{d(VC_i)}{dt} = q_j C_j^F - qC_j - k_q V \int_0^\infty G(L,t)n(L,t)L^2 dL \tag{18}$$

$$\frac{dV}{dt} = q^F - q \tag{19}$$

where: C_j^F is the concentration of species j in the j-th feed stream, C_j is the reactor concentration of species j, q_j^F is the feed flow rate of stream j, q^F is the total feed flow rate, q is the total outlet flow rate, V is the volume of contents in the reactor, k_a is the area factor, L is the characteristic particle size, $G(L, t)$ is the growth rate of particle, $n(L, t)$ is the particle size distribution (number of particles per volume of solvent per particle size), t is the time of reaction, j is equal to 1 for $Ca(OH)_2$ or 2 for Na_2CO_3.

The mass balance equation should be solved together with the population balance equation:

$$\frac{d[Vn(L,t)]}{dt} + V\frac{\partial[G(L,t)n(L,t)](Vn(L,t))}{\partial L} = VP(L, n, t) - qn(L.t) \tag{20}$$

where: $P(L.n.t)$ is a number of density
with corresponding initial condition $n(L,t_0) = n_{t0}(L)$ and boundary condition $n(L_0,t) = n_{L0}(t)$ (i.e. the number of nucleated particles), where L_0 is the nucleated particle size.

Next important equation needed is an equation describing nucleation rate. Typically, nucleation and growth rates of precipitation and crystallization processes are represented by semi-empirical power laws. A proper, nucleation model has to take into account the both primary nucleation induced by supersaturation without particles and secondary nucleation related to the existing particles in the reactor. Growth rate is a function of supersaturation and particle size and can be calculated from the following equations (Eek et al., 1995):

$$n_{L0}(t) = \frac{1}{G(0,t)} \int_0^\infty a_n C_s^{b_n} n L^{2,5} dL \qquad (21)$$

$$G_L(t) = a_t C_s^{b_t} \frac{1}{1-exp[-a_L(L-b_L)]} \qquad (22)$$

where: C_s is the supersaturation of the solute, which is defined as: $[(C_1C_2)^{0.5}-1]$ (expressed in terms of the normalized concentration) and an $a_n, b_n, a_t, b_t, a_L, b_L$ are the parameters.

The kinetic equations have strong nonlinearity due to the power terms what combined with the mass balance equation makes the problem difficult. However, the computationally demanding part of the precipitation reactor model is the population balance equation. In general, the population balance equation can be converting into a set of ordinary differential equations. Many, various forms of the finite element method and the finite difference method can be applied for this purpose. The details on the solution techniques can be found in a (Ramkrishna, 2000) book.

The population balance equation can be simplified in the case when the right-hand side of eq. (20) is a linear or an independent function of the density number. Then a closed-form of the solution can be obtained using the method of characteristics (Varma & Morbidelli, 1997). We can further simplify the model equations assuming that the aggregation and breakage are negligible and the growth rate takes a separable form of G_tG_L in eq. (22) (where: G_t is the time-dependent part of the growth rate and G_L is the size-dependent part). In this case we get the following equations for the particle size distribution:

$$n(L,t) = \frac{V(t_0)}{V(t)} n_{t0}(L_b) \frac{G(L_b)}{G(L)} \quad \text{for } L_b = L(L,t) \geq 0 \qquad (23)$$

$$n(L,t) = \frac{V(t_b)}{V(t)} n_{L0}(t_b) \frac{G(0)}{G(L)} \quad \text{for } t_b = L(L,t) > 0 \qquad (24)$$

where t_0 is the start time of growth reaction, L_b in the birth size and t_b is the birth time of the L size particle at time t, which can be obtained by solving the following equations:

$$\int_{L_b}^L \frac{1}{G(l)} dl = \int_{t_0}^t G_t(\tau) d\tau \qquad (25)$$

$$\int_0^L \frac{1}{G(l)} dl = \int_{t_b}^t G_t(\tau) d\tau \qquad (26)$$

In the case of size dependent growth, there are no general theoretical kinetics and the separable form is the exclusively used empirical form.

In order to simulate the precipitation reactor, the mass balance and the population balance equation should be solved together. They can be solved used an explicit integration method in which the algebraic equations are solved just once at the beginning of each integration step and held constant or finite element method (Kangwook et al., 2002).

The usefulness of this model for the calcium carbonate precipitation (both an explicit integration and finite element method gave almost the same results) has been checked successfully by (Kangwook et al., 2002) but is necessary to remember, that the assumption of negligible agglomeration and breakage (limits of the model) can be applied only for the reactor where the particle density is maintained on the low level (Kataki & Tsuge, 1990).

For the calcium carbonate precipitation in the batch reactor, breakage can be treated as a negligible phenomenon but the agglomeration is usually significant according to the high particle density in the reactor (Collier & Hounslow, 1999). So, if we want to avoid the aggregation and breakage phenomena in this reactor we have to operate the process in a special way, maintaining the low particle density.

Generally, the presented approach can be implemented for simple precipitation reaction and is, especially, very useful in the case of "run-to-run" or "on-line" controlling of the particle size distribution in a batch (or semi-batch) reactors (Kangwook et al., 2002).

2.3 Crystallization in tube

Every model describing crystallization in a tube has to take into account the fluid dynamics, the fluid flow through the tube and crystallization processes acting simultaneously. The simplest model describing crystallization from solution with feed concentration c_0, in a wall-cooled tube with a defined length and radius, where the supersaturation is generated by cooling of the solution by means of an energy withdrawal at the wall, can be derived making the following assumptions (Kulikov et al., 2005):

-   the system is considered to be quasi-homogeneous - it is assumed that the flow through the tube causes very well mixing of the fluid and solid (very small crystals) phases. So, instead of writing separate transport equations for the fluid and the solid phases, a single equation for the whole suspension is formulated. This results in assuming no slip and no particle drag which also implies no segregation of the particles,
-   mixture properties (density ρ, molecular viscosity v, specific heat capacity c_p, thermal conductivity λ) are assumed to be constant,
-   no heat of crystallization is released,
-   agglomeration and particle breakage are not considered.

The fluid dynamics of the homogeneous mixture can be described by the Reynolds-averaged Navier–Stokes equations consisting of the equations for mass and momentum conservation:

$$\nabla \tilde{V} = 0 \tag{27}$$

$$\frac{\partial \tilde{V}}{\partial t} + (\tilde{V} \cdot \nabla)\tilde{V} = -\frac{1}{\rho}\nabla p + (v + v_t) + \nabla^2 \tilde{V} + g \tag{28}$$

where: $\tilde{V}$ is the vector of Reynolds-averaged velocities, p is the static pressure, g is the gravitational acceleration, v and v_t are viscosity and the turbulent viscosity, respectively.

Using the introduced assumptions, a boundary condition was set for the flow at the tube inlet, no-slip condition was used at the wall and the standard k-ε model (Ferziger & Perić, 1996) has been used, as a turbulence model for a closure of the system. So, the energy balance can be expressed as:

$$\frac{\partial T}{\partial t} + \tilde{\mathbf{V}} \cdot \nabla T = \frac{\lambda}{c_p \rho} \nabla^2 T \tag{29}$$

where: T is the temperature and a boundary temperature condition is specified at the walls of the tube.

The population balance equation used in this model has been taken from (Marchisio et al., 2003). It contains (Kulikov et al., 2005) the accumulation term, the particle growth term, the convective transport term, terms reflecting molecular and turbulent diffusion of particles with the molecular diffusion coefficient D_m and the turbulent diffusion coefficient D_t, respectively, as well as particle birth b and death d terms and can be written in the following form:

$$\frac{\partial n}{\partial t} + \frac{\partial (Gn)}{\partial L} + \nabla (\tilde{\mathbf{V}} n) - (D_m + D_t) \nabla^2 n = b - d \tag{30}$$

where: $n(L, x, t)$ is a particle size distribution and L is a characteristic particle size.

In this case, it is assumed a simple kinetics with the growth term G obeying McCabe's law (size-independent growth) and being first order dependent on supersaturation:

$$G = k_1 [c - c_s (T)] \tag{31}$$

where: c and c_s is the solution concentration and the equilibrium concentration at saturation, respectively. k_1 is a constant.

As it assumed both birth term B and death term D are set to zero:

$$B = 0 \qquad D = 0 \tag{32}$$

and nucleation B_0 is accounted for as a left boundary condition as follows:

$$n(L = 0, x, t) = \frac{B_0}{G} \tag{33}$$

and expressed by a power law equation:

$$B_0 = (1 - \alpha) k_2 \exp \left(- \frac{k_3}{[c/c_s (T) - 1]^2} \right) \tag{34}$$

where: a is the volume fraction of solids and k_2 and k_3 are constants.

The initial condition for nucleation are given by the following equations;

$$n(L, t = 0) = 0 \qquad c(t = 0) = c_0 \tag{35}$$

and mass balance for the solute in the liquid phase by the following:

$$\frac{\partial n}{\partial t} + \nabla (\tilde{\mathbf{V}} c) - D_c \nabla^2 c = -3 \rho_{cr} k_v G \int_0^\infty n L^3 dL \tag{36}$$

where: D_c is the solute diffusivity, ρ_{cr} is the density of the crystals and k_v is are the shape factor of the particles.

The presented model specified in eqs (27)–(36) is a multidimensional dynamic problem containing partial differential equations formulated in spatial coordinates x, one internal particle size distribution coordinate L and the time coordinate t. The locally distributed velocities, temperatures, and particle size distribution are the unknown variables which cannot be calculated analytically and have to be obtained by a numerical simulation.

As it was mentioned before the numerical simulation can be done using two approaches. The first aims at the reduction of the complexity of the population balance discretization by selection a small number of variables characterizing the particle size distribution. It causes some loss of accuracy in the solution of the population balance, which is reformulated in terms of these variables. Transport equations are also reformulated for these variables and solved along with the CFD problem on the proper spatial grid. Usually, these variables are the moments of the distribution function i.e. the Quadrature Method of Moments (Marchisio et al., 2003). A main disadvantage of this approach is the inaccurate reconstruction of the particle size distribution when no a-priori information about its shape is available.

The second approach is based on the reduction of the spatial resolution for the population balance only. Most crystallization phenomena like growth, agglomeration, etc. do not change significantly on the resolution of the CFD grid and can be considered to act on larger scales. This allows for the representation of the population balance by collecting a set of CFD cells in an 'ideally-mixed' compartment. The population balance equations can then be solved in this compartment by a highly accurate discretization scheme. Set of such ideally-mixed compartments represents different regions of the crystallizer. This approach has been well described in the literature (Kramer et al., 2000).

It is difficult to claim the superiority of one of these approaches over the other. The proper selection of the approach very much depends on the application to which it is addressed. The compartmental approach better describes the major crystallization phenomena in a cooling crystallizer with complex breakage and aggregation behaviour while the reduced population balance approach better describes a high spatial fluctuation of supersaturation, e.g., in reactive crystallization.

2.4 Bubble column reactor

Bubble column reactor is an apparatus in which simplicity of design gives rise to extraordinary complexity in the physical and chemical phenomena. That is why modeling of the precipitation process in this reactor needs an integration of reaction kinetics, population balance and hydrodynamic principles.

Such successful modeling of the bubble column reactor applied for the precipitation of calcium carbonate by carbon dioxide absorption into lime has been done by (Rigopoulos. & Jones, 2003a). They used their own (Rigopoulos. & Jones, 2003b) finite element method for solving the time-dependent population balance equation with combined nucleation, growth, agglomeration, and breakage. The previous studies of gas-liquid precipitation (Rigopoulos. & Jones, 2001) which used the method of moments, took into account only a nucleation growth. However, experiment in both gas-liquid (Wachi & Jones, 1991) and liquid-liquid (Tai & Chen, 1995, Collier & Hounslow, 1999) precipitation of $CaCO_3$ have evidenced the presence of agglomeration and demonstrated its importance in determining of the product crystal size distribution.

The time-dependent population balance equation (Rigopoulos. & Jones, 2003b) that describes the evolution of the particle size distribution in a finite, spatially uniform domain, with particle volume as the "internal" coordinate, and including nucleation, growth, and agglomeration, can be written as follows:

$$\frac{dn(V,t)}{dt} = \frac{n_{in}(V,t) - n(V,t)}{\tau} - \frac{\partial}{\partial V}[G(V)n(V,t)] + B_0\delta(V - V_0) +$$

$$+ \frac{1}{2}\int_0^V \beta_a(V-V',V')n(V-V',t)n(V',t)dV' - n(V,t)\int_0^\infty \beta_a(V,V')n(V',t)dV' \tag{37}$$

where: $n(v,t)$ and $n_{in}(v,t)$ is the population density at the reactor and at the inlet, respectively; $G(v)$ is the volumetric growth rate; and B_0, β_a, and V_0 are the nucleation rate, agglomeration kernel, and volume of the nuclei, respectively, δ is a width of boundary layer.
The equation should be solved with the following initial and boundary conditions:

$$n(V,0) = n_0(V) \qquad \text{(initial distribution)} \tag{38}$$

$$n(0,t) = 0 \qquad \text{(no crystals zero of size)} \tag{39}$$

The mass balance equation is derived from the concept of penetration theory (Astarita, 1967) where mass transfer and chemical reactions at the interface are treated simultaneously. The interface and the bulk are considered as two separate dynamic reactors which operate independently and interact at discrete time intervals. Thus, the diffusion and reaction of chemical components is described by the following equations (c_i is the concentration of component i, superscripts I and B denotes variables at the interface and bulk, respectively):

$$\frac{\partial c_i^I}{\partial t} = D\frac{\partial^2 c_i^I}{\partial x^2} + \sum_{k=1}^K r_k(c_1^I, c_2^I, ..., c_n^I) \tag{40}$$

with initial and boundary conditions:

$$t = 0, \quad x > 0 \rightarrow c_i^I = c_i^B \tag{41}$$

$$x = 0, \quad t > 0 \rightarrow c_i^I = c_i^* \quad \text{(volatile species)} \tag{42}$$

$$x = \delta, \quad t > 0 \rightarrow \frac{dc_i^I}{dx} = 0 \quad \text{(non-volatile species)} \tag{43}$$

In the most cases of the bubble column the precipitation phenomena at the interface can be neglected because of the very short contact time between the reagents compared to the bulk. Generally. nonideal mixing should be considered in the column but for the relatively short height of the column and intense recirculation a full mixing in the bulk can be assumed. In this case the mass balance in the bulk can be described in the following way:

$$\frac{\partial c_i^B}{\partial t} = \sum_{k=1}^K r_k(c_1^B, c_2^B, ..., c_n^B, n_1, n_2, ..., n_m) \tag{44}$$

$$\frac{\partial n_j}{\partial t} = f_j\left(c_1^B, c_2^B, ..., c_n^B, n_1, n_2, ..., n_m\right) \tag{45}$$

where $f_j(c_i^B, n_j)$ is a function into which the original population balance is transformed via the finite element discretization. Initial conditions are calculated from the mixing of bulk and interface at the end of the previous contact time. The solution of the interface equations is obtained numerically with an implicit iterative scheme, while the bulk equations are calculated according to Adams method (Hindmarsh, 1983).

The reactions occurring during the process can be described in the following way:

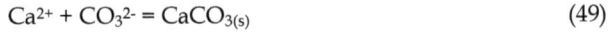

$$CO_{2(g)} = CO_{2(l)} \tag{46}$$

$$CO_{2(g)} + OH^- = HCO_3^- \tag{47}$$

$$HCO_3^- + OH^- = CO_3^{2-} + H_2O \tag{48}$$

$$Ca^{2+} + CO_3^{2-} = CaCO_{3(s)} \tag{49}$$

The first step (eg. (46)) is a CO_2 absorption in water at the gas-liquid equilibrium. The equilibrium can be described by Henry's law, taking into account that we deal with an ionic system, as follows:

$$log\left(\frac{H}{H_0}\right) = -\sum_i I_i h_i \tag{50}$$

$$h_i = h_+ + h_- + h_g \tag{51}$$

where: H and H_0 is a Henry's constant for an ionic and nonionic system, respectively, I_i is an ionic strength of component "i" and h_i, h_-, h_+, h_g is a component "i", anions, cations and gas contribution, respectively.

The kinetics of carbon dioxide absorption into alkali solutions are determined by the conversion of $CO_2(aq)$ into HCO_3^- (eq. (47)), which proceeds at a great, but finite rate. This reaction is followed by an instantaneous ionic reaction eq. (48). and the precipitation reaction eq. (49).

The rate of $CaCO_3(s)$ production is determined by a crystallization mechanism but always volumetric crystal growth is size-dependent even when the linear growth is size-independent (McCabe's law). To obtain the rate of change for the whole crystal mass is necessary to integrate the volumetric growth function over the whole range of crystal volumes:

$$\frac{dc_{CaCO3}}{dt} = \int_{V_0}^{V_\infty} G(V)n(V)dV\frac{\rho_{CaCo3}}{M_{CaCO3}} \tag{52}$$

where: ρ_{CaCO3} and M_{CaCO3} is calcium carbonate density and molar mass, respectively.

To estimate the rate of crystal mass production, which is coupled with the population balance, it is necessary to derive a complete kinetic model of precipitation taking into account the whole information concerning the crystal formation i.e. nucleation, crystal growth as well as agglomeration and breakage.

The growth rate kinetics is usually described by the linear growth rate (the increase in particle diameter or radius) G_l, by the following expression:

$$G_l = k_g (\lambda_s - 1)^2 \tag{53}$$

where: k_g is a kinetic constant and λ_s is the saturation ratio (Rigopoulos. & Jones, 2003b).
Nucleation process can have a variety of mechanisms (homogeneous, heterogeneous, secondary, etc). In the bubble column it can be assumed (Rigopoulos. & Jones, 2003b) that in the beginning of the process high supersaturation levels induce primary nucleation, but later, secondary nucleation causes the rise of crystal growth. So, the overall nucleation model consists of the sum of the two models: primary and secondary. The primary nucleation which depends mainly on supersaturation is usually described by a power law:

$$B_0 = k_{n1} (\lambda_s - 1)^{k_{n2}} \tag{54}$$

Secondary nucleation is induced by the existing crystals (Garside & Davey, 1980) and is a function of the crystal mass (M_c):

$$B_0 = k_{ns} (\lambda_s - 1)^{k_{n3}} M_c^{k_{n4}} \tag{55}$$

where: k_{n1}, k_{n2}, k_{n3}, k_{n4} and k_{ns} are the appropriate constants (Rigopoulos. & Jones, 2003b).
Thus, the overall nucleation model can be expressed as follows:

$$B_0 = k_{n1} (\lambda_s - 1)^{k_{n2}} + k_{ns} (\lambda_s - 1)^{k_{n3}} M_c^{k_{n4}} \tag{56}$$

It is necessary to point out that calcium carbonate can appear, during the precipitation process, in three different polymorphs where the most prevailing polymorph appears to be calcite. That is why, a kinetic model should account for their simultaneous presence in the solution (Chakraborty & Bhatia, 1996). Usually, because of the complexity and difficulty of such calculations, the considerations are limited only to calcite.
Agglomeration of crystals is a very complex and system-dependent process. Usually, it can be simplified and treated as a two-step process. The first step of agglomeration, i.e. the formation of flocculates through collisions and interparticle attraction, is similar to the phenomena occurring in colloids and aerosols. The second step is the growth of crystalline material between the clusters at so-called cementing sites (Hounslow et al., 2001). In the case of the bubble column (Rigopoulos. & Jones, 2003b) the agglomeration can be assumed to be roughly proportional to growth (Hounslow et al., 2001) and described as the second-order dependence on supersaturation, in the following way:

$$\beta_a (V', V - V') = k_a (\lambda_s - 1)^2 [(V')^{1/3} + (V - V')^{1/3}]^3 \tag{57}$$

where: k_a is the agglomeration constant (Rigopoulos. & Jones, 2003b).
Hydrodynamics of the gas-liquid precipitation strongly depends on the gas holdup which determines the rates of the chemical phenomena. The Eulerian-Eulerian multiphase CFD model (Rigopoulos, & Jones, 2001), where the turbulence in the liquid phase is calculated with k-ε model (Schwarz & Turner, 1988), can be used for its description. This model can be

successfully used for modeling large-scale equipment because it gives sufficiently accurate results with respect to averaged properties. However, it is less successful in reproducing fine details i.e. the radial phase distribution.

The above model very well (good agreement with the experiment) described the precipitation of $CaCO_3$ by CO_2 absorption into lime, in the bubble column. The conjunction of penetration theory and CFD predictions of the gas holdup seems to yield an adequate description of the reactor performance. Such integration of the population balance, reaction kinetics and hydrodynamic principles allowed for proper formulation of modeling approach for the gas-liquid precipitation process and the model can be used as a tool for the analysis and scale-up of industrial-class equipment.

2.5 Thin film reactor

The another approach (Kędra-Królik & Gierycz, 2010) is necessary when the precipitation goes in the thin film. It happens in the Rotating Disc Precipitation Reactor (Kędra-Królik & Gierycz, 2006, Kędra-Królik & Gierycz, 2009) used for calcium carbonate production. The reaction in liquid phase goes in contact with continuously flowing gaseous carbon dioxide in the thin film formed on the surface of the rotating disc (Kędra-Królik & Gierycz, 2006, Kędra-Królik & Gierycz, 2009). This creates a constant surface area of gas-liquid interface and the carbonation reaction of lime water involves gas, liquid and solid phase. The reactions occurring during the process are described by eqs. (46-49).

The model (Kędra-Królik & Gierycz, 2010) has taken into account not only kinetics of the multiphase reaction but also crystal growth rate. The film theory (Wachi & Jones, 1991, Danckwerts, 1970) describes the mass balance of reactants in these reactions as follows:

$$\frac{\partial c_{CO2}}{\partial t} = D_{CO2}\left(\frac{\partial^2 c_{CO2}}{\partial x^2}\right) - kc_{CO2}c_{OH} \tag{58}$$

$$\frac{\partial c_{OH}}{\partial t} = D_{OH}\left(\frac{\partial^2 c_{OH}}{\partial x^2}\right) - kc_{CO2}c_{OH} \tag{59}$$

$$\frac{\partial c_{CO3}}{\partial t} = D_{CO3}\left(\frac{\partial^2 c_{CO3}}{\partial x^2}\right) + kc_{CO2}c_{OH} - G' - B' \tag{60}$$

where: t is time, c_{CO2}, c_{OH}, c_{CO3} are the concentrations of gas reactant ($CO_{2(g)}$), liquid reactant (OH-) and the product (CO_3^{2-}), respectively, G', B' are rate of nucleation and crystal growth, respectively; k is second order chemical reaction constant; D_{CO2}, D_{OH}, D_{CO3} are diffusivity of ($CO_{2(g)}$), (OH-) and (CO_3^{2-}), respectively.

The component (CO_3^{2-}) is formed by reaction (48) and consumed by the precipitation reaction (49). It is assumed also that the concentration of (CO_3^{2-}) is constant across the diffusion layer. Thus the population balance of the precipitated particles is given by the following equation (Hill & Ng, 1995):

$$\frac{\partial N}{\partial t} + G\frac{\partial N}{\partial L} = D_P\left(\frac{\partial^2 N}{\partial x^2}\right) \tag{61}$$

where: N is a population density of particles, G is linear growth rate; L is a coordinate of particle dimension; D_P is the diffusivity of particles. Substituting: $N = P/L$, we get:

$$\frac{\partial P}{\partial t} + G\frac{\partial P}{\partial L} = D_P\left(\frac{\partial^2 P}{\partial x^2}\right) + \frac{G}{L}P \tag{62}$$

where: $P(x,L_i t)$ is a number of density discretized in L_i, L_i is a particle size coordinate, L_0 is an effective nucleic size, for newly nucleated particles.

In the case of the precipitation of CaCO₃ in the thin film the small crystals are obtained due to the very high nucleation rate compared to the crystal growth rate (Kędra-Królik & Gierycz, 2010). For such very small particles, the diffusivity of the crystals (D_P) within the liquid film can be described by the Stokes-Einstein equation (Hostomsky & Jones, 1991):

$$D_P = k_B T / (6\pi\mu r) \tag{63}$$

where: k_B is the Boltzmann constant, T is temperature, μ is viscosity and r is radius of particle

The number rate of nucleation (J_n) and linear crystal growth (G) can be expressed by the Nielsen equations (Hounslow, 1990):

$$J_n = k_n (c - c^*)^n \tag{64}$$

$$G = k_g (c - c^*)^g \tag{65}$$

where: n, g – the orders of nucleation and growth, respectively; c, c^* - the concentration and equilibrium saturation concentration, respectively; k_n, k_g – nucleation and growth rate constants, respectively.

The equations can be rewritten to the following forms:

$$J_n = k_n \left(\sqrt{c_{Ca}c_{CO3}} - \sqrt{K_{sp}}\right)^n \tag{66}$$

$$G = k_g \left(\sqrt{c_{Ca}c_{CO3}} - \sqrt{K_{sp}}\right)^g \tag{67}$$

where: c_{Ca} is the concentrations of (Ca²⁺), K_{sp} is solubility of the product (calcium carbonate). The corresponding mass based rate equations both of nucleation and growth can be expressed by the following equations (Wachi & Jones, 1991, Danckwerts, 1970):

$$B' = a\rho J_n L_0^3 \qquad (a = \pi/6 \text{ for the sphere}) \tag{68}$$

$$G' = \sum_{i=0}^{\infty} \beta\rho P(x, L_i)GL_i^2 \qquad (\beta = \pi \text{ for the sphere}) \tag{69}$$

where: ρ is crystal density.

The boundary conditions for the gas-liquid interface, assuming that except for the gaseous reactant ($CO_{2(g)}$) every component is non-volatile, are as follows:

$$\text{at } x = 0; t > 0 \rightarrow c_{CO_2} = c_{CO_2}^0, dc_{OH}/dx = 0, dc_{CO_3}/dx = 0, dP/dx = 0 \tag{70}$$

and for the film formed on the disk surface, assuming that newly nucleated particles have an effective nucleic size equal to L_0, as follows:

$$\text{at } 0 < x < \delta; t > 0; \rightarrow L = L_0; \frac{\partial P}{\partial t} + G \frac{\partial P}{\partial L} = D_P \left(\frac{\partial^2 P}{\partial x^2} \right) + J_n ; L = \infty; P = 0 \tag{71}$$

Solving the mass balance equations (eqs. 58-60) and population equation (eq. 62) with the boundary conditions we can calculate both the discretized density number of particles ($P(t,x,L_i)$) and discretized diameter L_i. The model describes properly the change of precipitation rate in the liquid film. The $Ca(OH)_2$ concentration decreases because of the very high nucleation rate. Higher supersaturation leads to smaller mean crystal size, since the nucleation rate is much more sensitive to the level of supersaturation than the growth rate. It agrees very well with the experiment (Kędra-Królik & Gierycz, 2006, Kędra-Królik & Gierycz, 2009) and is caused by the fact that the high level of supersaturation is accumulated within the liquid film due to the large diffusion resistance.

The proposed model very well describes the $CaCO_3$ crystals formation in the rotating disc reactor and can be used and recommended for accurate calculation of the particle size and distribution obtained by gas-liquid precipitation in the reactor. However, it is necessary to remember that the model has not taken into account agglomeration of the obtained crystals and cannot be used for calculation of the aggregation process in the reactor.

3. Conclusion

The aim of this paper was to present the different approaches to the proper and accurate modeling and simulation of $CaCO_3$ formation and growth in multiphase reaction. This very complex problem has been presented for most popular, different types of reactors, i.e. batch, tube and thin film reactor as well as bubble column.

The batch (or semi-batch) precipitation process has been described by closed-form solution of population balance equation, which has not taken into account aggregation and breakage, what simplifies the simulation. However, the presented strategy is general and can be applied to batch or semi-batch processes described by more complex types of population balance equations.

In the case of tube reactor integration of simulation of crystallization and fluid dynamics was successfully applied by means of the Method of Moments. The used method allowed for reconstructing the solids fraction profiles on the fine CFD grid, while preserving the full information on particle size distribution on the coarser compartment scale. The technique is well established and has moderate computational costs.

The thin film reactor has been described by the model which takes into account both kinetics of the multiphase reaction and crystals growth rate. Results of calculation agreed very well with the experiment and the model described properly the change of precipitation rate from bulk liquid to the film region and showed that the higher supersaturation leads to smaller mean crystal size, since the nucleation rate is more sensitive to the level of supersaturation than the growth rate.

The gas-liquid precipitation process in the bubble column was modeled by integration the population balance, reaction kinetics and hydrodynamic principles. The used model well-described the precipitation of $CaCO_3$ by CO_2 absorption into lime and can be recommended the analysis and scale-up of industrial-class equipment. It gave also some explanations for the experimental results. It showed that the crystal mean size increase after the pH drop is due to the disappearance of the smaller crystals by dissolution, the secondary nucleation take place because a new wave of nucleation-growth is induced by the existing crystals and crystal agglomeration starts to take place at relatively high pH and proceeds to a considerable extent because the aggregates are less frequently disrupted than in stirred tanks.

Moreover, a wide review of different methods and approaches to the accurate description of crystallization processes as well as main CFD problems has been presented in this chapter. It can serve as a basic material for formulation and implementation of new, accurate models describing not only multiphase crystallization processes but also any processes taking place in different chemical reactors.

Combined population balance and kinetic models, computational fluid dynamics and mixing theory enable well prediction and scale-up of crystallization and precipitation systems but it is necessary to remember that each process (performed in the well defined reactor) needs always its own modeling.

4. References

Astarita, G. (1967). *Mass Transfer with Chemical Reaction*; Elsevier Publishing, ISBN 66-25758,: Amsterdam, The Netherlands

Baldyga, J. & Bourne, J.R. (1984a). A fluid mechanical approach to turbulent mixing and chemical reaction. Part I: Inadequacies of available methods. *Chemical Engineering Communications*, Vol.28, No.4-6, pp. 231-241, ISSN 0098-6445

Baldyga, J. & Bourne, J.R. (1984b). A fluid mechanical approach to turbulent mixing and chemical reaction. Part II: Micromixing in the light of turbulence theory. *Chemical Engineering Communications*, Vol.28, No.4-6, pp. 243-258, ISSN 0098-6445

Baldyga, J. & Bourne, J.R. (1984c). A fluid mechanical approach to turbulent mixing and chemical reaction. Part III: Computational and experimental results for the new micromixing model. *Chemical Engineering Communications*, Vol.28, No.4-6, pp. 259-281, ISSN 0098-6445

Baldyga, J., Podgorska, W. & Pohorecki R. (1995), Mixing-precipitation model with application to double feed semibatch precipitation. *Chemical Engineering Science*, Vol.50, No.8, (April 1995), pp. 1281-1300, ISSN 0009-2509

Bandyopadhyaya, R.; Kumar, R. & Gandhi K.S. (2001). Modelling of $CaCO_3$ nanoparticle formation during overbasing of lubricating oil additive. *Langmuir*, Vol.17, No.4, (February 2001), pp. 1015-1029, ISSN 0743-7463

Cafiero, L.M.; Baffi, G.; Chianese, A. & Jachuck, R.J.J. (2002). Process intensification: precipitation of barium sulfate using a spinning disk reactor. *Industrial & Engineering Chemistry Research*, Vol.41, No.21, (October 2002), pp. 5240-5246, ISSN 0888-5885

Chakraborty, D. & Bhatia, S.K. (1996). Formation and aggregation of polymorphs in continuous precipitation. 2. Kinetics of CaCO₃ precipitation. *Industrial & Engineering Chemistry Research*, Vol.35, No.6, (June 1996), pp. 1995-2006, ISSN 0888-5885

Chen, J.F.; Wang, Y.H.; Guo, F.; Wang, X.M. & Zheng, Ch. (2000). Synthesis of nanoparticles with novel technology: High-gravity reactive precipitation. *Industrial & Engineering Chemistry Research*, Vol.39, No.4, (April 2000), pp. 948-954, ISSN 0888-5885

Cheng, B.; Lei, M.; Yu, J. & Zhao, X. (2004). Preparation of monodispersed cubic calcium carbonate particles via precipitation reaction. *Materials Letters*, Vol.58, No.10, (April 2004), pp. 1565-1570, ISSN 0167-577X

Colfen, H. & Antonietti, M. (2005). Mesocrystals: Inorganic superstructures made by highly parallel crystallization and controlled alignment. *Angewandte Chemie International Edition*, Vol.44, No.35, (September 2005), pp. 5576-5591, ISSN 1433-7851

Collier, A.P. & Hounslow, M. J. (1999). Growth and aggregation rates for calcite and calcium oxalate monohydrate. *American Institute of Chemical Engineering Journal*, Vol.45, No.11, (November 1999), pp. 2298–2305, ISSN 0001-1541

Danckwerts, P.V. (1970). *Gas-Liquid Reaction*, McGraw-Hill, ISBN 007015287X, New York, US

Dindore, V.Y.; Brilman, D.W.F. & Versteeg, G.F. (2005). Hollow fiber membrane contactor as a gas–liquid model contactor. *Chemical Engineering Science*, Vol.60, No.2, (January 2005), pp. 467-479, ISSN 0009-2509

Feng, B.; Yonga, A.K. & Ana, H. (2007). Effect of various factors on the particle size of calcium carbonate formed in a precipitation process. *Materials Science and Engineering A*, Vol.445-446, (February 2007), pp. 170-179, ISSN 0921-5093.

Ferziger, J.H. & Perić, M. (1996). *Computational Methods for Fluid Dynamics*, Springer-Verlag, ISBN 3-540-59434-5, Berlin, Germany

Gahn, C. & Mersmann, A. (1999). Brittle fracture in crystallization processes. Part A. Attrition and abrasion of brittle solids. *Chemical Engineering Science*, Vol.54, No.9, (May 1999), pp. 1273–1282, ISSN 0009-2509

Garside, J. & Davey, R. J. (1980). Invited review secondary contact nucleation: kinetics, growth and scale-up. *Chemical Engineering Communications*, Vol.4, No.4&5, pp. 393-424, ISSN 0098-6445

Hill, P.J. & Ng, K.M. (1995). New discretization procedure for the breakage equation. *American Institution of Chemical Engineers Journal*, Vol.41, No.5, (May 1995), pp. 1204–1217, ISSN 0001-1541

Hindmarsh, A.C. (1983). Odepack, a systematized collection of ode solvers. In: *Scientific Computing. Vol.1 of IMACS Transactions on Scientific Computation*, R. S. Stepleman, M. Carver, R. Peskin, W.F. Ames & R. Vichnevetsky, (Eds.), 55-64, IMACS/North-Holland, Amsterdam, The Netherlands

Hostomsky, J. & Jones, A.G. (1991). Calcium carbonate crystallization, agglomeration and form during continuous precipitation from solution. *Journal of Physics D: Applied Physics*, Vol.24, No.2, (February 1991), pp. 165-170, ISSN 0022-3727

Hounslow, M.J. (1990). A discretized population balance for continuous systems at steady state. *American Institution of Chemical Engineers Journal*, Vol.36, No.1, (January 1990), pp. 106-116, ISSN 0001-1541

Hounslow, M.J.; Ryall, R.L. & Marshall, V.R. (1988). A discretized population balance for nucleation, growth, and aggregation. *American Institution of Chemical Engineers Journal*, Vol.34, No.11, (November 1988), pp. 1821-1832, ISSN 0001-1541

Hounslow, M.J.; Mumtaz, H.S.; Collier, A.P.; Barrick, J.P. and Bramley, A.S. (2001). A micro-mechanical model for the rate of aggregation during precipitation from solution. *Chemical Engineering Science*, Vol.56, No.7, (April 2001), pp. 2543–2552, ISSN 0009-2509

Hulburt, H.M. & Katz, S. (1964). Some Problems in Particle Technology – Statistical Mechanical Formulation. *Chemical Engineering Science*, Vol.19, No.8, (August 1964), pp. 555-574, ISSN 0009-2509

Jones, A.G.; Rigopoulos, S. & Zauner, R. (2005). Crystallization and precipitation engineering. *Computers & Chemical Engineering*, Vol.29, No.6, (May 2005), pp. 1159-1166, ISSN 0098-1354

Judat, B. & Kind, M. (2004). Morphology and internal structure of barium sulfate—derivation of a new growth mechanism. *Journal of Colloid and Interface Science*, Vol.269, No.2, (January 2004), pp. 341-353, ISSN 0021-9797

Jung, T.; Kim, W.S. & Choi, Ch.K. (2005). Effect of monovalent salts on morphology of calcium carbonate crystallized in Couette-Taylor reactor. *Crystal Research and Technology*, Vol.40, No.6, (June 2005), pp. 586-592, ISSN 0232-1300

Kangwook L.; Jay, H.L.; Dae R.Y. & Mahoney, A.W. (2002). Integrated run-to-run and on-line model-based control of particle size distribution for a semi-batch precipitation reactor. *Computers and Chemical Engineering*, Vol.26, No.7-8, (August 2002), pp. 1117–1131, ISSN 0098-1354

Kataki, Y. & Tsuge, H. (1990). Reactive crystallization of calcium carbonate by gas–liquid and liquid–liquid reactions. *Canadian Journal of Chemical Engineering*, Vol.68, No.3, (June 1990), pp. 435–442, ISSN 0008-4034

Kędra-Królik, K. & Gierycz P. (2006). Obtaining calcium carbonate in a multiphase system by the use of new rotating disc precipitation reactor. *Journal of Thermal Analysis and Calorymetry*, Vol.83, No.3, pp. 579-582, ISSN 1388-6150

Kędra-Królik, K. & Gierycz P. (2009). Precipitation of nanostructured calcite in a controlled multiphase process. *Journal of Crystal Growth*, Vol.311, No.14, (July 2009), pp. 3674-3681, ISSN 0022-0248

Kędra-Królik, K. & Gierycz P. (2010). Simulation of nucleation and growing of CaCO₃ nanoparticles obtained in the rotating disk reactor. *Journal of Crystal Growth*, Vol.312, No.12-13, (June 2010), pp. 1945-1952, ISSN 0022-0248

Kitano, Y.; Park, K. & Hood, D.W. (1962). Pure aragonite synthesis. *Journal of Geophysical Research*, Vol.67, No.12, pp. 4873-4874, ISSN 0148-0227

Kramer, H.J.M.; Dijkstra, J.W.; Verheijen, P.J.T. & van Rosmalen, G.M. (2000). Modeling of industrial crystallizers for control and design purposes, *Powder Technology*, Vol.108, No.2-3, (March 2000), pp. 185–191, ISSN 0032-5910

Kulikov, V.; Briesen, H. & Marquardt, W. (2005). Scale integration for the coupled simulation of crystallization and fluid dynamics. *Chemical Engineering Research and Design*, Vol.83, No.6, (June 2005), pp. 706–717, ISSN 0263-8762

Kumar, S. & Ramkrishna, D. (1996). On the solution of population balance equations by discretization—II. A moving pivot technique. *Chemical Engineering Science*, Vol.51, No.8, (April 1996), pp. 1333-1342, ISSN 0009-2509

Malkaj, P.; Chrissanthopoulos, A. & Dalas, E. (2004). Understanding nucleation of calcium carbonate on gallium oxide using computer simulation. *Journal of Crystal Growth*, Vol.264, No.1-3, (March 2004), pp. 430-437, ISSN 0022-0248

Marchisio, D.L.; Vigil, R.D. and Fox, R.O. (2003). Implementation of quadrature method of moments in CFD codes for aggregation-breakage problems. *Chemical Engineering Science*, Vol.58, No.15, (August 2003), pp. 3337–3351, ISSN 0009-2509

Montes-Hernandez, G.; Renard, F.; Geoffroy, N.; Charlet, L. & Pironon, J. (2007). Calcite precipitation from CO_2–H_2O–$Ca(OH)_2$ slurry under high pressure of CO_2. *Journal of Crystal Growth*, Vol.308, No.1, (October 2007), pp. 228-236, ISSN 0022-0248

Mullin, J.W. (2001). *Crystallization*, Butterworth-Heinemann, ISBN 978-075-0648-33-2, Oxford, UK

Myerson, A.S. (1999). *Molecular modelling applications in crystallization*, Cambridge University Press, ISBN 0 521 55297 4, Cambridge, UK

Nicmanis, N. & Hounslow, M.J. (1998). Finite-element methods for steady-state population balance equations. *American Institution of Chemical Engineers Journal*, Vol.44, No.10, (October 1998), pp. 2258-2272, ISSN 0001-1541

Prasher, C.L. (1987). *Crushing & Grinding Process Handbook*, Wiley, ISBN 047110535X, New York, US

Quigley, D. & Roger, P.M. (2008). Free energy and structure of calcium carbonate nanoparticles during early stages of crystallization. *Journal of Chemical Physics*, Vol.128, No.22, (June 2008) pp. 2211011-2211014, ISSN 0021-9606

Ramkrishna, D. (2000). *Population Balances. Theory and Applications to Particulate Systems in Engineering*, Academic Press, ISBN 0-12-576970-9, San Diego, US

Randolph, A.D. & Larson, M.A. (1988). *Theory of Particulate Processes*, Academic Press, ISBN 0125796528, New York, US

Reddy, M.M. & Nancollas, G.H. (1976). The crystallization of calcium carbonate : IV. The effect of magnesium, strontium and sulfate ions. *Journal of Crystal Growth*, Vol.35, No.1, (August 1976), pp. 33-38, ISSN 0022-0248

Rigopoulos, S. & Jones, A.G. (2001). Dynamic modelling of a bubble column for particle formation via a gas-liquid reaction. *Chemical Engineering Science*, Vol.56, No.21-22, (November 2001), pp. 6177-6183, ISSN 0009-2509

Rigopoulos, S. & Jones, A.G. (2003a). Modeling of semibatch agglomerative gas–liquid precipitation of $CaCO_3$ in a bubble column reactor. *Industrial & Engineering Chemistry Research*, Vol.42, No.25, (December 2003), pp. 6567- 6575, ISSN 0888-5885

Rigopoulos, S. & Jones, A.G. (2003b) Finite element scheme for solution of the dynamic population balance. *American Institute of Chemical Engineering Journal*, Vol.49, No.5, (May 2003), pp. 1127-1139, ISSN 0001-1541

Schlomach, J.; Quarch, K. & Kind, M. (2006). Investigation of precipitation of calcium carbonate at high supersaturations. *Chemical Engineering & Technology*, Vol.29, No.2, (February 2006), pp. 215-220, ISSN 1521-4125

Schwarz, M.P. & Turner, W.J. (1988). Applicability of the standard k-ε turbulence model to gas-stirred baths. *Applied Mathematical Modelling*, Vol.12, No.3, (June 1988), pp. 273-279, ISSN 0307-904X

Sha, Z. & Palosaari, S. (2000). Mixing and crystallization in suspensions. *Chemical Engineering Science*, Vol.55, No.10, (May 2000), pp. 1797–1806, ISSN 0009-2509

Sohnel, O. & Mullin, J.W. (1982). Precipitation of calcium carbonate. *Journal of Crystal Growth*, Vol.60, No.2, (December 1982), pp. 239-250, ISSN 0022-0248

Spanos, N. & Koutsoukos, P.G. (1998). Kinetics of precipitation of calcium carbonate in alkaline pH at constant supersaturation. spontaneous and seeded growth. *Journal of Physical Chemistry B*, Vol.102, No.34, (August 1998), pp. 6679-6684, ISSN 1520-6106

Spiegelman M. (2004). *Myths & Methods in Modeling*, LDEO, Columbia University, New York, US

Tai, C.Y. & Chen, P.-C. (1995). Nucleation, agglomeration and crystal morphology of calcium carbonate. *American Institute of Chemical Engineering Journal*, Vol.41, No.1, (January 1995), pp. 68-77, ISSN 0001-1541

Tobias, J. & Klein, M.L. (1996). Molecular dynamics simulations of a calcium carbonate/calcium sulfonate reverse micelle. *Journal of Physical Chemistry B*, Vol.100, No.16, (April 1996), pp. 6637-6648, ISSN 1520-6106

Varma, A. & Morbidelli, M. (1997). *Mathematical Methods in Chemical Engineering*, Oxford University Press, ISBN 0-19-509821-8, New York, US

Villermaux, J. & Falk, L. (1994). A generalized mixing model for initial contacting of reactive fluids. *Chemical Engineering Science*, Vol.49, No.24(2), (December 1994), pp. 5127-5140, ISSN 0009-2509

Wachi, S. & Jones, A.G. (1991). Mass transfer with chemical reaction and precipitation. *Chemical Engineering Science*, Vol.46, No.4, pp. 1027-1033, ISSN 0009-2509

Wan, B. & Ring, T.A. (2006). Verification of SMOM and QMOM population balance modeling in CFD code using analytical solutions for batch particulate processes. *China Particuology*, Vol.4, No.5, (October 2006), pp. 243-249, ISSN 1672-2515

Wang, T.; Antonietti, M. & Colfen, H. (2006). Calcite mesocrystals: "Morphing" crystals by a polyelectrolyte. *Chemistry – A European Journal*, Vol.12, No.22, (July 2006), pp. 5722-5730, ISSN 0947-6539

Wei, H.Y. & Garside, J. (1997). Application of CFD modelling to precipitation systems. *Chemical Engineering Research and Design*, Vol.75, No.2, (February 1997), pp. 219-227, ISSN 0263-8762

Wojcik, J. & Jones, A.G. (1998). Dynamics and stability of continuous MSMPR agglomerative precipitation: Numerical analysis of the dual particle coordinate model. *Computers and Chemical Engineering*, Vol.22, No.4-5, pp. 535-545, ISSN 0098-1354

Wuklow, M.; Gerstlauer, A. & Nieken, U. (2001). Modeling and simulation of crystallization processes using parsival. *Chemical Engineering Science*, Vo.56., No.7, (April 2001), pp. 2575-2588, ISSN 0009-2509

Permissions

The contributors of this book come from diverse backgrounds, making this book a truly international effort. This book will bring forth new frontiers with its revolutionizing research information and detailed analysis of the nascent developments around the world.

We would like to thank Nikolai N. Kolesnikov, for lending his expertise to make the book truly unique. He has played a crucial role in the development of this book. Without his invaluable contribution this book wouldn't have been possible. He has made vital efforts to compile up to date information on the varied aspects of this subject to make this book a valuable addition to the collection of many professionals and students.

This book was conceptualized with the vision of imparting up-to-date information and advanced data in this field. To ensure the same, a matchless editorial board was set up. Every individual on the board went through rigorous rounds of assessment to prove their worth. After which they invested a large part of their time researching and compiling the most relevant data for our readers. Conferences and sessions were held from time to time between the editorial board and the contributing authors to present the data in the most comprehensible form. The editorial team has worked tirelessly to provide valuable and valid information to help people across the globe.

Every chapter published in this book has been scrutinized by our experts. Their significance has been extensively debated. The topics covered herein carry significant findings which will fuel the growth of the discipline. They may even be implemented as practical applications or may be referred to as a beginning point for another development. Chapters in this book were first published by InTech; hereby published with permission under the Creative Commons Attribution License or equivalent.

The editorial board has been involved in producing this book since its inception. They have spent rigorous hours researching and exploring the diverse topics which have resulted in the successful publishing of this book. They have passed on their knowledge of decades through this book. To expedite this challenging task, the publisher supported the team at every step. A small team of assistant editors was also appointed to further simplify the editing procedure and attain best results for the readers.

Our editorial team has been hand-picked from every corner of the world. Their multi-ethnicity adds dynamic inputs to the discussions which result in innovative outcomes. These outcomes are then further discussed with the researchers and contributors who give their valuable feedback and opinion regarding the same. The feedback is then collaborated with the researches and they are edited in a comprehensive manner to aid the understanding of the subject.

Apart from the editorial board, the designing team has also invested a significant amount of their time in understanding the subject and creating the most relevant covers. They scrutinized every image to scout for the most suitable representation of the subject and create an appropriate cover for the book.

The publishing team has been involved in this book since its early stages. They were actively engaged in every process, be it collecting the data, connecting with the contributors or procuring relevant information. The team has been an ardent support to the editorial, designing and production team. Their endless efforts to recruit the best for this project, has resulted in the accomplishment of this book. They are a veteran in the field of academics and their pool of knowledge is as vast as their experience in printing. Their expertise and guidance has proved useful at every step. Their uncompromising quality standards have made this book an exceptional effort. Their encouragement from time to time has been an inspiration for everyone.

The publisher and the editorial board hope that this book will prove to be a valuable piece of knowledge for researchers, students, practitioners and scholars across the globe.

List of Contributors

Manuel García-Méndez
Centro de Investigación en Ciencias Físico-Matemáticas, FCFM de la UANL Manuel L. Barragán S/N, Cd. Universitaria, México

H. R. Aghabozorg, S. Sadegh Hassani and F. Salehirad
Research Institute of Petroleum Industry, Iran

Hwisim Hwang, Yasutomo Uetsuji and Eiji Nakamachi
Doshisha University, Japan

Shadia J. Ikhmayies
Al Isra University, Faculty of Science and Information Technology, Amman, Jordan

Mallikarjuna N. Nadagouda
Water Supply and Water Resources Division, National Risk Management Research Laboratory, U. S. Environmental Protection Agency, Ohio, USA

Lisheng Huang
Department of Physics, National Cheng Kung University, Tainan, Taiwan
National Laboratory of Solid State Microstructures, Nanjing University, Nanjing, China
College of Sciences & College of Materials Science and Engineering, Nanjing University of Technology, Nanjing, China

Yinjie Su
National Laboratory of Solid State Microstructures, Nanjing University, Nanjing, China

Wanchuan Chen1
Department of Physics, National Cheng Kung University, Tainan, Taiwan

Satyawati S. Joshi
University of Pune, India

Igor Nederlof, Eric van Genderen, Jan Pieter Abrahams and Dilyana Georgieva
Leiden University, The Netherlands

Flip Hoedemaeker
Kabta Consultancy Ltd., The Netherlands

André Abts, Christian K. W. Schwarz, Britta Tschapek, Sander H. J. Smits and Lutz Schmitt
Institute of Biochemistry, Heinrich-Heine University, Düsseldorf, Germany

V. Borshchevskiy
Research-educational Centre "Bionanophysics", Moscow Institute of Physics and Technology, Russia
Institute of Complex Systems (ICS), ICS-5: Molecular Biophysics, Research Centre Juelich, Germany

V. Gordeliy
Laboratoire des Protéines Membranaires, Institut de Biology Structurale J.-P., France
Research-educational Centre "Bionanophysics", Moscow Institute of Physics and Technology, Russia
Institute of Complex Systems (ICS), ICS-5: Molecular Biophysics, Research Centre Juelich, Germany

A. Antony Joseph and C. Ramachandra Raja
Department of Physics, Government Arts College (Autonomous), Kumbakonam,Tamil Nadu, India

E. C. H. Ng and C. C. Wong
Singapore-MIT Alliance, Nanyang Technological University, Singapore

Y. K. Koh
DSO National Laboratories, Singapore

Pawel Gierycz
Faculty of Chemical and Process Engineering, Warsaw University of Technology, Institute of Physical Chemistry, Polish Academy of Sciences, Warsaw, Poland